Feature Engineering and Selection
A Practical Approach for Predictive Models

CHAPMAN & HALL/CRC DATA SCIENCE SERIES

Reflecting the interdisciplinary nature of the field, this book series brings together researchers, practitioners, and instructors from statistics, computer science, machine learning, and analytics. The series will publish cutting-edge research, industry applications, and textbooks in data science.

The inclusion of concrete examples, applications, and methods is highly encouraged. The scope of the series includes titles in the areas of machine learning, pattern recognition, predictive analytics, business analytics, Big Data, visualization, programming, software, learning analytics, data wrangling, interactive graphics, and reproducible research.

Published Titles

Feature Engineering and Selection: A Practical Approach for Predictive Models
Max Kuhn and Kjell Johnson

Probability and Statistics for Data Science: Math + R + Data
Norman Matloff

Feature Engineering and Selection

A Practical Approach for Predictive Models

Max Kuhn
Kjell Johnson

CRC Press
Taylor & Francis Group
Boca Raton London New York

CRC Press is an imprint of the
Taylor & Francis Group, an **informa** business

A CHAPMAN & HALL BOOK

Map tiles on front cover: © OpenStreetMap © CartoDB.

CRC Press
Taylor & Francis Group
6000 Broken Sound Parkway NW, Suite 300
Boca Raton, FL 33487-2742

© 2020 by Taylor & Francis Group, LLC
CRC Press is an imprint of Taylor & Francis Group, an Informa business

No claim to original U.S. Government works

Printed on acid-free paper

International Standard Book Number-13: 978-1-138-07922-9 (Hardback)

Visit the Taylor & Francis Web site at
http://www.taylorandfrancis.com

and the CRC Press Web site at
http://www.crcpress.com

To:
Maryann Kuhn
Valerie, Truman, and Louie

Contents

Preface

The goal of our previous work, *Applied Predictive Modeling*, was to elucidate a framework for constructing models that generate accurate predictions for future, yet-to-be-seen data. This framework includes pre-processing the data, splitting the data into training and testing sets, selecting an approach for identifying optimal tuning parameters, building models, and estimating predictive performance. This approach protects from overfitting to the training data and helps models to identify truly predictive patterns that are generalizable to future data, thus enabling good predictions for that data. Authors and modelers have successfully used this framework to derive models that have won Kaggle competitions (Raimondi, 2010), have been implemented in diagnostic tools (Jahani and Mahdavi, 2016; Luo, 2016), are being used as the backbone of investment algorithms (Stanković et al., 2015), and are being used as a screening tool to assess the safety of new pharmaceutical products (Thomson et al., 2011).

In addition to having a good approach to the modeling process, building an effective predictive model requires other good practices. These practices include garnering expert knowledge about the process being modeled, collecting the appropriate data to answer the desired question, understanding the inherent variation in the response and taking steps, if possible, to minimize this variation, ensuring that the predictors collected are relevant for the problem, and utilizing a range of model types to have the best chance of uncovering relationships among the predictors and the response.

Despite our attempts to follow these good practices, we are sometimes frustrated to find that the best models have less-than-anticipated, less-than-useful useful predictive performance. This lack of performance may be due to a simple to explain, but difficult to pinpoint, cause: relevant predictors that were collected are represented in a way that models have trouble achieving good performance. Key relationships that are not directly available as predictors may be between the response and:

- a transformation of a predictor,
- an interaction of two or more predictors such as a product or ratio,
- a functional relationship among predictors, or
- an equivalent re-representation of a predictor.

Adjusting and reworking the predictors to enable models to better uncover predictor-response relationships has been termed *feature engineering*. The engineering connotation implies that we know the steps to take to fix poor performance and to guide predictive improvement. However, we often do not know the best re-representation of the predictors to improve model performance. Instead, the re-working of predictors is more of an *art*, requiring the right tools and experience to find better predictor representations. Moreover, we may need to search many alternative predictor representations to improve model performance. This process, too, can lead to overfitting due to the vast number of alternative predictor representations. So appropriate care must be taken to avoid overfitting during the predictor creation process.

The goals of *Feature Engineering and Selection* are to provide tools for re-representing predictors, to place these tools in the context of a good predictive modeling framework, and to convey our experience of utilizing these tools in practice. In the end, we hope that these tools and our experience will help you generate better models. When we started writing this book, we could not find any comprehensive references that described and illustrated the types of tactics and strategies that can be used to improve models by focusing on the predictor representations (that were not solely focused on images and text).

Like in *Applied Predictive Modeling*, we have used R as the computational engine for this text. There are a few reasons for that. First, while not the only good option, R has been shown to be popular and effective in modern data analysis. Second, R is free and open-source. You can install it anywhere, modify the code, and have the ability to see exactly how computations are performed. Third, it has excellent community support via the canonical R mailing lists and, more importantly, with Twitter,[1] StackOverflow,[2] and RStudio Community.[3] Anyone who asks a reasonable, reproducible question has a pretty good chance of getting an answer.

Also, as in our previous effort, it is critically important to us that all the software and *data* are freely available. This allows everyone to reproduce our work, find bugs/errors, and to even extend our approaches. The data sets and R code are available in the GitHub repository `https://github.com/topepo/FES`. An HTML version of this text can be found at `https://bookdown.org/max/FES`.

We'd like to thank everyone that contributed feedback, typos, or discussions while the book was being written. GitHub contributors included, as of June 21, 2019, the following: @alexpghayes, @AllardJM, @AndrewKostandy, @bashhwu, @btlois, @cdr6934, @danielwo, @davft, @draben, @eddelbuettel, @endore, @feinmann, @gtesei, @ifellows, @JohnMount, @jonimatix, @jrfiedler, @juliasilge, @jwillage, @kaliszp, @KevinBretonnelCohen, @kieroneil, @KnightAdz, @kransom14, @LG-1, @LluisRamon, @LoweCoryr, @lpatruno, @mlduarte, @monogenea, @mpettis, @Nathan-Furnal, @nazareno, @PedramNavid, @r0f1, @Ronen4321, @shinhongwu, @stecaron, @StefanZaaiman, @treysp, @uwesterr, and @van1991. Hadley Wickham also provided excellent feedback and suggestions. Max would also like to thank RStudio for provid-

---

[1]https://twitter.com/hashtag/rstats?src=hash
[2]https://stackoverflow.com/questions/tagged/r
[3]https://community.rstudio.com/

ing the time and support for this activity as well as the wonderful `#rstats` Twitter community. We would also like to thank the reviewers for their work, which helped improve the text. Our excellent editor, John Kimmel, has been a breath of fresh air for us. Most importantly, we would like to thank our families for their support: Louie, Truman, Stefan, Dan, and Valerie.

Author Bios

Max Kuhn, Ph.D., is a software engineer at RStudio. He worked in 18 years in drug discovery and medical diagnostics applying predictive models to real data. He has authored numerous R packages for predictive modeling and machine learning.

Kjell Johnson, Ph.D., is the owner and founder of Stat Tenacity, a firm that provides statistical and predictive modeling consulting services. He has taught short courses on predictive modeling for the American Society for Quality, American Chemical Society, International Biometric Society, and for many corporations.

Kuhn and Johnson have also authored *Applied Predictive Modeling*, which is a comprehensive, practical guide to the process of building a predictive model. The text won the 2014 *Technometrics* Ziegel Prize for Outstanding Book.

1 | Introduction

Statistical models have gained importance as they have become ubiquitous in modern society. They enable us by generating various types of predictions in our daily lives. For example, doctors rely on general rules derived from models that tell them which specific cohorts of patients have an increased risk of a particular ailment or event. A numeric prediction of a flight's arrival time can help understand if our airplane is likely to be delayed. In other cases, models are effective at telling us what is important or concrete. For example, a lawyer might utilize a statistical model to quantify the likelihood that potential hiring bias is occurring by chance or whether it is likely to be a systematic problem.

In each of these cases, models are created by taking existing data and finding a mathematical representation that has acceptable fidelity to the data. From such a model, important statistics can be estimated. In the case of airline delays, a prediction of the outcome (arrival time) is the quantity of interest while the estimate of a possible hiring bias might be revealed through a specific model parameter. In the latter case, the hiring bias estimate is usually compared to the estimated uncertainty (i.e., noise) in the data and a determination is made based on how uncommon such a result would be relative to the noise–a concept usually referred to as "statistical significance." This type of model is generally thought of as being *inferential*: a conclusion is reached for the purpose of understanding the state of nature. In contrast, the prediction of a particular value (such as arrival time) reflects an *estimation problem* where our goal is not necessarily to understand if a trend or fact is genuine but is focused on having the most accurate determination of that value. The uncertainty in the prediction is another important quantity, especially to gauge the trustworthiness of the value generated by the model.

Whether the model will be used for inference or estimation (or in rare occasions, both), there are important characteristics to consider. *Parsimony* (or simplicity) is a key consideration. Simple models are generally preferable to complex models, especially when inference is the goal. For example, it is easier to specify realistic distributional assumptions in models with fewer parameters. Parsimony also leads to a higher capacity to interpret a model. For example, an economist might be interested in quantifying the benefit of postgraduate education on salaries. A simple model might represent this relationship between years of education and job salary linearly. This parameterization would easily facilitate statistical inferences on the potential benefit of such education. But suppose that the relationship differs substantially between occupations and/or is not linear. A more complex model would do a better job at capturing the data patterns but would be much less interpretable.

The problem, however, is that **accuracy should not be seriously sacrificed for the sake of simplicity**. A simple model might be easy to interpret but would not succeed if it does not maintain acceptable level of faithfulness to the data; if a model is only 50% accurate, should it be used to make inferences or predictions? Complexity is usually the solution to poor accuracy. By using additional parameters or by using a model that is inherently nonlinear, we might improve accuracy but interpretability will likely suffer greatly. This trade-off is a key consideration for model building.

Thus far the discussion has been focused on aspects of the model. However, the variables that go into the model (and how they are represented) are just as critical to success. It is impossible to talk about modeling without discussing models, but one of goal this book is to increase the emphasis on the predictors in a model.

In terms of nomenclature, the quantity that is being modeled or predicted is referred to as either: the *outcome*, response, or dependent variable. The variables that are used to model the outcome are called the *predictors*, *features*, or independent variables (depending on the context). For example, when modeling the sale price of a house (the outcome), the characteristics of a property (e.g., square footage, number of bed rooms and bath rooms) could be used as predictors (the term features would also be suitable). However, consider artificial model terms that are composites of one or more variables, such as the number of bedrooms *per* bathroom. This type of variable might be more appropriately called a feature (or a derived feature). In any case, features and predictors are used to explain the outcome in a model.[4]

As one might expect, there are good and bad ways of entering predictors into a model. In many cases, there are multiple ways that an underlying piece of information can be represented or encoded. Consider a model for the sale price of a property. The location is likely to be crucial and can be represented in different ways. Figure 1.1 shows locations of properties in and around Ames, Iowa, that were sold between 2006 and 2010. In this image, the colors represent the reported neighborhood of residence. There are 28 neighborhoods represented here and the number of properties per neighborhoods range from a single property in Landmark, to 443 in North Ames. A second representation of location in the data is longitude and latitude. A realtor might suggest using ZIP Code^{TM} as a predictor in the model as a proxy for school district since this can be an important consideration for buyers with children. But from an information theory point of view, longitude and latitude offer the most specificity for measuring physical location and one might make an argument that this representation has higher information content (assuming that this particular information is predictive).

---

[4]Also, to some extent, the choice of these terms is driven by whether a person is more computer science-centric or statistics-centric.

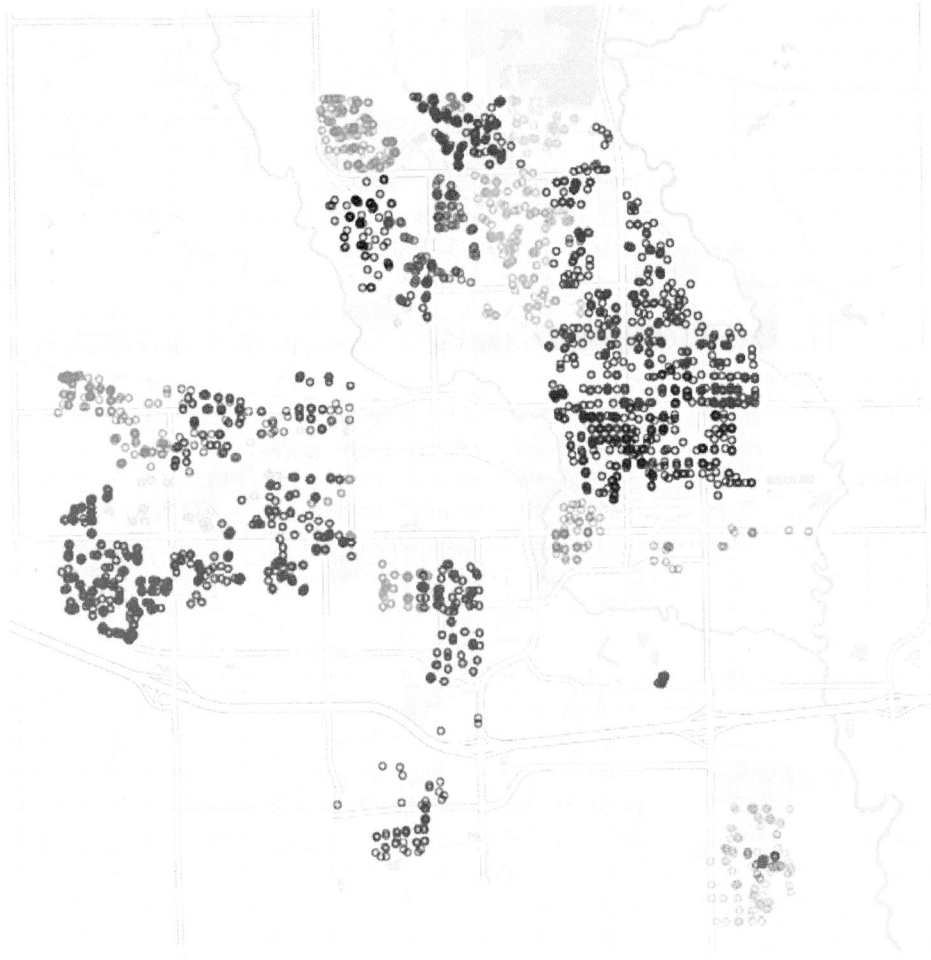

Figure 1.1: Houses for sale in Ames, Iowa, colored by neighborhood. Leaflet | ©
OpenStreetMap © CartoDB

The idea that there are different ways to represent predictors in a model, and that
some of these representations are better than others, leads to the idea of **feature
engineering**—the process of creating representations of data that increase the
effectiveness of a model.

Note that model effectiveness is influenced by many things. Obviously, if the predictor
has no relationship to the outcome then its representation is irrelevant. However, it
is very important to realize that there are a multitude of types of models and that
each has its own sensitivities and needs. For example:

- Some models cannot tolerate predictors that measure the same underlying
  quantity (i.e., multicollinearity or correlation between predictors).
- Many models cannot use samples with any missing values.
- Some models are severely compromised when irrelevant predictors are in the
  data.

Feature engineering and variable selection can help mitigate many of these issues. ***The goal of this book is to help practitioners build better models by focusing on the predictors***. "Better" depends on the context of the problem but most likely involves the following factors: accuracy, simplicity, and robustness. To achieve these characteristics, or to make good trade-offs between them, it is critical to understand the interplay between predictors used in a model and the type of model. Accuracy and/or simplicity can sometimes be improved by representing data in ways that are more palatable to the model or by reducing the number of variables used. To demonstrate this point, a simple example with two predictors is shown in the next section. Additionally, a more substantial example is discussed in Section 1.3 that more closely resembles the modeling process in practice.

1.1   A Simple Example

As a simple example of how feature engineering can affect models, consider Figure 1.2a that shows a plot of two correlated predictor variables (labeled as A and B). The data points are colored by their outcome, a discrete variable with two possible values ("PS" and "WS"). These data originate from an experiment from Hill et al. (2007) which includes a larger predictor set. For their task, a model would require a high degree of accuracy but would not need to be used for inference. For this illustration, only these two predictors will be considered. In this figure, there is clearly a diagonal separation between the two classes. A simple logistic regression model (Hosmer and Lemeshow, 2000) will be used here to create a prediction equation from these two variables. That model uses the following equation:

$$log(p/(1-p)) = \beta_0 + \beta_1 A + \beta_2 B$$

where p is the probability that a sample is the "PS" class and the β values are the model parameters that need to be estimated from the data.

A standard procedure (maximum likelihood estimation) is used to estimate the three regression parameters from the data. The authors used 1009 data points to estimate the parameters (i.e., a *training set*) and reserved 1010 samples strictly for estimating performance (a *test set*).[5] Using the training set, the parameters were estimated to be $\hat{\beta}_0 = 1.73$, $\hat{\beta}_1 = 0.003$, and $\hat{\beta}_2 = -0.064$.

To evaluate the model, predictions are made on the test set. Logistic regression naturally produces class probabilities that give an indication of likelihood for each class. While it is common to use a 50% cutoff to make hard class predictions, the performance derived from this default might be misleading. To avoid applying a probability cutoff, a technique called the receiver operating characteristic (ROC) curve is used here.[6] The ROC curve evaluates the results on all possible cutoffs and plots the true positive rate versus the false positive rate. The curve for this

---

[5] These types of data sets are discussed more in Section 3.3.

[6] This technique is discussed in more detail in Section 3.2

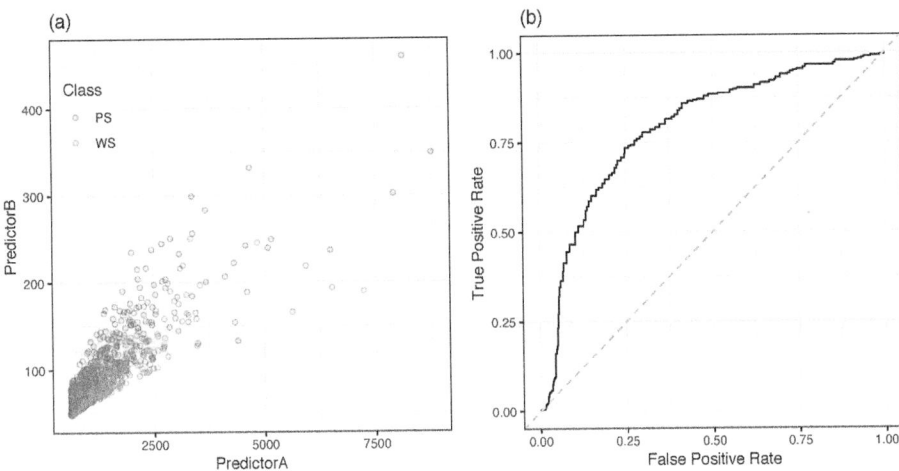

Figure 1.2: (a) An example data set and (b) the ROC curve from a simple logistic regression model.

example is shown in Figure 1.2b. The best possible curve is one that is shifted as close as possible to the upper left corner while a ineffective model will stay along the dashed diagonal line. A common summary value for this technique is to use the area under the ROC curve where a value of 1.0 corresponds a perfect model while values near 0.5 are indicative of a model with no predictive ability. For the current logistic regression model, the area under the ROC curve is 0.794 which indicates moderate accuracy in classifying the response.

Given these two predictor variables, it would make sense to try different transformations and encodings of these data in an attempt to increase the area under the ROC curve. Since the predictors are both greater than zero and appear to have right-skewed distributions, one might be inclined to take the ratio A/B and enter only this term in the model. Alternatively, we could also evaluate if simple transformations of each predictor would be helpful. One method is the Box-Cox transformation[7] which uses a separate estimation procedure prior to the logistic regression model that can put the predictors on a new scale. Using this methodology, the Box-Cox estimation procedure recommended that both predictors should be used on the inverse scale (i.e., $1/A$ instead of A). This representation of the data is shown in Figure 1.3a. When these transformed values were entered into the logistic regression model in lieu of the original values, the area under the ROC curve changed from 0.794 to 0.848, which is a substantial increase. Figure 1.3b shows both curves. In this case the ROC curve corresponding to the transformed predictors is uniformly better than the original result.

This example demonstrates how an alteration of the predictors, in this case a simple transformation, can lead to improvements to the effectiveness of the model. When comparing the data in Figures 1.2a and Figure 1.3a, it is easier to visually discriminate

---

[7]Discussed in Section 6.1.

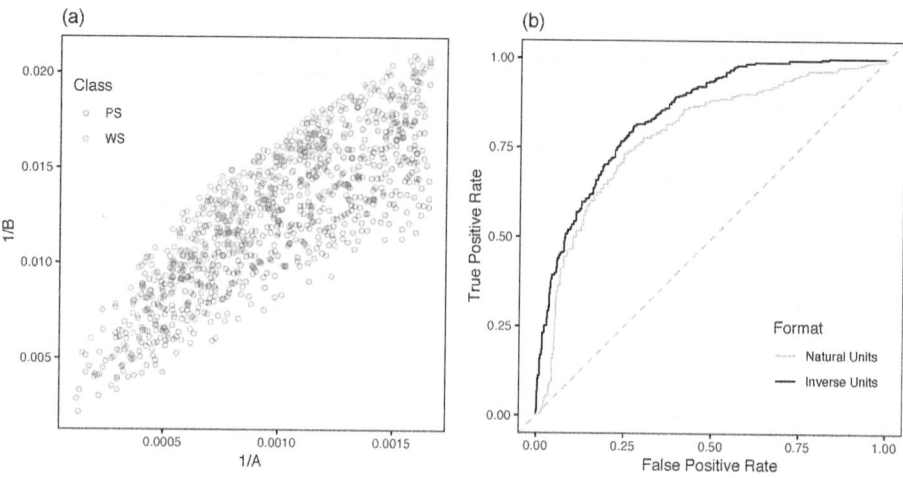

Figure 1.3: (a) The transformed predictors and (b) the ROC curve from both logistic regression models.

the two groups of data. In transforming the data individually, we enabled the logistic regression model to do a better job of separating the classes. Depending on the nature of the predictors, using the inverse of the original data might make inferential analysis more difficult.

However, different models have different requirements of the data. If the skewness of the original predictors was the issue affecting the logistic regression model, other models exist that do not have the same sensitivity to this characteristic. For example, a neural network can also be used to fit these data without using the inverse transformation of the predictors.[8] This model was able to achieve an area under the ROC curve of 0.848 which is roughly equivalent to the improved logistic model results. One might conclude that the neural network model is inherently always better than logistic regression since neural network was not susceptible to the distributional aspects of the predictors. But we should not jump to a blanket conclusion like this due to the "No Free Lunch" theorem (see Section 1.2.3). Additionally, the neural network model has its own drawbacks: it is completely uninterpretable and requires extensive parameter tuning to achieve good results. Depending on how the model will be utilized, one of these models might be more favorable than the other for these data.

A more comprehensive summary of how different features can influence models in different ways is given shortly in Section 1.3. Before this example, the next section discusses several key concepts that will be used throughout this text.

---

[8]The predictor values were normalized to have mean zero and a standard deviation of one, as is needed for this model. However, this does not affect the skewness of the data.

1.2 | Important Concepts

Before proceeding to specific strategies and methods, there are some key concepts that should be discussed. These concepts involve theoretical aspects of modeling as well as the practice of creating a model. A number of these aspects are discussed here and additional details are provided in later chapters and references are given throughout this work.

1.2.1 | Overfitting

Overfitting is the situation where a model fits very well to the current data but fails when predicting new samples. It typically occurs when the model has relied too heavily on patterns and trends in the current data set that do not occur otherwise. Since the model only has access to the current data set, it has no ability to understand that such patterns are anomalous. For example, in the housing data shown in Figure 1.1, one could determine that properties that had square footage between 1,267.5 and 1,277 and contained three bedrooms could have their sale prices predicted within $1,207 of the true values. However, the accuracy on other houses (not in this data set) that satisfy these conditions would be much worse. This is an example of a trend that does not *generalize* to new data.

Often, models that are very flexible (called "low-bias models" in Section 1.2.5) have a higher likelihood of overfitting the data. It is not difficult for these models to do extremely well on the data set used to create the model and, without some preventative mechanism, can easily fail to generalize to new data. As will be seen in the coming chapters, especially Section 3.5, overfitting is one of the primary risks in modeling and should be a concern for practitioners.

While models can overfit to the *data points*, such as with the housing data shown above, feature selection techniques can overfit to the *predictors*. This occurs when a variable appears relevant in the current data set but shows no real relationship with the outcome once new data are collected. The risk of this type of overfitting is especially dangerous when the number of data points, denoted as n, is small and the number of potential predictors (p) is very large. As with overfitting to the data points, this problem can be mitigated using a methodology that will show a warning when this is occurring.

1.2.2 | Supervised and Unsupervised Procedures

Supervised data analysis involves identifying patterns between predictors and an identified *outcome* that is to be modeled or predicted, while unsupervised techniques are focused solely on identifying patterns among the predictors.

Both types of analyses would typically involve some amount of exploration. Exploratory data analysis (EDA) (Tukey, 1977) is used to understand the major characteristics of the predictors and outcome so that any particular challenges associated with the data can be discovered prior to modeling. This can include investigations of correlation structures in the variables, patterns of missing data, and/or anomalous motifs in the data that might challenge the initial expectations of the modeler.

Obviously, predictive models are strictly supervised since there is a direct focus on finding relationships between the predictors and the outcome. Unsupervised analyses include methods such as cluster analysis, principal component analysis, and similar tools for discovering patterns in data.

Both supervised and unsupervised analyses are susceptible to *overfitting* but supervised are particularly inclined to discovering erroneous patterns in the data for predicting the outcome. In short, we can use these techniques to create a *self-fulfilling predictive prophecy*. For example, it is not uncommon for an analyst to conduct a supervised analysis of data to detect which predictors are significantly associated with the outcome. These significant predictors are then used in a visualization (such as a heatmap or cluster analysis) on the same data. Not surprisingly, the visualization reliably demonstrates clear patterns between the outcomes and predictors and appears to provide evidence of their importance. However, since the same data are shown, the visualization is essentially *cherry picking* the results that are only true for these data and which are unlikely to generalize to new data. This issue is discussed at various points in this text but most pointedly in Section 3.8.

1.2.3 | No Free Lunch

The "No Free Lunch" Theorem (Wolpert, 1996) is the idea that, without any specific knowledge of the problem or data at hand, no one predictive model can be said to be the best. There are many models that are optimized for some data characteristics (such as missing values or collinear predictors). In these situations, it might be reasonable to assume that they would do better than other models (all other things being equal). In practice, things are not so simple. One model that is optimized for collinear predictors might be constrained to model linear trends in the data and is sensitive to missingness in the data. It is very difficult to predict the best model especially before the data are in hand.

There have been experiments to judge which models tend to do better than others *on average*, notably Demsar (2006) and Fernandez-Delgado et al. (2014). These analyses show that some models have a tendency to produce the most accurate models but the rate of "winning" is not high enough to enact a strategy of "always use model *X*."

In practice, it is wise to try a number of disparate types of models to probe which ones will work well with your particular data set.

1.2.4 | The Model versus the Modeling Process

The process of developing an effective model is both iterative and heuristic. It is difficult to know the needs of any data set prior to working with it and it is common for many approaches to be evaluated and modified before a model can be finalized. Many books and resources solely focus on the modeling technique but this activity is often a small part of the overall process. Figure 1.4 shows an illustration of the overall process for creating a model for a typical problem.

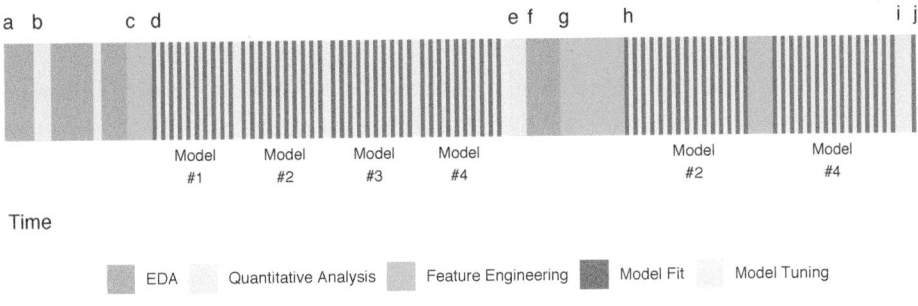

Figure 1.4: A schematic for the typical modeling process.

The initial activity begins[9] at marker (*a*) where exploratory data analysis is used to investigate the data. After initial explorations, marker (*b*) indicates where early data analysis might take place. This could include evaluating simple summary measures or identifying predictors that have strong correlations with the outcome. The process might iterate between visualization and analysis until the modeler feels confident that the data are well understood. At milestone (*c*), the first draft for how the predictors will be represented in the models is created based on the previous analysis.

At this point, several different modeling methods might be evaluated with the initial feature set. However, many models can contain *hyperparameters* that require tuning.[10] This is represented at marker (*d*) where four clusters of models are shown as thin red marks. This represents four distinct models that are being evaluated but each one is evaluated multiple times over a set of candidate hyperparameter values. This model tuning process is discussed in Section 3.6 and is illustrated several times in later chapters. Once the four models have been tuned, they are numerically evaluated on the data to understand their performance characteristics (*e*). Summary measures for each model, such as model accuracy, are used to understand the level of difficulty for the problem and to determine which models appear to best suit the data. Based on these results, more EDA can be conducted on the model results (*f*), such as residual analysis. For the previous example of predicting the sale prices of houses, the properties that are poorly predicted can be examined to understand if there is any systematic issues with the model. As an example, there may be particular ZIP Codes that are difficult to accurately assess. Consequently, another round of feature engineering (*g*) might be used to compensate for these obstacles. By this point, it may be apparent which models tend to work best for the problem at hand and another, more extensive, round of model tuning can be conducted on fewer models

---

[9]This assumes that the data have been sufficiently *cleaned* and that no erroneous values are present. Data cleaning can easily take an extended amount of time depending on the source of the data.

[10]Discussed more in Section 3.6.

(*h*). After more tuning and modification of the predictor representation, the two candidate models (#2 and #4) have been finalized. These models can be evaluated on an external test set as a final "bake off" between the models (*i*). The final model is then chosen (*j*) and this fitted model will be used going forward to predict new samples or to make inferences.

The point of this schematic is to illustrate there are far more activities in the process than simply fitting a single mathematical model. For most problems, it is common to have feedback loops that evaluate and reevaluate how well any model/feature set combination performs.

1.2.5 │ Model Bias and Variance

Variance is a well-understood concept. When used in regard to data, it describes the degree in which the values can fluctuate. If the same object is measured multiple times, the observed measurements will be different to some degree. In statistics, *bias* is generally thought of as the degree in which something deviates from its true underlying value. For example, when trying to estimate public opinion on a topic, a poll could be systematically biased if the people surveyed over-represent a particular demographic. The bias would occur as a result of the poll incorrectly estimating the desired target.

Models can also be evaluated in terms of variance and bias (Geman et al., 1992). A model has high variance if small changes to the underlying data used to estimate the parameters cause a sizable change in those parameters (or in the structure of the model). For example, the sample mean of a set of data points has higher variance than the sample median. The latter uses only the values in the center of the data distribution and, for this reason, it is insensitive to moderate changes in the values. A few examples of models with *low variance* are linear regression, logistic regression, and partial least squares. High-variance models include those that strongly rely on individual data points to define their parameters such as classification or regression trees, nearest neighbor models, and neural networks. To contrast low-variance and high-variance models, consider linear regression and, alternatively, nearest neighbor models. Linear regression uses all of the data to estimate slope parameters and, while it can be sensitive to outliers, it is much less sensitive than a nearest neighbor model.

Model bias reflects the ability of a model to conform to the underlying theoretical structure of the data. A low-bias model is one that can be highly flexible and has the capacity to fit a variety of different shapes and patterns. A high-bias model would be unable to estimate values close to their true theoretical counterparts. Linear methods often have high bias since, without modification, they cannot describe nonlinear patterns in the predictor variables. Tree-based models, support vector machines, neural networks, and others can be very adaptable to the data and have low bias.

As one might expect, model bias and variance can often be in opposition to one another; in order to achieve low bias, models tend to demonstrate high variance (and

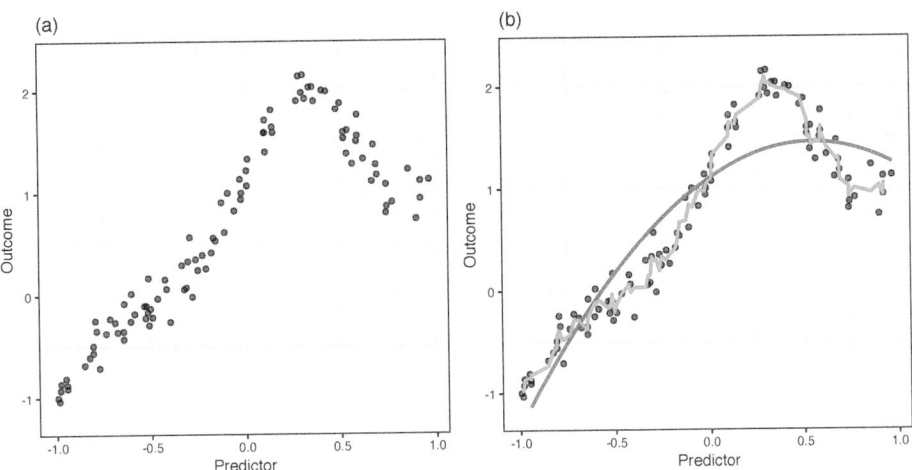

Figure 1.5: A simulated data set and model fits for a 3-point moving average (green) and quadratic regression (purple).

vice versa). The *variance-bias trade-off* is a common theme in statistics. In many cases, models have parameters that control the flexibility of the model and thus affect the variance and bias properties of the results. Consider a simple sequence of data points such as a daily stock price. A moving average model would estimate the stock price on a given day by the average of the data points within a certain window of the day. The size of the window can modulate the variance and bias here. For a small window, the average is much more responsive to the data and has a high potential to match the underlying trend. However, it also inherits a high degree of sensitivity to those data in the window and this increases variance. Widening the window will average more points and will reduce the variance in the model but will also desensitize the model fit potential by risking over-smoothing the data (and thus increasing bias).

Consider the example in Figure 1.5a that contains a single predictor and outcome where their relationship is nonlinear. The right-hand panel (b) shows two model fits. First, a simple three-point moving average is used (in green). This trend line is bumpy but does a good job of tracking the nonlinear trend in the data. The purple line shows the results of a standard linear regression model that includes a term for the predictor value and a term for the square of the predictor value. Linear regression is linear *in the model parameters* and adding polynomial terms to the model can be an effective way of allowing the model to identify nonlinear patterns. Since the data points start low on the y-axis, reach an apex near a predictor value of 0.3 then decrease, a quadratic regression model would be a reasonable first attempt at modeling these data. This model is very smooth (showing low variance) but does not do a very good job of fitting the nonlinear trend seen in the data (i.e., high bias).

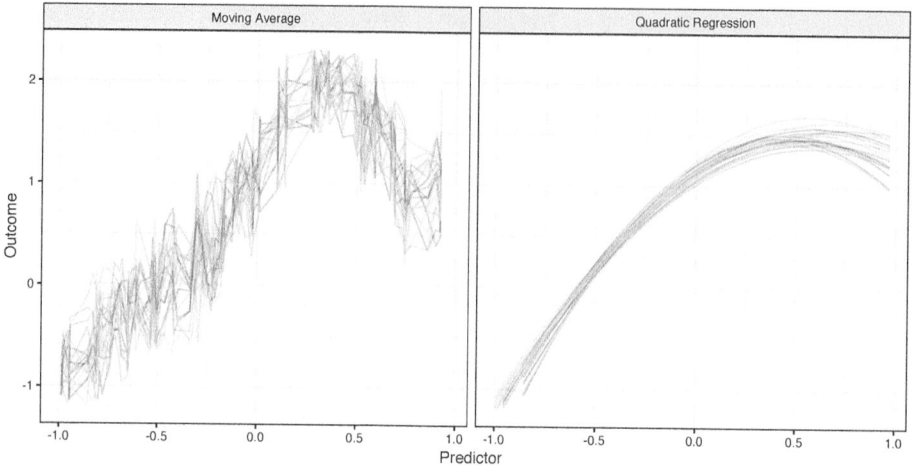

Figure 1.6: Model fits for twenty jittered versions of the data set.

To accentuate this point further, the original data were "jittered" multiple times by adding small amounts of random noise to their values. This was done twenty times and, for each version of the data, the same two models were fit to the jittered data. The fitted curves are shown in Figure 1.6. The moving average shows a significant degree of noise in the regression predictions but, on average, manages to track the data patterns well. The quadratic model was not confused by the extra noise and generated very similar (although inaccurate) model fits.

The notions of model bias and variance are central to the ideas in this text. As previously described, simplicity is an important characteristic of a model. One method of creating a *low-variance, low-bias* model is to augment a low-variance model with appropriate representations of the data to decrease the bias. The previous example in Section 1.1 is a simple example of this process; a logistic regression (high bias, low variance) was improved by modifying the predictor variables and was able to show results on par with a neural networks model (low bias). As another example, the data in Figure 1.5(a) were generated using the following equation

$$y = x^3 + \left[\beta_1 \, exp(\beta_2 \, (x - \beta_3)^2) \right] + \epsilon.$$

Theoretically, if this functional form could be determined from the data, then the best possible model would be a nonlinear regression model (low variance, low bias). We revisit the variance-bias relationship in Section 3.4.6 in the context of measuring performance using resampling.

In a similar manner, models can have reduced performance due to irrelevant predictors causing excess model variation. Feature selection techniques improve models by reducing the unwanted noise of extra variables.

1.2.6 | Experience-Driven Modeling and Empirically Driven Modeling

Projects may arise where no modeling has previously been applied to the data. For example, suppose that a new customer database becomes available and this database contains a large number of fields that are potential predictors. Subject matter experts may have a good sense of what features should be in the model based on previous experience. This knowledge allows experts to be prescriptive about exactly which variables are to be used and how they are represented. Their reasoning should be strongly considered given their expertise. However, since the models estimate parameters from the data, there can be a strong desire to be *data-driven* rather than *experience-driven*.

Many types of models have the ability to empirically discern which predictors should be in the model and can derive the representation of the predictors that can maximize performance (based on the available data). The perceived (and often real) danger in this approach is twofold. First, as previously discussed, data-driven approaches run the risk of overfitting to false patterns in the data. Second, they might yield models that are highly complex and may not have any obvious rational explanation. In the latter case a circular argument may arise where practitioners only accept models that quantify what they already know but expect better results than what a human's manual assessment can provide. For example, if an unexpected, novel predictor is found that has a strong relationship with the outcome, this may challenge the current conventional wisdom and be viewed with suspicion.

It is common to have some conflict between experience-driven modeling and empirically driven modeling. Each approach has its advantages and disadvantages. In practice, we have found that a combination of the two approaches works best as long as both sides see the value in the contrasting approaches. The subject matter expert may have more confidence in a novel model feature if they feel that the methodology used to discover the feature is rigorous enough to avoid spurious results. Also an empirical modeler might find benefit in an expert's recommendations to initially whittle down a large number of predictors or at least to help prioritize them in the modeling process. Also, the process of feature engineering requires some level of expertise related to what is being modeled. It is difficult to make recommendations on how predictors should be represented in a vacuum or without knowing the context of the project. For example, in the simple example in Section 1.1, the inverse transformation for the predictors might have seemed obvious to an experienced practitioner.

1.2.7 | Big Data

The definition of Big Data is somewhat nebulous. Typically, this term implies a large number of data points (as opposed to variables) and it is worth noting that the *effective sample size* might be smaller than the actual data size. For example, if there is a severe class imbalance or rare event rate, the number of events in the data might be fairly small. Click-through rate on online ads is a good example of this.

Another example is when one particular region of the predictor space is abundantly sampled. Suppose a data set had billions of records but most correspond to white males within a certain age range. The number of distinct samples might be low, resulting in a data set that is not diverse.

One situation where large datasets probably doesn't help is when samples are added within the mainstream of the data. This simply increases the granularity of the distribution of the variables and, after a certain point, may not help in the data analysis. More rows of data can be helpful when new areas of the population are being accrued. In other words, big data does not necessarily mean *better data*.

While the benefits of big data have been widely espoused, there are some potential drawbacks. First, it simply might not solve problems being encountered in the analysis. Big data cannot automatically induce a relationship between the predictors and outcome when none exists. Second, there are often computational ramifications to having large amounts of data. Many high-variance/low-bias models tend to be very complex and computationally demanding and the time to fit these models can increase with data size and, in some cases, the increase can be nonlinear. Adding more data allows these models to more accurately reflect the complexity of the data but would require specialized solutions to be feasible. This, in itself, is not problematic unless the solutions have the effect of restricting the types of models that can be utilized. It is better for the problem to dictate the type of model that is needed.

Additionally, not all models can exploit large data volumes. For high-bias, low-variance models, big data tends to simply drive down the standard errors of the parameter estimates. For example, in a linear regression created on a million data records, doubling or tripling the amount of training data is unlikely to improve the parameter estimates to any practical degree (all other things being equal).

However, there are models that can effectively leverage large data sets. In some domains, there can be large amounts of *unlabeled* data where the outcome is unknown but the predictors have been measured or computed. The classic examples are images and text but unlabeled data can occur in other situations. For example, pharmaceutical companies have large databases of chemical compounds that have been designed but their important characteristics have not been measured (which can be expensive). Other examples include public governmental databases where there is an abundance of data that have not been connected to a specific outcome.

Unlabeled data can be used to solve some specific modeling problems. For models that require formal probability specifications, determining multivariate distributions can be extremely difficult. Copious amounts of predictor data can help estimate or specify these distributions. Autoencoders, discussed in Section 6.3.2, are models that can denoise or smooth the predictor values. The outcome is not required to create an autoencoder so unlabeled data can potentially improve the situation.

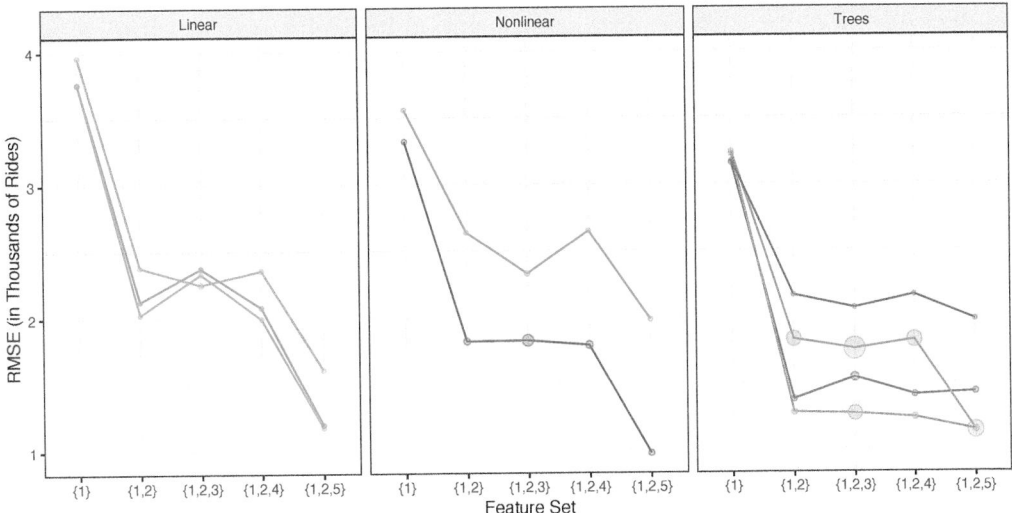

Figure 1.7: A series of model and feature set combinations for modeling train ridership in Chicago. The size of the points is proportional to the time required to tune and train the model.

Overall, when encountering (or being offered) large amounts of data, one might think to ask:

- What are you using it for? Does it solve some unmet need?
- Will it get in the way?

1.3 | A More Complex Example

To illustrate the interplay between models and features we present a more realistic example.[11] The case study discussed here involves predicting the ridership on Chicago "L" trains (i.e., the number of people entering a particular station on a daily basis). If a sufficiently predictive model can be built, the Chicago Transit Authority could use this model to appropriately staff trains and number of cars required per line. This data set is discussed in more detail in Section 4.1 but this section describes a series of models that were evaluated when the data were originally analyzed.

To begin, a simple set of four predictors was considered. These initial predictors, labeled as "Set 1", were developed because they are simple to calculate and visualizations showed strong relationships with ridership (the outcome). A variety of different models were evaluated and the root mean squared error (RMSE) was estimated using resampling methods. Figure 1.7 shows the results for several different types of models (e.g., tree-based models, linear models, etc.). RMSE values for the initial feature set ranged between 3151 and 3960 daily rides.[12] With the same feature set, tree-based models had the best performance while linear models had the worst

---

[11]This example will be analyzed at length in later chapters.

[12]A RMSE value of 3000 can correspond to R^2 values of between 0.80 and 0.90 in these data. However, as discussed in Section 3.2.1, R^2 can be misleading here due to the nature of these data.

results. Additionally, there is very little variation in RMSE results within a model type (i.e., the linear model results tend to be similar to each other).

In an effort to improve model performance, some time was spent deriving a second set of predictors that might be used to augment the original group of four. From this, 128 numeric predictors were identified that were lagged versions of the ridership at different stations. For example, to predict the ridership one week in the future, today's ridership would be used as a predictor (i.e., a seven-day lag). This second set of predictors had a beneficial effect overall but were especially helpful to linear models (see the x-axis value of {1, 2} in Figure 1.7). However, the benefit varied between models and model types.

Since the lag variables were important for predicting the outcome, more lag variables were created using lags between 8 and 14 days. Many of these variables show a strong correlation to the other predictors. However, models with predictor sets 1, 2, and 3 did not show much meaningful improvement above and beyond the previous set of models and, for some, the results were worse. One particular linear model suffered since this expanded set had a high degree of between-variable correlation. This situation is generally known as *multicollinearity* and can be particularly troubling for some models. Because this expanded group of lagged variables didn't now show much benefit overall, it was not considered further.

When brainstorming which predictors could be added next, it seemed reasonable to think that weather conditions might affect ridership. To evaluate this conjecture, a fourth set of 18 predictors was calculated and used in models with the first two sets (labeled as {1, 2, 4}). Like the third set, the weather did not show any relevance to predicting train ridership.

After conducting exploratory data analysis of residual plots associated with models with sets 1 and 2, a fifth set of 49 binary predictors were developed to address days where the current best models did poorly. These predictors resulted in a substantial drop in model error and were retained (see Figures 3.12 and 4.19). Note that the improvement affected models differently and that, with feature sets 1, 2, and 5, the simple linear models yielded results that are on par with more complex modeling techniques.

The overall points that should be understood from this demonstration are:

1. When modeling data, there is almost never a single model fit or feature set that will immediately solve the problem. The process is more likely to be a *campaign* of trial and error to achieve the best results.
2. The effect of feature sets *can* be much larger than the effect of different models.
3. The interplay between models and features is complex and somewhat unpredictable.
4. With the right set of predictors, is it common that many different types of models can achieve the same level of performance. Initially, the linear models had the worst performance but, in the end, showed some of the best performance.

Techniques for discovering, representing, adding, and subtracting are discussed in subsequent chapters.

1.4 | Feature Selection

In the previous example, new sets of features were derived sequentially to improve performance of the model. These sets were developed, added to the model, and then resampling was used to evaluate their utility. The new predictors were not prospectively filtered for statistical significance prior to adding them to the model. This would be a *supervised* procedure and care must be taken to make sure that overfitting is not occurring.

In that example, it was demonstrated that some of the predictors have enough underlying information to adequately predict the outcome (such as sets 1, 2, and 5). However, this collection of predictors might very well contain non-informative variables and this might impact performance to some extent. To whittle the predictor set to a smaller set that contains only the informative predictors, a supervised feature selection technique might be used. Additionally, there is the possibility that there are a small number of important predictors in sets 3 and 4 whose utility was not discovered because of all of the non-informative variables in these sets.

In other cases, all of the raw predictors are known and available at the beginning of the modeling process. In this case, a less sequential approach might be used by simply using a feature selection routine to attempt to sort out the best and worst predictors.

There are a number of different strategies for supervised feature selection that can be applied and these are discussed in Chapters 10 through 12. A distinction between search methods are how subsets are derived:

- *Wrapper* methods use an *external* search procedure to choose different subsets of the whole predictor set to evaluate in a model. This approach separates the feature search process from the model fitting process. Examples of this approach would be backwards or stepwise selection as well as genetic algorithms.
- *Embedded methods* are models where the feature selection procedure occurs naturally in the course of the model fitting process. Here an example would be a simple decision tree where variables are selected when the model uses them in a split. If a predictor is never used in a split, the prediction equation is functionally independent of this variable and it has been selected out.

As with model fitting, the main concern during feature selection is overfitting. This is especially true when wrapper methods are used and/or if the number of data points in the training set is small relative to the number of predictors.

Finally, *unsupervised* selection methods can have a very positive effect on model performance. Recall the Ames housing data. A property's neighborhood might be a useful predictor in the model. Since most models require predictors to be represented as numbers, it is common to encode such data as *dummy* or *indicator variables*. In this case, the single neighborhood predictor, with 28 possible values, is converted to a set of 27 binary variables that have a value of one when a property is in that neighborhood and zero otherwise. While this is a well-known and common approach, here it leads to cases where 2 of the neighborhoods have only one or two properties in these data, which is less than 1% of the overall set. With such a low frequency, such a predictor might have a detrimental effect on some models (such as linear regression) and removing them prior to building the model might be advisable.

When conducting a search for a subset of variables, it is important to realize that there may not be a unique set of predictors that will produce the best performance. There is often a compensatory effect where, when one seemingly important variable is removed, the model adjusts using the remaining variables. This is especially true when there is some degree of correlation between the explanatory variables or when a low-bias model is used. For this reason, feature selection should not be used as a formal method of determining feature *significance*. More traditional inferential statistical approaches are a better solution for appraising the contribution of a predictor to the underlying model or to the data set.

1.5   An Outline of the Book

The goal of this text is to provide effective tools for uncovering relevant and predictively useful representations of predictors. These tools will be the bookends of the predictive modeling process. At the beginning of the process we will explore techniques for augmenting the predictor set. Then at the end of the process we will provide methods for filtering the enhanced predictor set to ultimately produce better models. These concepts will be detailed in Chapters 2-12 as follows.

We begin by providing a short illustration of the interplay between the modeling and feature engineering process (Chapter 2). In this example, we use feature engineering and feature selection methods to improve the ability of a model to predict the risk of ischemic stroke.

In Chapter 3 we provide a review of the process for developing predictive models, which will include an illustration of the steps of data splitting, validation approach selection, model tuning, and performance estimation for future predictions. This chapter will also include guidance on how to use feedback loops when cycling through the model building process across multiple models.

Exploratory visualizations of the data are crucial for understanding relationships among predictors and between predictors and the response, especially for high-dimensional data. In addition, visualizations can be used to assist in understanding the nature of individual predictors including predictors' skewness and missing data patterns. Chapter 4 will illustrate useful visualization techniques to explore relationships within and between predictors. Graphical methods for evaluating model lack-of-fit will also be presented.

Chapter 5 will focus on approaches for encoding discrete, or categorical, predictors. Here we will summarize standard techniques for representing categorical predictors. Feature engineering methods for categorical predictors such as feature hashing are introduced as a method for using existing information to create new predictors that better uncover meaningful relationships. This chapter will also provide guidance on practical issues such as how to handle rare levels within a categorical predictor and the impact of creating dummy variables for tree and rule-based models. Date-based predictors are present in many data sets and can be viewed as categorical predictors.

Engineering numeric predictors will be discussed in Chapter 6. As mentioned above, numeric predictors as collected in the original data may not be optimal for predicting the response. Univariate and multivariate transformations are a first step to finding better forms of numeric predictors. A more advanced approach is to use basis expansions (i.e., splines) to create better representations of the original predictors. In certain situations, transforming continuous predictors to categorical or ordinal bins reduces variation and helps to improve predictive model performance. Caveats to binning numerical predictors will also be provided.

Up to this point in the book, a feature has been considered as one of the observed predictors in the data. In Chapter 7 we will illustrate that important features for a predictive model could also be the interaction between two or more of the original predictors. Quantitative tools for determining which predictors interact with one another will be explored along with graphical methods to evaluate the importance of these types of effects. This chapter will also discuss the concept of estimability of interactions.

Every practitioner working with real-world data will encounter missing data at some point. While some predictive models (e.g., trees) have novel ways of handling missing data, other models do not and require complete data. Chapter 8 explores mechanisms that cause missing data and provides visualization tools for investigating missing data patterns. Traditional and modern tools for removing or imputing missing data are provided. In addition, the imputation methods are evaluated for continuous and categorical predictors.

Working with profile data, such as time series (longitudinal), cellular-to-wellular, and image data will be addressed in Chapter 9. These kinds of data are normally collected in the fields of finance, pharmaceutical, intelligence, transportation, and weather forecasting, and this particular data structure generates unique challenges to many models. Some modern predictive modeling tools such as partial least squares can naturally handle data in this format. But many other powerful modeling techniques do not have direct ways of working with this kind of data. These models require that profile data be summarized or collapsed prior to modeling. This chapter will illustrate techniques for working with profile data in ways that strive to preserve the predictive information while creating a format that can be used across predictive models.

The feature engineering process as described in Chapters 5-9 can lead to many more predictors than what was contained in the original data. While some of the additional predictors will likely enhance model performance, not all of the original and new predictors will likely be useful for prediction. The final chapters will discuss feature selection as an overall strategy for improving model predictive performance. Important aspects include: the goals of feature selection, consequences of irrelevant predictors, comparisons with selection via regularization, and how to avoid overfitting (in the feature selection process).

1.6 | Computing

The website `http://bit.ly/fes-intro` contains R programs for reproducing these analyses.

2 | Illustrative Example: Predicting Risk of Ischemic Stroke

As a primer to feature engineering, an abbreviated example is presented with a modeling process similar to the one shown in Figure 1.4. For the purpose of illustration, this example will focus on exploration, analysis fit, and feature engineering, through the lens of a single model (logistic regression).

To illustrate the value of feature engineering for enhancing model performance, consider the application of trying to better predict patient risk for ischemic stroke (Schoepf et al., ress). Historically, the degree arterial stenosis (blockage) has been used to identify patients who are at risk for stroke (Lian et al., 2012). To reduce the risk of stroke, patients with sufficient blockage ($> 70\%$) are generally recommended for surgical intervention to remove the blockage (Levinson and Rodriguez, 1998). However, historical evidence suggests that the degree of blockage alone is actually a poor predictor of future stroke (Meier et al., 2010). This is likely due to the theory that while blockages may be of the same size, the composition of the plaque blockage is also relevant to the risk of stroke outcome. Plaques that are large, yet stable and unlikely to be disrupted, may pose less stroke risk than plaques that are smaller, yet less stable.

To study this hypothesis, a historical set of patients with a range of carotid artery blockages were selected. The data consisted of 126 patients, 44 of which had blockages greater than 70%. All patients had undergone Computed Tomography Angiography (CTA) to generate a detailed three-dimensional visualization and characterization of the blockage. These images were then analyzed by Elucid Bioimaging's vascuCAP (TM) software which generates anatomic structure estimates such as percent stenosis, arterial wall thickness, and tissue characteristics such as lipid-rich necrotic core and calcification. As an example, consider Figure 2.1 (a) which represents a carotid artery with severe stenosis as represented by the tiny tube-like opening running through the middle of the artery. Using the image, the software can calculate the maximal cross-sectional stenosis by area (MaxStenosisByArea) and by diameter (MaxStenosisByDiameter). In addition, The gray area in this figure represents

(a)   (b)   (c)

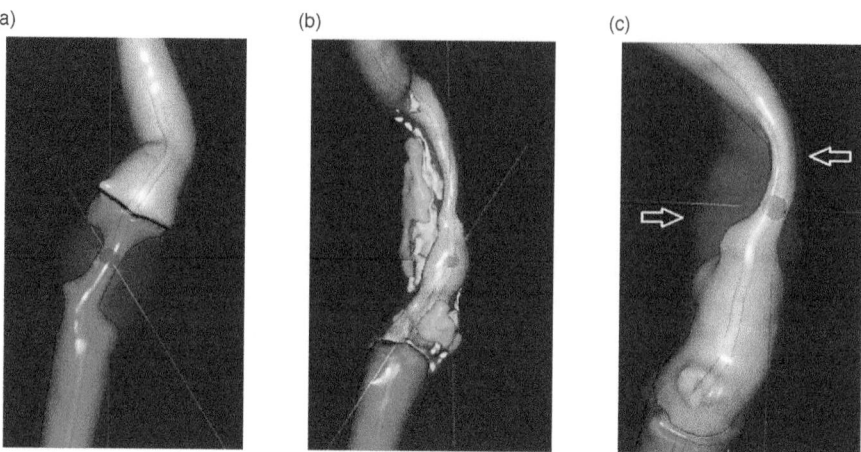

Figure 2.1: Three illustrations of the vascuCAP software applied to different carotid arteries and examples of the imaging predictors the software generates. (a) A carotid artery with severe stenosis, represented by the tiny opening through the middle. (b) A carotid artery with calcified plaque (green) and a lipid-rich necrotic core (yellow). (c) A carotid artery with severe stenosis and positive remodeling plaque.

macromolecules (such as collagen, elastin, glycoproteins, and proteoglycans) that provide structural support to the arterial wall. This structural region can be quantified by its area (MATXAREA). Figure 2.1 (b) illustrates an artery with severe stenosis and calcified plaque (green) and a lipid-rich necrotic core (yellow). Both plaque and lipid-rich necrotic core are thought to contribute to stroke risk, and these regions can be quantified by their volume (CALCVOL and LRNCVOL) and maximal cross-sectional area (MaxCALCAREA and MaxLRNCAREA). The artery presented in 2.1 (c) displays severe stenosis and outward arterial wall growth. The top arrow depicts the cross-section of greatest stenosis (MAXSTENOSISBYDIAMETER) and the bottom arrow depicts the cross-section of greatest positive wall remodeling (MAXREMODELINGRATIO). Remodeling ratio is a measure of the arterial wall where ratios less than 1 indicate a wall shrinkage and ratios greater than 1 indicate wall growth. This metric is likely important because coronary arteries with large ratios like the one displayed here have been associated with rupture (Cilla et al., 2013; Abdeldayem et al., 2015). The MAXREMODELINGRATIO metric captures the region of maximum ratio (between the two white arrows) in the three-dimensional artery image. A number of other imaging predictors are generated based on physiologically meaningful characterizations of the artery.

The group of patients in this study also had follow-up information on whether or not a stroke occurred at a subsequent point in time. The association between blockage categorization and stroke outcome is provided in Table 2.1. For these patients, the association is not statistically significant based on a chi-squared test of association (p = 0.42), indicating that blockage categorization alone is likely not a good predictor of stroke outcome.

Table 2.1: Association between blockage categorization and stroke outcome.

	Stroke=No	Stroke=Yes
Blockage < 70%	43	39
Blockage > 70%	19	25

If plaque characteristics are indeed important for assessing stroke risk, then measurements of the plaque characteristics provided by the imaging software could help to improve stroke prediction. By improving the ability to predict stroke, physicians may have more actionable data to make better patient management or clinical intervention decisions. Specifically, patients who have large (blockages > 70%), but stable plaques may not need surgical intervention, saving themselves from an invasive procedure while continuing on medical therapy. Alternatively, patients with smaller, but less stable plaques may indeed need surgical intervention or more aggressive medical therapy to reduce stroke risk.

The data for each patient also included commonly collected clinical characteristics for risk of stroke such as whether or not the patient had atrial fibrillation, coronary artery disease, and a history of smoking. Demographics of gender and age were included as well. These readily available risk factors can be thought of as another potentially useful predictor set that can be evaluated. In fact, this set of predictors should be evaluated first to assess their ability to predict stroke since these predictors are easy to collect, are acquired at patient presentation, and do not require an expensive imaging technique.

To assess each set's predictive ability, we will train models using the risk predictors, imaging predictors, and combinations of the two. We will also explore other representations of these features to extract beneficial predictive information.

2.1 Splitting

Before building these models, we will split the data into one set that will be used to develop models, preprocess the predictors, and explore relationships among the predictors and the response (the training set) and another that will be the final arbiter of the predictor set/model combination performance (the test set). To partition the data, the splitting of the original data set will be done in a *stratified* manner by making random splits in each of the outcome classes. This will keep the proportion of stroke patients approximately the same (Table 2.2). In the splitting, 70% of the data were allocated to the training set.

2.2 Preprocessing

One of the first steps of the modeling process is to understand important predictor characteristics such as their individual distributions, the degree of missingness within

Table 2.2: Distribution of stroke outcome by training and test split.

Data Set	Stroke = Yes (n)	Stroke = No (n)
Train	51% (45)	49% (44)
Test	51% (19)	49% (18)

each predictor, potentially unusual values within predictors, relationships between predictors, and the relationship between each predictor and the response and so on. Undoubtedly, as the number of predictors increases, our ability to carefully curate each individual predictor rapidly declines. But automated tools and visualizations are available that implement good practices for working through the initial exploration process such as Kuhn (2008) and Wickham and Grolemund (2016).

For these data, there were only 4 missing values across all subjects and predictors. Many models cannot tolerate any missing values. Therefore, we must take action to eliminate missingness to build a variety of models. Imputation techniques replace the missing values with a rational value, and these techniques are discussed in Chapter 8. Here we will replace each missing value with the median value of the predictor, which is a simple, unbiased approach and is adequate for a relatively small amount of missingness (but nowhere near optimal).

This data set is small enough to manually explore, and the univariate exploration of the imaging predictors uncovered many interesting characteristics. First, imaging predictors were mean centered and scaled to unit variance to enable direct visual comparisons. Second, many of the imaging predictors had distributions with long tails, also known as positively skewed distributions. As an example, consider the distribution of the maximal cross-sectional area (mm^2) of lipid-rich necrotic core (MaxLRNCArea, displayed in Figure 2.2a). MaxLRNCArea is a measurement of the mixture of lipid pools and necrotic cellular debris for a cross-section of the stenosis. Initially, we may believe that the skewness and a couple of unusually high measurements are due to a small subset of patients. We may be tempted to remove these unusual values out of fear that these will negatively impact a model's ability to identify predictive signal. While our intuition is correct for many models, skewness as illustrated here is often due to the underlying distribution of the data. The distribution, instead, is where we should focus our attention. A simple log-transformation, or more complex Box-Cox or Yeo-Johnson transformation (Section 6.1), can be used to place the data on a scale where the distribution is approximately symmetric, thus removing the appearance of outliers in the data (Figure 2.2b). This kind of transformation makes sense for measurements that increase exponentially. Here, the lipid area naturally grows multiplicatively by definition of how area is calculated.

Next, we will remove predictors that are highly correlated ($R^2 > 0.9$) with other predictors. The correlation among the imaging predictors can be seen in the heatmap in Figure 2.3 where the order of the columns and rows are determined by a clustering

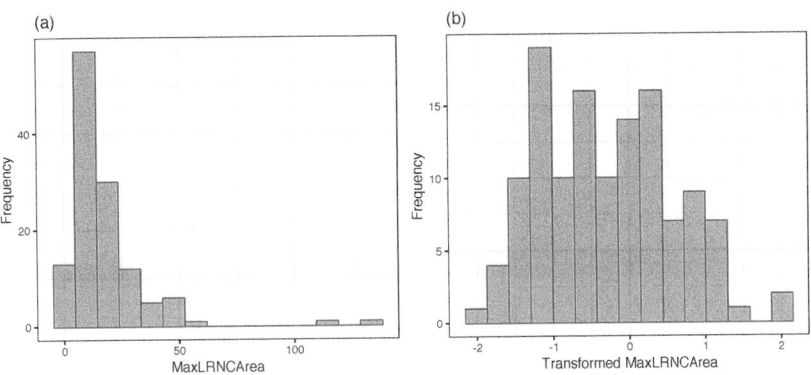

Figure 2.2: (a) Distribution of maximal cross-sectional area of lipid-rich necrotic core. (b) The distribution of the Yeo-Johnson transformed maximal cross-sectional area of lipid-rich necrotic core.

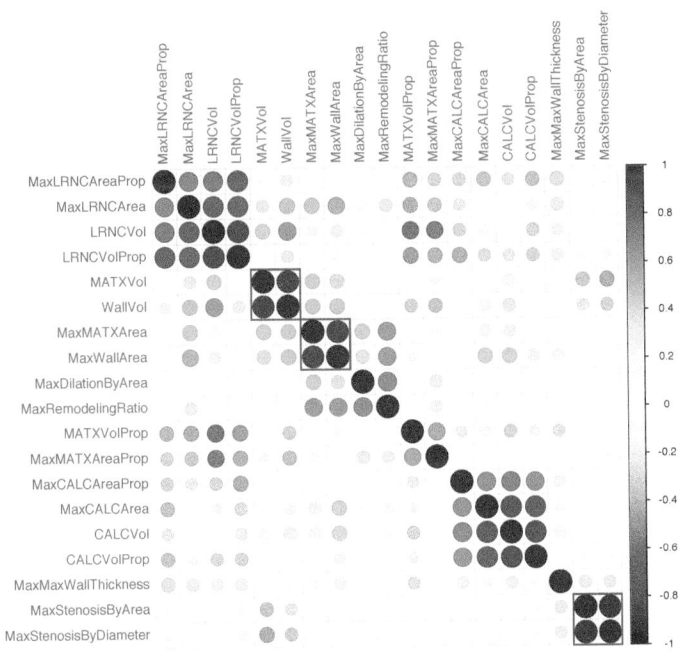

Figure 2.3: Visualization of the imaging predictor correlation matrix.

algorithm. Here, there are three pairs of predictors that show unacceptably high correlations:

- vessel wall volume in mm^3 (WALLVOL) and matrix volume (MATXVOL),

- maximum cross-sectional wall area in mm^2 (MAXWALLAREA) and maximum matrix area (MAXMATXAREA)

- maximum cross-sectional stenosis based on area (MAXSTENOSISBYAREA), and maximum cross-sectional stenosis based on diameter (MAXSTENOSISBYDIAMETER).

These three pairs are highlighted in red boxes along the diagonal of the corrleation matrix in Figure 2.3. It is easy to understand why this third pair has high correlation, since the calculation for area is a function of diameter. Hence, we only need one of these representations for modeling. While only 3 predictors cross the high correlation threshold, there are several other pockets of predictors that approach the threshold. For example, calcified volume (CALCVOL) and maximum cross-sectional calcified area in mm^2 (MAXCALCAREA) (r = 0.87), and maximal cross-sectional area of lipid-rich necrotic core (MAXLRNCAREA) and volume of lipid-rich necrotic core (LRNCVOL) ($r = 0.8$) have moderately strong positive correlations but do not cross the threshold. These can be seen in a large block of blue points along the diagonal. The correlation threshold is arbitrary and may need to be raised or lowered depending on the problem and the models to be used. Chapter 3 contains more details on this approach.

2.3   Exploration

The next step is to explore potential predictive relationships between individual predictors and the response and between pairs of predictors and the response.

However, before proceeding to examine the data, a mechanism is needed to make sure that any trends seen in this small data set are not over-interpreted. For this reason, a *resampling* technique will be used. A resampling scheme described in Section 3.4 called repeated 10-fold cross-validation is used. Here, 50 variations of the training set are created and, whenever an analysis is conducted on the training set data, it will be evaluated on each of these resamples. While not infallible, it does offer some protection against overfitting.

For example, one initial question is, which of the predictors have simple associations with the outcome? Traditionally, a statistical hypothesis test would be generated on the entire data set and the predictors that show statistical significance would then be slated for inclusion in a model. The problem with this approach is that it uses all of the data to make this determination and these same data points will be used in the eventual model. The potential for overfitting is strong because of the wholesale reuse of the entire data set for two purposes. Additionally, if predictive performance is the focus, the results of a formal hypothesis test may not be aligned with predictivity.

As an alternative, when we want to compare two models (M_1 and M_2), the following procedure, discussed more in Section 3.7, will be used:

Algorithm 2.1: Model comparisons using resampling-based performance metrics.

1 **for** *each resample* **do**
2 | Use the resample's 90% to fit models M_1 and M_2
3 | Predict the remaining 10% for both models
4 | Compute the area under the ROC curve for M_1 and M_2
5 | Determine the difference in the two AUC values
6 **end**
7 Use a one sided t-test on the differences to test that M_2 is better than M_1.

The 90% and 10% numbers are not always used and are a consequence of using 10-fold cross-validation (Section 3.4.1).

Using this algorithm, the potential improvements in the second model are evaluated over many versions of the training set and the improvements are quantified in a relevant metric.[13]

To illustrate this algorithm, two logistic regression models were considered. The simple model, analogous to the statistical "null model," contains only an intercept term, while the more complex model has a single term for an individual predictor from the risk set. These results are presented in Table 2.3 and order the risk predictors from most significant to least significant in terms of improvement in ROC. For the training data, several predictors provide marginal, but significant improvement in predicting stroke outcome as measured by the improvement in ROC. Based on these results, our intuition would lead us to believe that the significant risk set predictors will likely be integral to the final predictive model.

Similarly, relationships between the continuous, imaging predictors and stroke outcome can be explored. As with the risk predictors, the predictive performance of the intercept-only logistic regression model is compared to the model with each of the imaging predictors. Figure 2.4 presents the scatter of each predictor by stroke outcome with the p-value (top center) of the test to compare the improvement in ROC between the null model and model with each predictor. For these data, the thickest wall across all cross-sections of the target (MaxMaxWallThickness), and the maximum cross-sectional wall remodeling ratio (MaxRemodelingRatio) have the strongest associations with stroke outcome. Let's consider the results for MaxRemodelingRatio, which indicates that there is a significant shift in the average value between the stroke categories. The scatter of the distributions of this predictor between stroke categories still has considerable overlap. To get a sense of how well

---

[13]As noted at the end of this chapter, as well as in Section 7.3, there is a deficiency in the application of this method.

Table 2.3: Improvement in ROC over null model of the risk set for predicting stroke outcome. The null model AUC values were approximately 0.50.

Predictor	Improvement	Pvalue	ROC
CoronaryArteryDisease	0.079	0.0003	0.579
DiabetesHistory	0.066	0.0003	0.566
HypertensionHistory	0.065	0.0004	0.565
Age	0.083	0.0011	0.583
AtrialFibrillation	0.044	0.0013	0.544
SmokingHistory	-0.010	0.6521	0.490
Sex	-0.034	0.9288	0.466
HypercholesterolemiaHistory	-0.102	1.0000	0.398

MAXREMODELINGRATIO separates patients by stroke category, the ROC curve for this predictor for the training set data is created (Figure 2.5). The curve indicates that there is some signal for this predictor, but the signal may not be sufficient for using as a prognostic tool.

At this point, one could move straight to tuning predictive models on the preprocessed and filtered risk and/or the imaging predictors for the training set to see how well the predictors together identify stroke outcome. This often is the next step that practitioners take to understand and quantify model performance. However, there are more exploratory steps can be taken to identify other relevant and useful constructions of predictors that improve a model's ability to predict. In this case, the stroke data in its original form does not contain direct representations of *interactions* between predictors (Chapter 7). Pairwise interactions between predictors are prime candidates for exploration and may contain valuable predictive relationships with the response.

For data that has a small number of predictors, all pair-wise interactions can be created. For numeric predictors, the interactions are simply generated by multiplying the values of each predictor. These new terms can be added to the data and passed in as predictors to a model. This approach can be practically challenging as the number of predictors increases, and other approaches to sift through the vast number of possible pairwise interactions might be required (see Chapter 7 for details). For this example, an interaction term was created for every possible pair of original imaging predictors (171 potentially new terms). For each interaction term, the same resampling algorithm was used to quantify the cross-validated ROC from a model with only the two main effects and a model with the main effects and the interaction term. The improvement in ROC as well as a p-value of the interaction model versus the main effects model was calculated. Figure 2.6 displays the relationship between the improvement in ROC due to the interaction term (on the x-axis) and the negative log10 p-value of the improvement (y-axis where larger is more significant). The size of the points symbolizes the baseline area under the ROC curve from the main effects

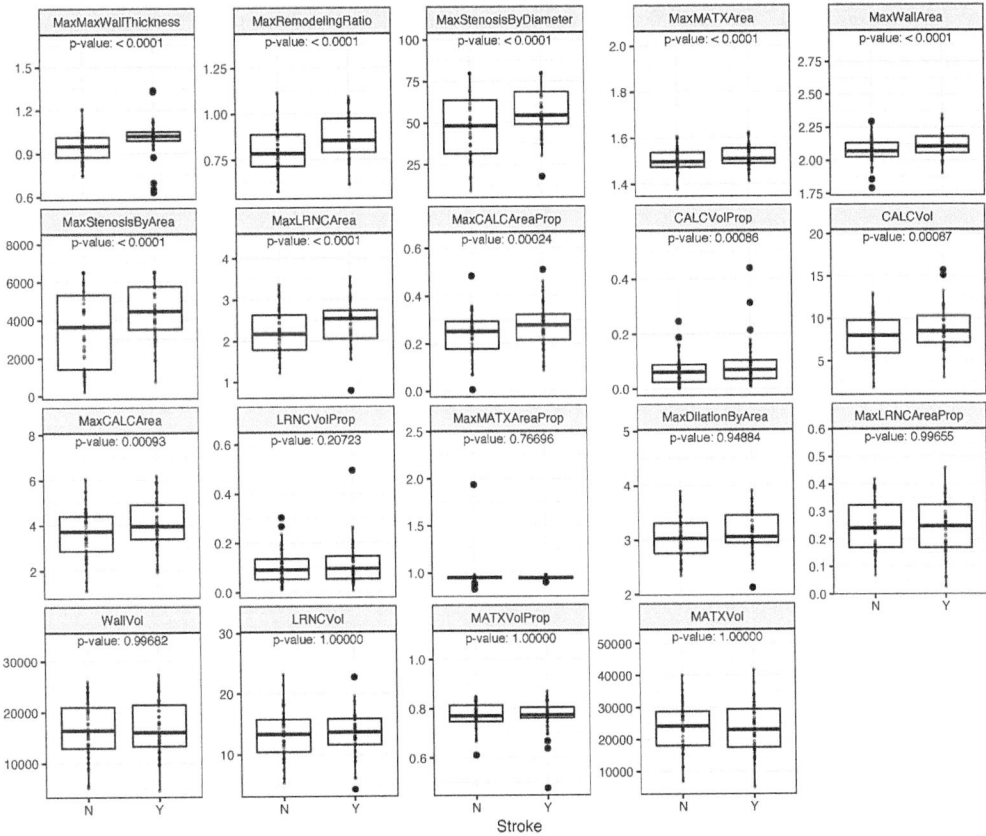

Figure 2.4: Univariate associations between imaging predictors and stroke outcome. The p-value of the improvement in ROC for each predictor over the intercept-only logistic regression model is listed in the top center of each facet.

models. Points that have relatively small symbols indicate whether the improvement was to a model that had already showed good performance.

Of the 171 interaction terms, the filtering process indicated that 18 provided an improvement over the main effects alone when filtered on p-values less than 0.2. Figure 2.7 illustrates the relationship between one of these interaction terms. Panel (a) is a scatter plot of the two predictors, where the training set data are colored by stroke outcome. The contour lines represent equivalent product values between the two predictors, which helps to highlight the characteristic that patients that did not have a stroke outcome generally have lower product values of these two predictors. Alternatively, patients with a stroke outcome generally have higher product outcomes. Practically, this means that a significant blockage combined with outward wall growth in the vessel increases risk for stroke. The boxplot of this interaction in (b) demonstrates that the separation between stroke outcome categories is stronger than for either predictor alone.

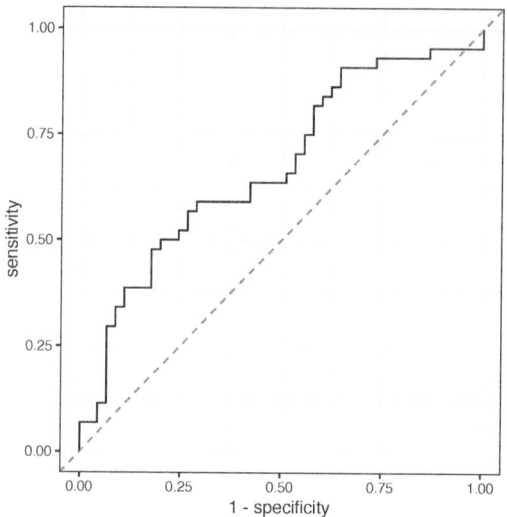

Figure 2.5: ROC curve for the maximum cross-sectional wall remodeling ratio (MaxRemodelingRatio) as a standalone predictor of stroke outcome based on the training data.

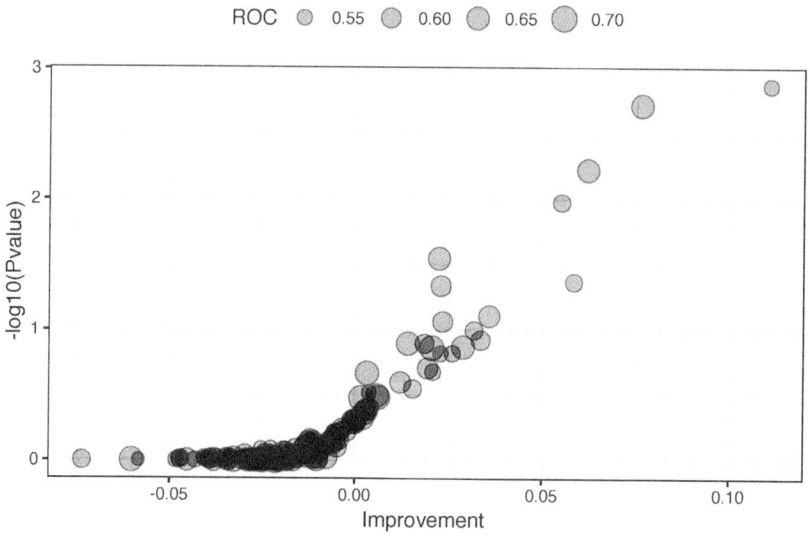

Figure 2.6: A comparison of the improvement in area under the ROC curve to the p-value of the importance of the interaction term for the imaging predictors.

2.4   Predictive Modeling across Sets

At this point there are at least five progressive combinations of predictor sets that could be explored for their predictive ability: original risk set alone, imaging predictors alone, risk and imaging predictors together, imaging predictors and interactions

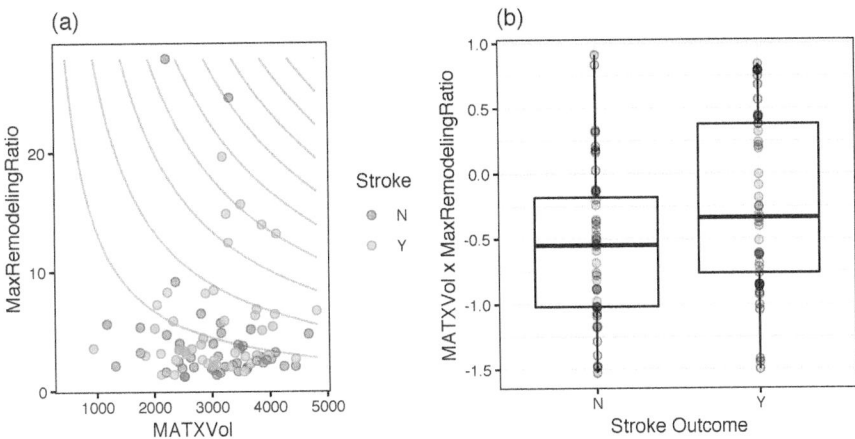

Figure 2.7: (a) A bivariate scatter plot of the two predictors (MATXVol and MaxRemodelingRatio) on the original scales of measurement. (b) The distribution of the multiplicative interaction between stroke outcome categories after preprocessing.

of imaging predictors, and risk, imaging predictors, and interactions of imaging predictors. We'll consider modeling several of these sets of data in turn.

Physicians have a strong preference towards logistic regression due to its inherent interpretability. However, it is well known that logistic regression is a high-bias, low-variance model which has a tendency to yield lower predictivity than other low-bias, high-variance models. The predictive performance of logistic regression is also degraded by the inclusion of correlated, non-informative predictors. In order to find the most predictive logistic regression model, the most relevant predictors should be identified to find the best subset for predicting stroke risk.

With these specific issues in mind for these data, a recursive feature elimination (RFE) routine (Chapters 10 and 11) was used to determine if fewer predictors would be advantageous. RFE is a simple backwards selection procedure where the largest model is used initially and, from this model, each predictor is ranked in importance. For logistic regression, there are several methods for determining importance, and we will use the simple absolute value of the regression coefficient for each model term (after the predictors have been centered and scaled). The RFE procedure begins to remove the least important predictors, refits the model, and evaluates performance. At each model fit, the predictors are preprocessed by an initial Yeo-Johnson transformation as well as centering and scaling.

As will be discussed in later chapters, correlation between the predictors can cause instability in the logistic regression coefficients. While there are more sophisticated approaches, an additional variable filter will be used on the data to remove the minimum set of predictors such that no pairwise correlations between predictors are greater than 0.75. The data preprocessing will be conducted with and without this step to show the potential effects on the feature selection procedure.

Our previous resampling scheme was used in conjunction with the RFE process. This means that the backwards selection was performed 50 different times on 90% of the training set and the remaining 10% was used to evaluate the effect of removing the predictors. The optimal subset size is determined using these results and the final RFE execution is one the entire training set and stops at the optimal size. Again, the goal of this resampling process is to reduce the risk of overfitting in this small data set. Additionally, all of the preprocessing steps are conducted within these resampling steps so that there is the potential that the correlation filter may select different variables for each resample. This is deliberate and is the only way to measure the effects and variability of the preprocessing steps on the modeling process.

The RFE procedure was applied to:

- The small risk set of 8 predictors. Because this is not a large set, all 28 pairwise interactions are generated. When the correlation filter is applied, the number of model terms might be substantially reduced.

- The set of 19 imaging predictors. The interaction effects derived earlier in the chapter are also considered for these data.

- The entire set of predictors. The imaging interactions are also combined with these variables.

Figure 2.8 shows the results. When only main effects are considered for the risk set, the full set of 8 predictors is favored. When the full set of 28 pairwise interactions are added, model performance was hurt by the extra interactions. Based on resampling, a set of 13 predictors was optimal (11 of which were interactions). When a correlation filter was applied, the main effect model was unaffected while the interaction model has, at most, 18 predictors. Overall, the filter did not help this predictor set.

For the imaging predictor set, the data set preferred a model with none of the previously discovered interactions and a correlation filter. This may seem counter-intuitive, but understand that the interactions were discovered in the absence of other predictors or interactions. The non-interaction terms appear to compensate or replace the information supplied by the most important interaction terms. A reasonable hypothesis as to why the correlation filter improved these predictors is that they tend to have higher correlations with one another (compared to the risker predictors). The best model so far is based on the filtered set of 7 imaging main effects.

When combining the two predictor sets, model performance without a correlation filter was middle-of-the-road and there was no real difference between the interaction and main effects models. Once the filter was applied, the data strongly favored the main effects model (with all 10 predictors that survived the correlation filter).

Using this training set, we estimated that the filtered predictor set of 7 imaging predictors was our best bet. The final predictor set was MaxLRNCArea, MaxLRNCAreaProp, MaxMaxWallThickness, MaxRemodelingRatio, MaxCALCAreaProp,

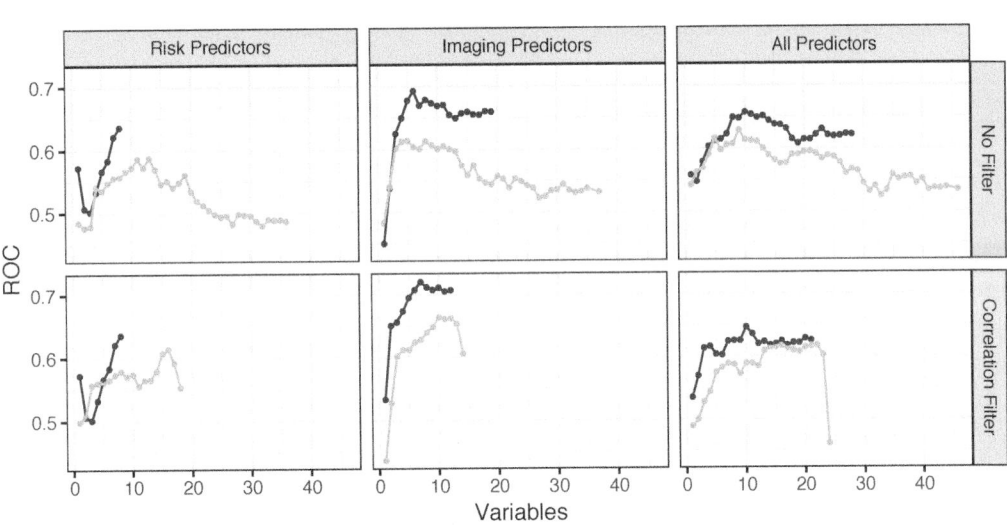

Figure 2.8: Recursive feature elimination results under various conditions.

MAXSTENOSISBYAREA, and MAXMATXAREA. To understand the variation in the selection process, Table 2.4 shows the frequency of the predictors that were selected in the 7 variable models across all 50 resamples. The selection results were fairly consistent especially for a training set this small.

Table 2.4: Consistency of RFE selection over resamples.

Predictor	Number of Times Selected	In Final Model?
MAXLRNCAREAPROP	49	Yes
MAXMAXWALLTHICKNESS	49	Yes
MAXCALCAREAPROP	47	Yes
MAXREMODELINGRATIO	45	Yes
MAXSTENOSISBYAREA	41	Yes
MAXLRNCAREA	37	Yes
MAXMATXAREA	24	Yes
MATXVOLPROP	18	No
CALCVOLPROP	12	No
MAXCALCAREA	7	No
MAXMATXAREAPROP	7	No
MAXDILATIONBYAREA	6	No
MATXVOL	4	No
CALCVOL	2	No
LRNCVOL	2	No

How well did this predictor set do on the test set? The test set area under the ROC curve was estimated to be 0.69. This is less than the resampled estimate of 0.72 but is greater than the estimated 90% lower bound on this number (0.674).

2.5 | Other Considerations

The approach presented here is not the only approach that could have been taken with these data. For example, if logistic regression is the model being evaluated, the `glmnet` model (Hastie et al., 2015) is a model that incorporates feature selection into the logistic regression fitting process.[14] Also, you may be wondering why we chose to preprocess only the imaging predictors, why we did not explore interactions among risk predictors or between risk predictors and imaging predictors, why we constructed interaction terms on the original predictors and not on the preprocessed predictors, or why we did not employ a different modeling technique or feature selection routine. And you would be right to ask these questions. In fact, there may be a different preprocessing approach, a different combination of predictors, or a different modeling technique that would lead to a better predictivity. Our primary point in this short tour is to illustrate that spending a little more time (and sometimes a lot more time) investigating predictors and relationships among predictors can help to improve model predictivity. This is especially true when marginal gains in predictive performance can have significant benefits.

2.6 | Computing

The website `http://bit.ly/fes-stroke` contains R programs for reproducing these analyses.

---

[14]This model will be discussed more in Section 7.3 and Chapter 10.

3 | A Review of the Predictive Modeling Process

Before diving into specific methodologies and techniques for modeling, there are necessary topics that should first be discussed and defined. These topics are fairly general with regard to empirical modeling and include metrics for measuring performance for regression and classification problems, approaches for optimal data usage (including data splitting and resampling), best practices for model tuning, and recommendations for comparing model performance.

There are two data sets used in this chapter to illustrate the techniques. First is the Ames housing price data first introduced in Chapter 1. The second data set focuses on the classification of a person's profession based on the information from an online dating site. These data are discussed in the next section.

3.1 | Illustrative Example: OkCupid Profile Data

OkCupid is an online dating site that serves international users. Kim and Escobedo-Land (2015) describe a data set where over 50,000 profiles from the San Francisco area were made available[15] and the data can be found in a `GitHub` repository.[16] The data contain several types of variables:

- open text essays related to an individual's interests and personal descriptions,
- single-choice type fields such as profession, diet, and education, and
- multiple-choice fields such as languages spoken and fluency in programming languages.

In their original form, almost all of the raw data fields are discrete in nature; only age and the last time since login were numeric. The categorical predictors were converted to dummy variables (Chapter 5) prior to fitting models. For the analyses of these data in this chapter, the open text data will be ignored but will be probed later

---

[15]While there have been instances where online dating information has been obtained without authorization, these data were made available with permission from OkCupid president and co-founder Christian Rudder. For more information, see the original publication. In these data, no user names or images were made available.

[16]github.com/rudeboybert/JSE_OkCupid

(Section 5.6). Of the 212 predictors that were used, there were clusters of variables for geographic location (i.e., town, $p = 3$), religious affiliation ($p = 13$), astrological sign ($p = 15$), children ($p = 15$), pets ($p = 15$), income ($p = 12$), education ($p = 32$), diet ($p = 17$), and over 50 variables related to spoken languages. For more information on this data set and how it was processed, see the book's GitHub repository.

For this demonstration, the goal will be to predict whether a person's profession is in the STEM fields (science, technology, engineering, and math). There is a moderate class imbalance in these data; only 18.5% of profiles work in these areas. While the imbalance has a significant impact on the analysis, the illustration presented here will mostly side-step this issue by *down-sampling* the instances such that the number of profiles in each class are equal. See Chapter 16 of Kuhn and Johnson (2013) for a detailed description of techniques for dealing with infrequent classes.

3.2 | Measuring Performance

While often overlooked, the metric used to assess the effectiveness of a model to predict the outcome is very important and can influence the conclusions. The metric we select to evaluate model performance depends on the outcome, and the subsections below describe the main statistics that are used.

3.2.1 | Regression Metrics

When the outcome is a number, the most common metric is the root mean squared error (RMSE). To calculate this value, a model is built and then it is used to predict the outcome. The *residuals* are the difference between the observed outcome and predicted outcome values. To get the RMSE for a model, the average of the squared residuals is computed, then the square root of this value is taken. Taking the square root puts the metric back into the original measurement units. We can think of RMSE as the average distance of a sample from its observed value to its predicted value. Simply put, the lower the RMSE, the better a model can predict samples' outcomes.

Another popular metric is the coefficient of determination, usually known as R^2. There are several formulas for computing this value (Kvalseth, 1985), but the most conceptually simple one finds the standard correlation between the observed and predicted values (a.k.a. R) and squares it. The benefit of this statistic is, for linear models, that it has a straightforward interpretation: R^2 is the proportion of the total variability in the outcome that can be explained by the model. A value near 1.0 indicates an almost perfect fit while values near zero result from a model where the predictions have no linear association with the outcome. One other advantage of this number is that it makes comparisons between different outcomes easy since it is unitless.

Unfortunately, R^2 can be a deceiving metric. The main problem is that it is a measure of correlation and not accuracy. When assessing the predictive ability of a model, we need to know how well the observed and predicted values *agree*. It is possible, and not unusual, that a model could produce predicted values that have a strong linear relationship with the observed values but the predicted values do *not* conform to the 45-degree line of agreement (where the observed and predicted values are equal). One example of this phenomenon occurs when a model under-predicts at one extreme of the outcome and overpredicts at the other extreme of the outcome. Tree-based ensemble methods (e.g., random forest, boosted trees, etc.) are notorious for these kinds of predictions. A second problem with using R^2 as a performance metric is that it can show very optimistic results when the outcome has large variance. Finally, R^2 can be misleading if there are a handful of outcome values that are far away from the overall scatter of the observed and predicted values. In this case the handful of points can artificially increase R^2.

To illustrate the problems with R^2, let's look at the results of one particular model of the Chicago train ridership data. For this model R^2 was estimated to be 0.9; at face value we may conclude that this is an extremely good model. However, the high value is mostly due to the inherent nature of the ridership numbers which are high during the workweek and correspondingly low on the weekends. The bimodal nature of the outcome inflates the outcome variance and, in turn, the R^2. We can see the impacts of the bi-modal outcome in Figure 3.1 (a). Part (b) of the figure displays a histogram of the residuals, some of which are greater than 10K rides. The RMSE for this model is 3,853 rides, which is somewhat large relative to the observed ridership values.

A second illustration of the problem of using R^2 can be seen by examining the blue and black lines in Figure 3.1(a). The blue line is the linear regression fit between the observed and predicted values, while the black line represents the line of agreement. Here we can see that the model under-predicts the smaller observed values (left) and over-predicts the larger observed values (right). In this case, the offset is not huge but it does illustrate how the RMSE and R^2 metrics can produce discordant results. For these reasons, we advise using RMSE instead of R^2.

To address the problem that the correlation coefficient is overly optimistic when the data illustrates correlation but not agreement, Lawrence and Lin (1989) developed the concordance correlation coefficient (CCC). This metric provides a measure of correlation relative to the line of agreement and is defined as the product of the usual correlation coefficient and a measure of bias from the line of agreement. The bias coefficient ranges from 0 to 1, where a value of 1 indicates that the data falls on the line of agreement. The further the data deviates from the line of agreement, the smaller the bias coefficient. Therefore, the CCC can be thought of as a penalized version of the correlation coefficient. The penalty will apply if the data exhibits poor correlation between the observed and predicted values or if the relationship between the observed and predicted values is far from the line of agreement.

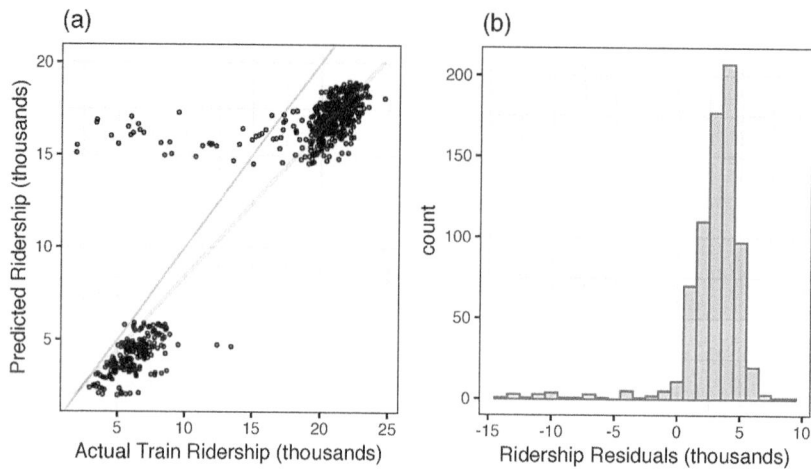

Figure 3.1: Observed and predicted values of Chicago train ridership (in thousands, panel (a)) along with a histogram of the corresponding model residuals (observed minus predicted, panel (b). In panel (a), the blue line corresponds to the linear regression line between the observed and predicted values, while the black line represents the line of agreement.

Both RMSE and R^2 are very sensitive to extreme values because each is based on the squared value of the individual sample's residuals. Therefore, a sample with a large residual will have an inordinately large effect on the resulting summary measure. In general, this type of situation makes the model performance metric appear worse that what it would be without the sample. Depending on the problem at hand, this vulnerability is not necessarily a vice but could be a virtue. For example, if the goal of the modeling problem is to rank-order new data points (e.g., the highest spending customers), then the size of the residual is not an issue so long as the most extreme values are predicted to be the most extreme. However, it is more often the case that we are interested in predicting the actual response value rather than just the rank. In this case, we need metrics that are not skewed by one or just a handful of extreme values. The field of robustness was developed to study the effects of extreme values (i.e., outliers) on commonly used statistical metrics and to derive alternative metrics that achieved the same purpose but were *insensitive* to the impact of outliers (Hampel et al., 1972). As a broad description, robust techniques seek to find numerical summaries for the majority of the data. To lessen the impact of extreme values, robust approaches down-weight the extreme samples or they transform the original values in a way that brings the extreme samples closer to the majority of the data. Rank-ordering the samples is one type of transformation that reduces the impact of extreme values. In the hypothetical case of predicting customers' spending, *rank correlation* might be a better choice of metric for the model since it measures how well the predictions rank order with their true values. This statistic computes the ranks of the data (e.g., 1, 2, etc.) and computes the standard correlation statistic from these values. Other robust measures for regression

Table 3.1: A confusion matrix for an OkCupid model. The columns are the true classes and the rows correspond to the predictions.

	stem	other
stem	5134	6385
other	2033	25257

are the median absolute deviation (MAD) (Rousseeuw and Croux, 1993) and the absolute error.

3.2.2 │ Classification Metrics

When the outcome is a discrete set of values (i.e., qualitative data), there are two different types of performance metrics that can be utilized. The first type described below is based on qualitative class prediction (e.g., stem or other) while the second type uses the predicted class probabilities to measure model effectiveness (e.g., $Pr[\texttt{stem}] = 0.254$).

Given a set of predicted classes, the first step in understanding how well the model is working is to create a *confusion matrix*, which is a simple cross-tabulation of the observed and predicted classes. For the OkCupid data, a simple logistic regression model was built using the predictor set mentioned above and Table 3.1 shows the resulting confusion matrix.[17]

The samples that were correctly predicted sit on the diagonal of the table. The STEM profiles mistakenly predicted as non-STEM are shown in the bottom left of the table ($n = 2033$), while the non-STEM profiles that were erroneously predicted are in the upper right cell ($n = 6385$). The most widely utilized metric is classification accuracy, which is simply the proportion of the outcome that was correctly predicted. In this example, the accuracy is $0.78 = (5134 + 25257)/(5134 + 6385 + 2033 + 25257)$. There is an implicit tendency to assess model performance by comparing the observed accuracy value to $1/C$, where C is the number of classes. In this case, 0.78 is much greater than 0.5. However, this comparison should be made only when there are nearly the same number of samples in each class. When there is an imbalance between the classes, as there is in these data, accuracy can be a quite deceiving measure of model performance since a value of 0.82 can be achieved by predicting all profiles as non-STEM.

As an alternative to accuracy, another statistic called Cohen's Kappa (Agresti, 2012) can be used to account for class imbalances. This metric normalizes the error rate to what would be expected by chance. Kappa takes on values between -1 and 1 where a value of 1 indicates complete concordance between the observed and predicted

---

[17]Note that these values were *not* obtained by simply re-predicting the data set. The values in this table are the set of "assessment" sets generated during the cross-validation procedure defined in Section 3.4.1.

values (and thus perfect accuracy). A value of -1 is complete discordance and is rarely seen.[18] Values near zero indicate that there is no relationship between the model predictions and the true results. The Kappa statistic can also be generalized to problems that have more than two groups.

A visualization technique that can be used for confusion matrices is the mosaic plot (see Figure 3.3). In these plots, each cell of the table is represented as a rectangle whose area is proportional to the number of values in the cell. These plots can be rendered in a number of different ways and for tables of many sizes. See Friendly and Meyer (2015) for more examples.

There are also specialized sets of classification metrics when the outcome has two classes. To use them, one of the class values must be designated as the *event of interest*. This is somewhat subjective. In some cases, this value might be the worst-case scenario (i.e., death) but the designated event should be the value that one is most interested in predicting.

The first paradigm of classification metrics focuses on false positives and false negatives and is most useful when there is interest in comparing the two types of errors. The *sensitivity* metric is simply the proportion of the events that were predicted correctly and is the *true positive* rate in the data. For our example,

$$sensitivity = \frac{\#\ \text{truly STEM predicted correctly}}{\#\ \text{truly STEM}}$$
$$= 5134/7167 = 0.716.$$

The *false positive* rate is associated with the *specificity*, which is

$$specificity = \frac{\#\ \text{truly non-STEM predicted correctly}}{\#\ \text{truly non-STEM}}$$
$$= 25257/31642 = 0.798.$$

The false positive rate is 1 - specificity (0.202 in this example).

The other paradigm for the two-class system is rooted in the field of *information retrieval* where the goal is to find the events. In this case, the metrics commonly used are *precision* and *recall*. Recall is equivalent to sensitivity and focuses on the number of true events found by the model. Precision is the proportion of events that are predicted correctly out of the total number of predicted events, or

$$precision = \frac{\#\ \text{truly STEM predicted correctly}}{\#\ \text{predicted STEM}}$$
$$= 5134/11519 = 0.446.$$

---

[18]Values close to -1 are rarely seen in predictive modeling since the models are seeking to find predicted values that are similar to the observed values. We have found that a predictive model that has difficulty finding a relationship between the predictors and the response has a Kappa value slightly below or near 0.

One facet of sensitivity, specificity, and precision that is worth understanding is that they are *conditional* statistics. For example, sensitivity reflects the probability that an event is correctly predicted *given that a sample is truly an event*. The latter part of this sentence shows the conditional nature of the metric. Of course, the true class is usually unknown and, if it were known, a model would not be needed. In any case, if Y denotes the true class and P denotes the prediction, we could write sensitivity as $\Pr[P = \text{STEM} \mid Y = \text{STEM}]$.

The question that one really wants to know is: "if my value was predicted to be an event, what are the chances that it is truly is an event?" or $\Pr[Y = \text{STEM} \mid P = \text{STEM}]$. Thankfully, the field of Bayesian analysis (McElreath, 2015) has an answer to this question. In this context, Bayes' Rule states that

$$Pr[Y|P] = \frac{Pr[Y] \times Pr[P|Y]}{Pr[P]} = \frac{Prior \times Likelihood}{Evidence}.$$

Sensitivity (or specificity, depending on one's point of view) is the "likelihood" part of this equation. The *prior probability*, or *prevalence*, is the overall rate that we see events in the wild (which may be different from what was observed in our training set). Usually, one would specify the overall event rate before data are collected and use it in the computations to determine the unconditional statistics. For sensitivity, its unconditional analog is called the *positive predictive value* (PPV):

$$PPV = \frac{sensitivity \times prevalence}{(sensitivity \times prevalence) + ((1 - specificity) \times (1 - prevalence))}.$$

The *negative predictive value* (NPV) is the analog to specificity and can be computed as

$$NPV = \frac{specificity \times (1 - prevalence)}{((1 - sensitivity) \times prevalence) + (specificity \times (1 - prevalence))}.$$

See Altman and Bland (1994b) for a clear and concise discussion of these measures. Also, simplified versions of these formulas are often shown for these statistics that assume the prevalence to be 0.50. These formulas, while correct when prevalence is 0.50, can produce very misleading results if the prevalence is different from this number.

For the OkCupid data, the difference in the sensitivity and PPV are:

- *sensitivity*: if the profile is truly STEM, what is the probability that it is correctly predicted?
- *PPV*: if the profile was predicted as STEM, what is the probability that it is STEM?

The positive and negative predictive values are not often used to measure performance. This is partly due to the nature of the prevalence. If the outcome is not well

Table 3.2: A comparison of typical probability-based measures used for classification models. The calculations presented here assume that Class 1 is the true class.

	Class 1	Class 2	Log-Likelihood	Gini	Entropy
Equivocal Model	0.5	0.5	-0.693	0.25	1.000
Good Model	0.8	0.2	-0.223	0.16	0.722
Bad Model	0.2	0.8	-1.609	0.16	0.722

understood, it is very difficult to provide a value (even when asking experts). When there is a sufficient amount of data, the prevalence is typically estimated by the proportion of the outcome data that correspond to the event of interest. Also, in other situations, the prevalence may depend on certain factors. For example, the proportion of STEM profiles in the San Francisco area can be estimated from the training set to be 0.18. Using this value as the prevalence, our estimates are PPV = 0.45 and NPV = 0.93. The PPV is significantly smaller than the sensitivity due to the model missing almost 28% of the true STEM profiles and the fact that the overall likelihood of being in the STEM fields is already fairly low.

The prevalence of people in STEM professions in San Francisco is likely to be larger than in other parts of the country. If we thought that the overall STEM prevalence in the United States was about 5%, then our estimates would change to PPV = 0.16 and NPV = 0.98. These computations only differ by the prevalence estimates and demonstrate how the smaller prevalence affects the unconditional probabilities of the results.

The metrics discussed so far depend on having a *hard* prediction (e.g., STEM or other). Most classification models can produce class probabilities as *soft* predictions that can be converted to a definitive class by choosing the class with the largest probability. There are a number of metrics that can be created using the probabilities.

For a two-class problem, an example metric is the binomial log-likelihood statistic. To illustrate this statistic, let i represent the index of the samples where $i = 1, 2, \ldots, n$, and let j represent the numeric value of the number of outcome classes where $j = 1, 2$. Next, we will use y_{ij} to represent the indicator of the true class of the ith sample. That is, $y_{ij} = 1$ if the ith sample is in the jth class and 0 otherwise. Finally, let p_{ij} represent the predicted probability of the ith sample in the jth class. Then the log-likelihood is calculated as

$$\log \ell = \sum_{i=1}^{n} \sum_{j=1}^{C} y_{ij} \log(p_{ij}),$$

where $C = 2$ for the two-class problem. In general, we want to maximize the log-likelihood. This value will be maximized if all samples are predicted with high probability to be in the correct class.

Two other metrics that are commonly computed on class probabilities are the Gini criterion (Breiman et al., 1984)

$$G = \sum_{i=1}^{n} \sum_{j \neq j'} p_{ij} p_{ij'}$$

and entropy (H) (MacKay, 2003):

$$H = -\sum_{i=1}^{n} \sum_{j=1}^{C} p_{ij} \log_2 p_{ij}$$

Unlike the log-likelihood statistic, both of these metrics are measures of *variance* or *impurity* in the class probabilities[19] and should be *minimized.*

Of these three metrics, it is important to note that the likelihood statistic is the only one to use the true class information. Because of this, it penalizes poor models in a supervised manner. The Gini and entropy statistics would only penalize models that are equivocal (i.e., produce roughly equal class probabilities). For example, Table 3.2 shows a two-class example. If the true outcome was the first class, the model results shown in the second row would be best. The likelihood statistic only takes into account the column called "Class 1" since that is the only column where $y_{ij} = 1$. In terms of the likelihood statistic, the equivocal model does better than the model that confidently predicts the wrong class. When considering Gini and entropy, the equivocal model does worst while the good and bad models are equivalent.[20]

When there are two classes, one advantage that the log-likelihood has over metrics based on a hard prediction is that it sidesteps the issue of the appropriateness of the probability cutoff. For example, when discussing accuracy, sensitivity, specificity, and other measures, there is the implicit assumption that the probability cutoff used to go between a soft prediction to a hard prediction is valid. This can often not be the case, especially when the data have a severe class imbalance.[21] Consider the OkCupid data and the logistic regression model that was previously discussed. The class probability estimates that were used to make definitive predictions that were contained in Table 3.1 are shown in Figure 3.2 where the top panel contains the profiles that were truly STEM and the bottom panel has the class probability distribution for the other profiles.[22] The common 50% cutoff was used to create the original table of observed by predicted classes. Table 3.1 can also be visualized using a mosaic plot such as the one shown in Figure 3.3(b) where the sizes of the blocks are proportional to the amount of data in each cell. What would happen to this table if we were more permissive about the level of evidence needed to call a profile

---

[19]In fact, the Gini statistic is equivalent to the binomial variance when there are two classes.

[20]While not helpful for comparing models, these two statistics are widely used in the process of creating decision trees. See Breiman et al. (1984) and Quinlan (1993) for examples. In that context, these metrics enable tree-based algorithms to create effective models.

[21]See Chapter 16 of Kuhn and Johnson (2013).

[22]As with the confusion matrix in Table 3.1, these data were created during 10-fold cross-validation.

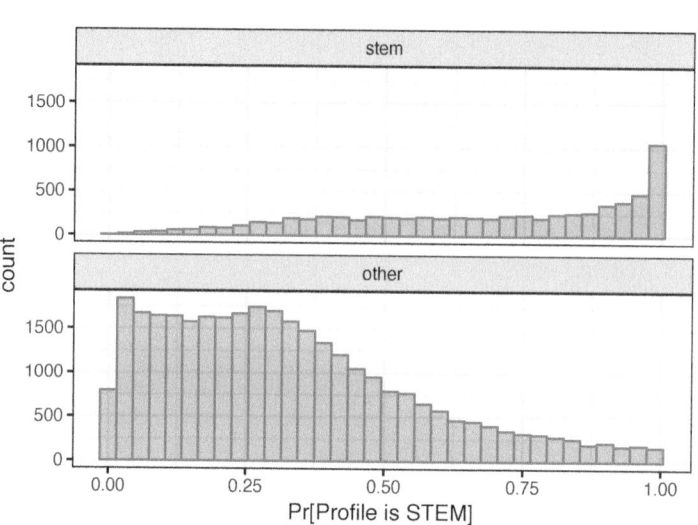

Figure 3.2: The class probability distributions for the OkCupid data. The top panel corresponds to the true STEM profiles while the bottom shows the probabilities for the non-STEM test set samples.

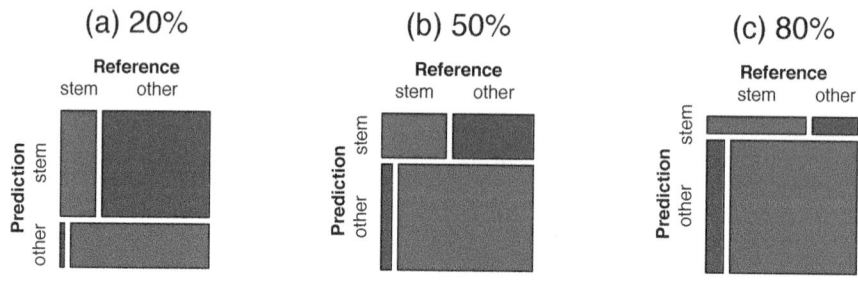

Figure 3.3: Mosaic plots of a confusion matrix under different cutoffs.

STEM? Instead of using a 50% cutoff, we might *lower* the threshold for the event to 20%. In this instance, more profiles would be called STEM. This might raise sensitivity since the true STEM profiles are more likely to be correctly predicted, but the cost is to increase the number of false positives. The mosaic plot for this confusion matrix is shown in Figure 3.3(a) where the blue block in the upper left becomes larger. But there is also an increase in the red block in the upper right. In doing so, the sensitivity increases from 0.72 to 0.96, but the specificity drops from 0.8 to 0.35. Increasing the level of evidence needed to predict a STEM profile to 80%, has the opposite effect as shown in Figure 3.3(c). Here, specificity improves but sensitivity is undermined.

The question then becomes "what probability cutoff should be used?" This depends on a number of things, including which error (false positive or false negative) hurts the most. However, if both types of errors are equally bad, there may be cutoffs that do better than the default.

The receiver operating characteristic (ROC) (Altman and Bland, 1994a) curve can be used to alleviate this issue. It considers *all possible cutoffs* and tracks the changes in sensitivity and specificity. The curve is composed by plotting the false positive rate (1 - specificity) versus the true positive rate. The ROC curve for the OkCupid data is shown in Figure 3.4(a). The best model is one that hugs the y-axis and directly proceeds to the upper left corner (where neither type of error is made) while a completely ineffective model's curve would track along the diagonal line shown in gray. This curve allows the user to do two important tasks. First, an appropriate cutoff can be determined based on one's expectations regarding the importance of either sensitivity or specificity. This cutoff can then be used to make the qualitative predictions. Secondly, and perhaps more importantly, it allows a model to be assessed without having to identify the best cutoff. Commonly, the area under the ROC curve (AUC) is used to evaluate models. If the best model immediately proceeds to the upper left corner, the area under this curve would be one in which the poor model would produce an AUC in the neighborhood of 0.50. Caution should be used, though, since two curves for two different models may cross; this indicates that there are areas where one model does better than the other. Used as a summary measure, the AUC annihilates any subtleties that can be seen in the curves. For the curve in Figure 3.4(a), the AUC was 0.839, indicating a moderately good fit.

From the information retrieval point of view, the precision-recall curve is more appropriate (Christopher et al., 2008). This is similar to the ROC curve in that the two statistics are calculated over every possible cutoff in the data. For the OkCupid data, the curve is shown in Figure 3.4(b). A poor model would result in a precision-recall curve that is in the vicinity of the horizontal gray line that is at the value of the observed prevalence (0.18 here). The area under the curve is used to summarize model performance. The best possible value is 1.0, while the worst is the prevalence. The area under this curve is 0.603.

During the initial phase of model building, a good strategy for data sets with two classes is to focus on the AUC statistics from these curves instead of metrics based on hard class predictions. Once a reasonable model is found, the ROC or precision-recall curves can be carefully examined to find a reasonable cutoff for the data and then qualitative prediction metrics can be used.

3.2.3 | Context-Specific Metrics

While the metrics discussed previously can be used to develop effective models, they may not answer the underlying question of interest. As an example, consider a scenario where we have collected data on customer characteristics and whether or not the customers clicked on an ad. Our goal may be to relate customer characteristics

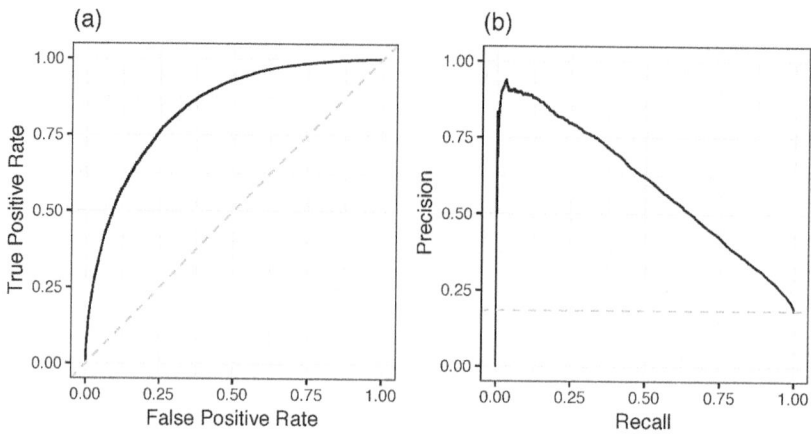

Figure 3.4: Examples of an ROC curve (a) and a precision-recall curve (b).

to the probability of a customer clicking on an ad. Several of the metrics described above would enable us to assess model performance if this was the goal. Alternatively, we may be more interested in answering "how much money will my company make if this model is used to predict who will click on an ad?" In another context, we may be interested in building a model to answer the question "what is my expected profit when the model is used to determine if this customer will repay a loan?" These questions are very context specific and do not directly fit into the previously described metrics.

Take the loan example. If a loan is requested for $\$M$, can we compute the expected profit (or loss)? Let's assume that our model is created on an appropriate data set and can produce a class probability P_r that the loan will be paid on time. Given these quantities, the interest rate, fees, and other known factors, the gross return on the loan can be computed for each data point and the average can then be used to optimize the model.

Therefore, we should let the question of interest lead us to an appropriate metric for assessing a model's ability to answer the question. It may be possible to use common, existing metrics to address the question. Or the problem may require development of custom metrics for the context. See Chapter 16 of Kuhn and Johnson (2013) for an additional discussion.

3.3 | Data Splitting

One of the first decisions to make when starting a modeling project is how to utilize the existing data. One common technique is to split the data into two groups typically referred to as the *training* and *testing* sets.[23] The training set is used to develop models and feature sets; it is the substrate for estimating parameters, comparing

---

[23]There are other types of subsets, such as a *validation* set. This is discussed in Section 3.4.5.

models, and all of the other activities required to reach a final model. The test set is used only at the conclusion of these activities for estimating a final, unbiased assessment of the model's performance. It is critical that the test set not be used prior to this point. Looking at the test set results would bias the outcomes since the testing data will have become part of the model development process.

How much data should be set aside for testing? It is extremely difficult to make a uniform guideline. The proportion of data can be driven by many factors, including the size of the original pool of samples and the total number of predictors. With a large pool of samples, the criticality of this decision is reduced once "enough" samples are included in the training set. Also, in this case, alternatives to a simple initial split of the data might be a good idea; see Section 3.4.7 below for additional details. The ratio of the number of samples (n) to the number of predictors (p) is important to consider, too. We will have much more flexibility in splitting the data when n is much greater than p. However, when n is less than p, then we can run into modeling difficulties even if n is seemingly large.

There are a number of ways to split the data into training and testing sets. The most common approach is to use some version of random sampling. Completely random sampling is a straightforward strategy to implement and usually protects the process from being biased towards any characteristic of the data. However, this approach can be problematic when the response is not evenly distributed across the outcome. A less risky splitting strategy would be to use a *stratified* random sample based on the outcome. For classification models, this is accomplished by selecting samples at random *within* each class. This approach ensures that the frequency distribution of the outcome is approximately equal within the training and test sets. When the outcome is numeric, artificial strata can be constructed based on the quartiles of the data. For example, in the Ames housing price data, the quartiles of the outcome distribution would break the data into four artificial groups containing roughly 230 houses. The training/test split would then be conducted within these four groups and the four different training set portions are pooled together (and the same for the test set).

Non-random sampling can also be used when there is a good reason. One such case would be when there is an important temporal aspect to the data. Here it may be prudent to use the most recent data as the test set. This is the approach used in the Chicago transit data discussed in Section 4.1.

3.4 | Resampling

As previously discussed, there are times where there is the need to understand the effectiveness of the model without resorting to the test set. Simply repredicting the training set is problematic so a procedure is needed to get an appraisal using the training set. Resampling methods will be used for this purpose.

Resampling methods can generate different versions of our training set that can be used to simulate how well models would perform on new data. These techniques

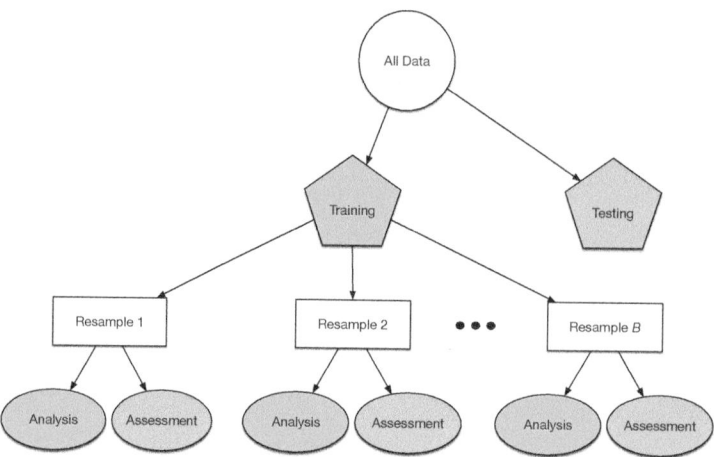

Figure 3.5: A diagram of typical data usage with B resamples of the training data.

differ in terms of how the resampled versions of the data are created and how many iterations of the simulation process are conducted. In each case, a resampling scheme generates a subset of the data to be used for modeling and another that is used for measuring performance. Here, we will refer to the former as the "analysis set" and the latter as the "assessment set". They are roughly analogous to the training and test sets described at the beginning of the chapter.[24] A graphic of an example data hierarchy with an arbitrary number of resamples is shown in Figure 3.5.

Before proceeding, we should be aware of *what* is being resampled. The *independent experimental unit* is the unit of data that is as statistically independent as possible from the other data. For example, for the Ames data, it would be reasonable to consider each house to be independent of the other houses. Since houses are contained in rows of the data, each row is allocated to either the analysis or assessment sets. However, consider a situation where a company's customer is the independent unit but the data set contains multiple rows per customer. In this case, each customer would be allotted to the analysis or assessment sets and all of their corresponding rows move with them. Such non-standard arrangements are discussed more in Chapter 9.

There are a number of different flavors of resampling that will be described in the next four sections.

3.4.1   *V*-Fold Cross-Validation and Its Variants

Simple *V*-fold cross-validation creates *V* different versions of the original training set that have the same approximate size. Each of the *V* assessment sets contains $1/V$ of the training set and each of these exclude different data points. The analysis sets contain the remainder (typically called the "folds") . Suppose $V = 10$, then there

---

[24]In fact, many people use the terms "training" and "testing" to describe the splits of the data produced during resampling. We avoid that here because 1) resampling is only ever conducted on the training set samples and 2) the terminology can be confusing since the same term is being used for different versions of the original data.

are 10 different versions of 90% of the data and also 10 versions of the remaining 10% for each corresponding resample.

To use V-fold cross-validation, a model is created on the first fold (analysis set) and the corresponding assessment set is predicted by the model. The assessment set is summarized using the chosen performance measures (e.g., RMSE, the area under the ROC curve, etc.) and these statistics are saved. This process proceeds in a round-robin fashion so that, in the end, there are V estimates of performance for the model and each was calculated on a different assessment set. The cross-validation estimate of performance is computed by averaging the V individual metrics.

Figure 3.6 shows a diagram of 10-fold cross-validation for a hypothetical data set with 20 training set samples. For each resample, two different training set data points are held out for the assessment set. Note that the assessment sets are mutually exclusive and contain different instances.

When the outcome is categorical, *stratified* splitting techniques can also be applied here to make sure that the analysis and assessment sets produce the same frequency distribution of the outcome. Again, this is a good idea when a continuous outcome is skewed or a categorical outcome is imbalanced, but is unlikely to be problematic otherwise.

For example, for the OkCupid data, stratified 10-fold cross-validation was used. The training set consists of 38,809 profiles and each of the 10 assessment sets contains 3,880 different profiles. The area under the ROC curve was used to measure performance of the logistic regression model previously mentioned. The 10 areas under the curve ranged from 0.83 to 0.854 and their average value was 0.839. Without using the test set, we can use this statistic to forecast how this model would perform on new data.

As will be discussed in Section 3.4.6, resampling methods have different characteristics. One downside to basic V-fold cross-validation is that it is *relatively* noisier (i.e., has more variability) than other resampling schemes. One way to compensate for this is to conduct *repeated* V-fold cross-validation. If R repeats are used, V resamples are created R separate times and, in the end, RV resamples are averaged to estimate performance. Since more data are being averaged, the reduction in the variance of the final average would decease by $\sqrt{R}$ (using a Gaussian approximation[25]). Again, the noisiness of this procedure is relative and, as one might expect, is driven by the amount of data in the assessment set. For the OkCupid data, the area under the ROC curve was computed from 3,880 profiles and is likely to yield sufficiently precise estimates (even if we only expect about 716 of them to be STEM profiles).

The assessment sets can be used for model validation and diagnosis. Table 3.1 and Figure 3.2 use these holdout predictions to visualize model performance. Also, Section 4.4 has a more extensive description of how the assessment data sets can be used to drive improvements to models.

---

[25] This is based on the idea that the standard error of the mean has $\sqrt{R}$ in the denominator.

One other variation, *leave-one-out* cross-validation, has V equal to the size of the training set. This is a somewhat deprecated technique and may only be useful when the training set size is extremely small (Shao, 1993).

In the previous discussion related to the independent experimental unit, an example was given where customers should be resampled and all of the rows associated with each customer would go into *either* the analysis or assessment sets. To do this, V-fold cross-validation would be used with the customer identifiers. In general, this is referred to as either "grouped V-fold cross-validation" or "leave-group-out cross-validation" depending on the value of V.

3.4.2 | Monte Carlo Cross-Validation

V-fold cross-validation produced V sets of splits with mutually exclusive assessment sets. Monte Carlo resampling produces splits that are likely to contain overlap. For each resample, a random sample is taken with π proportion of the training set going into the analysis set and the remaining samples allocated to the assessment set. Like the previous procedure, a model is created on the analysis set and the assessment set is used to evaluate the model. This splitting procedure is conducted B times and the average of the B results are used to estimate future performance. B is chosen to be large enough so that the average of the B values has an acceptable amount of precision.

Figure 3.6 also shows Monte Carlo fold cross-validation with 10 resamples and $\pi = 0.90$. Note that, unlike 10-fold cross-validation, some of the same data points are used in different assessment sets.

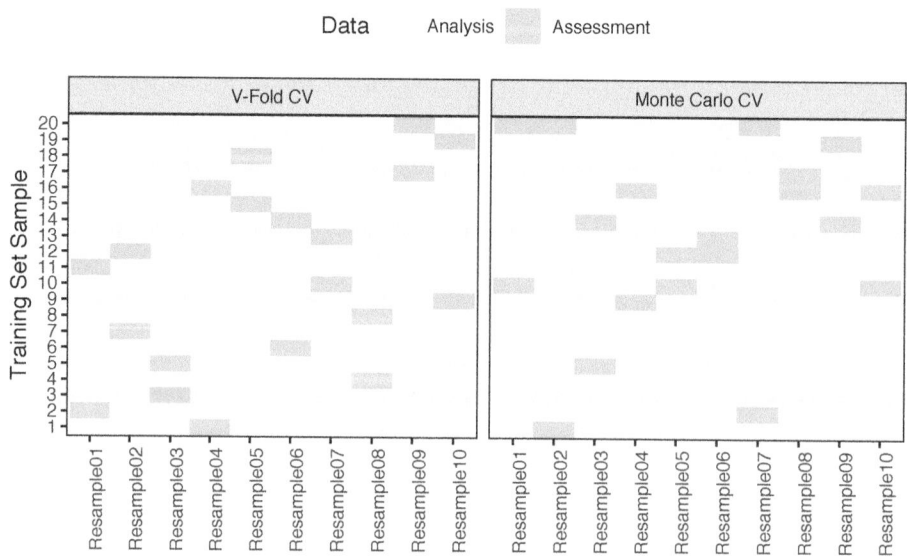

Figure 3.6: A diagram of two types of cross-validation for a training set containing 20 samples.

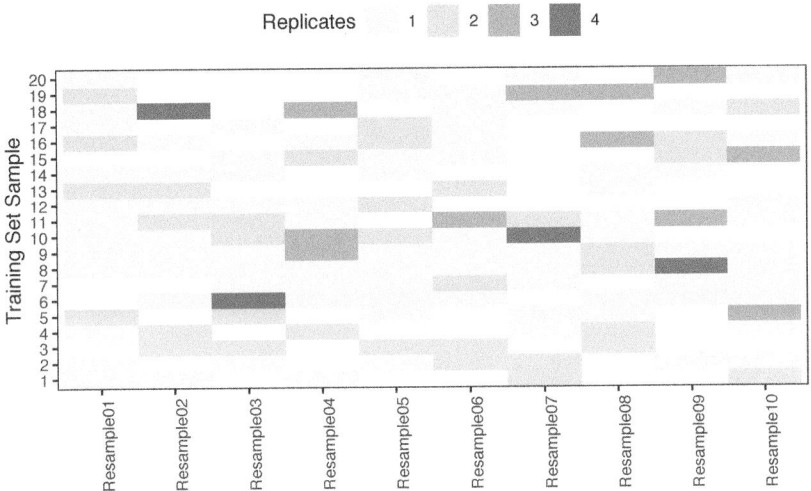

Figure 3.7: A diagram of bootstrap resampling for a training set containing 20 samples. The colors represent how many times a data point is replicated in the analysis set.

3.4.3 | The Bootstrap

A bootstrap resample of the data is defined to be a simple random sample that is the same size as the training set where the data are sampled *with replacement* (Davison and Hinkley, 1997). This means that when a bootstrap resample is created there is a 63.2% chance that any training set member is included in the bootstrap sample at least once. The bootstrap resample is used as the analysis set and the assessment set, sometimes known as the out-of-bag sample, consists of the members of the training set not included in the bootstrap sample. As before, bootstrap sampling is conducted B times and the same modeling/evaluation procedure is followed to produce a bootstrap estimate of performance that is the mean of B results.

Figure 3.7 shows an illustration of ten bootstrap samples created from a 20-sample data set. The colors show that several training set points are selected multiple times for the analysis set. The assessment set would consist of the rows that have no color.

3.4.4 | Rolling Origin Forecasting

This procedure is specific to time-series data or any data set with a strong temporal component (Hyndman and Athanasopoulos, 2013). If there are seasonal or other chronic trends in the data, random splitting of data between the analysis and assessment sets may disrupt the model's ability to estimate these patterns.

In this scheme, the first analysis set consists of the first M training set points, assuming that the training set is ordered by time or other temporal component. The assessment set would consist of the next N training set samples. The second

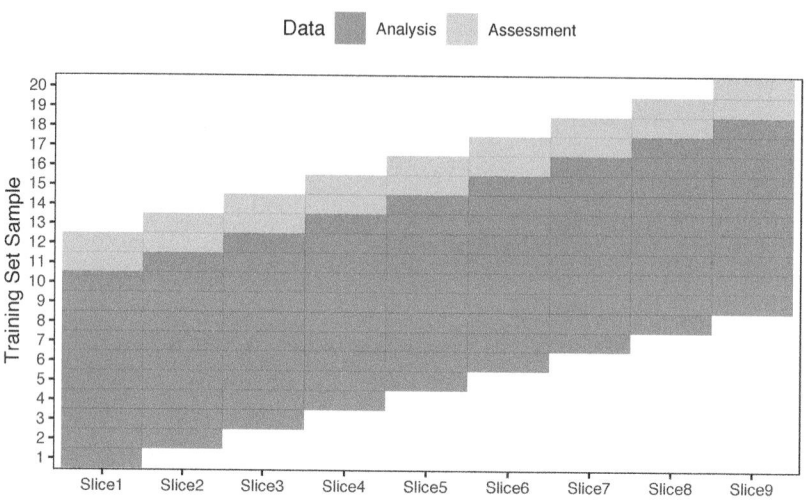

Figure 3.8: A diagram of a rolling origin forcasting resampling with $M = 10$ and $N = 2$.

resample keeps the data set sizes the same but increments the analysis set to use samples 2 through $M + 1$ while the assessment contains samples $M + 2$ to $M + N + 1$. The splitting scheme proceeds until there is no more data to produce the same data set sizes. Supposing that this results in B splits of the data, the same process is used for modeling and evaluation and, in the end, there are B estimates of performance generated from each of the assessment sets. A simple method for estimating performance is to again use simple averaging of these data. However, it should be understood that, since there is significant overlap in the rolling assessment sets, the B samples themselves constitute a time-series and might also display seasonal or other temporal effects.

Figure 3.8 shows this type of resampling where 10 data points are used for analysis and the subsequent two training set samples are used for assessment.

There are a number of variations of the procedure:

- The analysis sets need not be the same size. They can cumulatively grow as the moving window proceeds along the training set. In other words, the first analysis set would contain M data points, the second would contain $M + 1$ and so on. This is the approach taken with the Chicago train data modeling and is described in Chapter 4.
- The splitting procedure could *skip* iterations to produce fewer resamples. For example, in the Chicago data, there are daily measurements from 2001 to 2016. Incrementing by one day would produce an excessive value of B. For these data, 13 samples were skipped so that the splitting window moves in two-week blocks instead of by individual day.
- If the training data are unevenly sampled, the same procedure can be used but

moves over time increments rather than data set row increments. For example, the window could move over 12-hour periods for the analysis sets and 2-hour periods for the assessment sets.

This resampling method differs from the previous ones in at least two ways. The splits are not random and the assessment data set is not the remainder of the training set data once the analysis set was removed.

3.4.5 | Validation Sets

A validation set is something in-between the training and test sets. Historically, in the neural network literature, there was the recognition that using the same data to estimate parameters and measure performance was problematic. Often, a small set of data was allocated to determine the error rate of the neural network for each iteration of training (also known as a training epoch in that patois). This validation set would be used to estimate the performance as the model is being trained to determine when it begins to overfit. Often, the validation set is a random subset of the training set so that it changes for each iteration of the numerical optimizer. In resampling this type of model, it would be reasonable to use a small, random portion of the analysis set to serve as a within-resample validation set.

Does this replace the test set (or, analogously, the assessment set)? No. Since the validation data are guiding the training process, they can't be used for a fair assessment for how well the modeling process is working.

3.4.6 | Variance and Bias in Resampling

In Section 1.2.5, variance and bias properties of *models* were discussed. Resampling methods have the same properties but their effects manifest in different ways. Variance is more straightforward to conceptualize. If you were to conduct 10-fold cross-validation many times on the same data set, the variation in the resampling scheme could be measured by determining the spread of the resulting averages. This variation could be compared to a different scheme that was repeated the same number of times (and so on) to get relative comparisons of the amount of noise in each scheme.

Bias is the ability of a particular resampling scheme to be able to hit the true underlying performance *parameter* (that we will never truly know). Generally speaking, as the amount of data in the analysis set shrinks, the resampling estimate's bias increases. In other words, the bias in 10-fold cross-validation is smaller than the bias in 5-fold cross-validation. For Monte Carlo resampling, this obviously depends on the value of π. However, through simulations, one can see that 10-fold cross-validation has less bias than Monte Carlo cross-validation when $\pi = 0.10$ and $B = 10$ are used. Leave-one-out cross-validation would have very low bias since its analysis set is only one sample apart from the training set.

Figure 3.9 contains a graphical representation of variance and bias in resampling schemes where the curves represent the distribution of the resampling statistics if

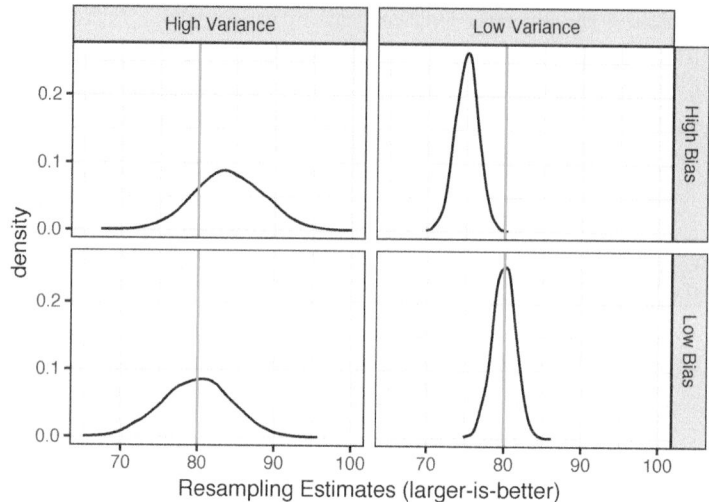

Figure 3.9: An illustration of the variance-bias differences in resampling schemes.

the same procedure was conducted on the same data set many times. Four possible variance/bias cases are represented. We will assume that the model metric being measured here is better when the value is large (such as R^2 or sensitivity) and that the true value is represented by the green vertical line. The upper right panel demonstrates a pessimistic bias since the values tend to be smaller than the true value while the panel below in the lower right shows a resampling scheme that has relatively low variance and the center of its distribution is on target with the true value (so that it nearly unbiased).

In general, for a fixed training set size and number of resamples, simple V-fold cross-validation is generally believed to be the noisiest of the methods discussed here and the bootstrap is the least variable.[26] The bootstrap is understood to be the most biased (since about 36.8% of the training set is selected for assessment) and its bias is generally pessimistic (i.e., likely to show worse model performance than the true underlying value). There have been a few attempts at correcting the bootstrap's bias such as Efron (1983) and Efron and Tibshirani (1997).

While V-fold cross-validation does have inflated variance, its bias is fairly low when V is 10 or more. When the training set is not large, we recommend using five or so repeats of 10-fold cross-validation, depending on the required precision, the training set size, and other factors.

3.4.7 | What Should Be Included Inside of Resampling?

In the preceding descriptions of resampling schemes, we have said that the assessment set is used to "build the model". This is somewhat of a simplification. In order for any resampling scheme to produce performance estimates that *generalize* to new data,

---

[26]This is a general trend; there are many factors that affect these comparisons.

it must contain all of the steps in the *modeling process* that could significantly affect the model's effectiveness. For example, in Section 1.1, a transformation was used to modify the predictor variables and this resulted in an improvement in performance. During resampling, this step should be included in the resampling loop. Other preprocessing or filtering steps (such as PCA signal extraction, predictor correlation filters, feature selection methods) must be part of the resampling process in order to understand how well the model is doing and to measure when the modeling process begins to *overfit*.

There are some operations that can be exempted. For example, in Chapter 2, only a handful of patients had missing values and these were imputed using the median. For such a small modification, we did not include these steps inside of resampling. In general though, imputation can have a substantial impact on performance and its variability should be propagated into the resample results. Centering and scaling can also be exempted from resampling, all other things being equal.

As another example, the OkCupid training data were downsampled so that the class proportions were equal. This is a substantial data processing step and it is important to propagate the effects of this procedure through the resampling results. For this reason, the downsampling procedure is executed on the analysis set of every resample and then again on the entire training set when the final model is created.

One other aspect of resampling is related to the concept of *information leakage* which is where the test set data are used (directly or indirectly) during the training process. This can lead to overly optimistic results that do not replicate on future data points and can occur in subtle ways.

For example, suppose a data set of 120 samples has a single predictor and is sequential in nature, such as a time series. If the predictor is noisy, one method for preprocessing the data is to apply a moving average to the predictor data and replace the original data with the smoothed values. Suppose a 3-point average is used and that the first and last data points retain their original values. Our inclination would be to smooth the whole sequence of 120 data points with the moving average, split the data into training and test sets (say with the first 100 rows in training). The subtle issue here is that the 100th data point, the last in the training set, uses the first in the test set to compute its 3-point average. Often the most recent data are most related to the next value in the time series. Therefore, including the most recent point will likely bias model performance (in a small way).

To provide a solid methodology, we should constrain ourselves to developing the list of preprocessing techniques, estimate them only in the presence of the training data points, and then apply the techniques to future data (including the test set). Arguably, the moving average issue cited above is most likely minor in terms of consequences, but illustrates how easily the test data can creep into the modeling process. The approach to applying this preprocessing technique would be to split the data, then apply the moving average smoothers to the training and test sets independently.

Another, more overt path to information leakage, can sometimes be seen in machine learning competitions where the training and test set data are given at the same time. While the test set data often have the outcome data blinded, it is possible to "train to the test" by only using the training set samples that are most similar to the test set data. This may very well improve the model's performance scores for this particular test set but might ruin the model for predicting on a broader data set.

Finally, with large amounts of data, alternative data usage schemes might be a good idea. Instead of a simple training/test split, multiple splits can be created for specific purposes. For example, a specific split of the data can be used to determine the relevant predictors for the model prior to model building using the training set. This would reduce the need to include the expensive feature selection steps inside of resampling. The same approach could be used for post-processing activities, such as determining an appropriate probability cutoff from a receiver operating characteristic curve.

3.5 │ Tuning Parameters and Overfitting

Many models include parameters that, while important, cannot be directly estimated from the data. The *tuning parameters* (sometimes called *hyperparameters*[27]) are important since they often control the complexity of the model and thus also affect any variance-bias trade-off that can be made.

As an example, the K-nearest neighbor model stores the training set data and, when predicting new samples, locates the K training set points that are in the closest proximity to the new sample. Using the training set outcomes for the neighbors, a prediction is made. The number of neighbors controls the complexity and, in a manner very similar to the moving average discussion in Section 1.2.5, controls the variance and bias of the models. When K is very small, there is the most potential for overfitting since only few values are used for prediction and are most susceptible to changes in the data. However, if K is too large, too many potentially irrelevant data points are used for prediction resulting in an *underfit* model.

To illustrate this, consider Figure 3.10 where a single test set sample is shown as the blue circle and its five closest (geographic) neighbors from the training set are shown in red. The test set sample's sale price is $282K and the neighbor's prices, from closest to farthest, are: $320K, $248K, $286K, $274K, $320K. Using $K = 1$, the model would miss the true house price by $37.5K. This illustrates the concept of *overfitting* introduced in Section 1.2.1; the model is too aggressively using patterns in the training set to make predictions on new data points. For this model, increasing the number of neighbors might help alleviate the issue. Averaging all $K = 5$ points to make a prediction substantially cuts the error to $7.3K.

This example illustrates the effect that the tuning parameter can have on the quality of the models. In some models, there could be more than one parameter. Again,

---

[27]Not to be confused with the hyperparameters of a prior distribution in Bayesian analysis.

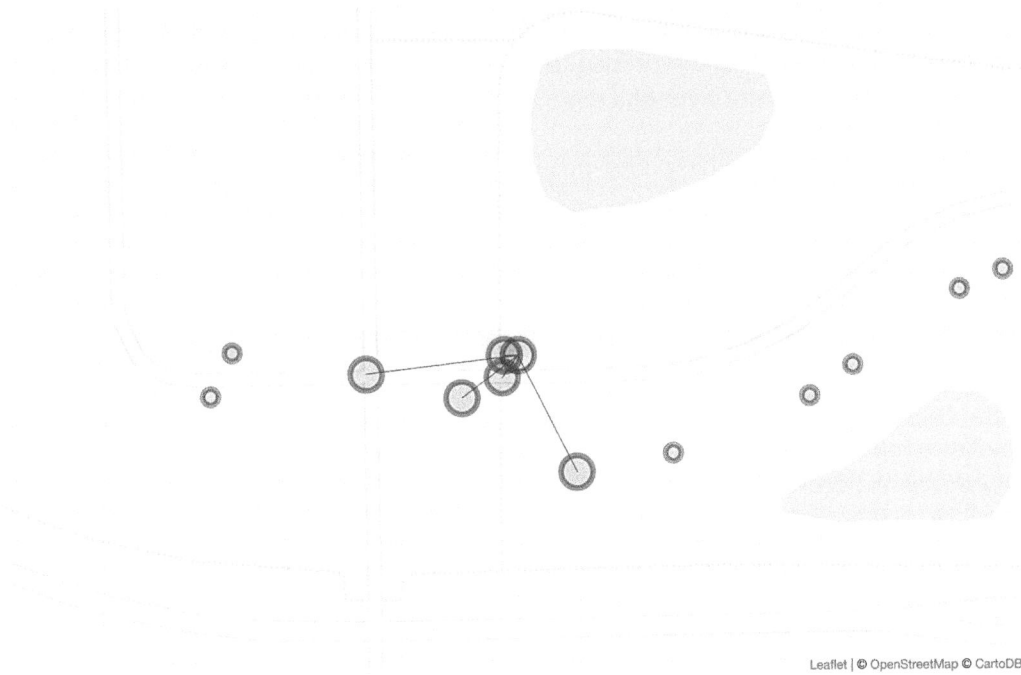

Figure 3.10: A house in Ames in the test set (blue) along with its five closest neighbors from the training set (red).

for the nearest neighbor model, a different distance metric could be used as well as difference schemes for weighting the neighbors so that more distant points have less of an effect on the prediction.

To make sure that proper values of the tuning parameters are used, some sort of search procedure is required along with a method for obtaining good, generalizable measures of performance. For the latter, repeatedly using the test set for these questions is problematic since it would lose its impartiality. Instead, resampling is commonly used. The next section will describe a few methods for determining optimal values of these types of parameters.

3.6   Model Optimization and Tuning

The search for the best tuning parameter values can be done in many ways but most fall into two main categories: those that predefine which values to evaluate and those that incrementally determine the values. In the first case, the most well-known procedure is *grid search*. Here, a set of candidate tuning parameter values are specified and then evaluated. In some cases, the model will have more than one tuning parameter and, in this case, a candidate parameter combination is multidimensional. We recommend using resampling to evaluate each distinct parameter value combination to get good estimates of how well each candidate performs. Once the results are calculated, the "best" tuning parameter combination is chosen and the final model is fit to the entire training set with this value. The best combination can be determined in various ways but the most common approach is to pick the candidate with the empirically best results.

As an example, a simple K-nearest neighbors model requires the number of neighbors. For the OkCupid data, this model will be used to predict the profession of the profile. In this context, the predictors contain many different types of profile characteristics, so that a "nearest neighbor" is really a similar profile based on many characteristics.

We will predefine the candidate set to be $K = 1, 3, \ldots 201$.[28] When combined with the same 10-fold cross-validation process, a total of 380 temporary models will be used to determine a good value of K. Once the best value of K is chosen, one final model is created using the optimal number of neighbors. The resampling profile is shown in Figure 3.11. Each black point shown in this graph is the average of performance for ten different models estimated using a distinct 90% of the training set. The configuration with the largest area under the ROC curve used 201 neighbors with a corresponding AUC of 0.806. Figure 3.11 shows the individual resampling profiles for each of the ten resamples. The reduced amount of variation in these data is mostly due to the size of the training set.[29]

When there are many tuning parameters associated with a model, there are several ways to proceed. First, a multidimensional grid search can be conducted where candidate parameter combinations and the grid of combinations are evaluated. In some cases, this can be very inefficient. Another approach is to define a range of possible values for each parameter and to randomly sample the multidimensional space enough times to cover a reasonable amount (Bergstra and Bengio, 2012). This *random search* grid can then be resampled in the same way as a more traditional grid. This procedure can be very beneficial when there are a large number of tuning parameters and there is no *a priori* notion of which values should be used. A large grid may be inefficient to search, especially if the profile has a fairly stable pattern with little change over some range of the parameter. Neural networks, gradient boosting machines, and other models can effectively be tuned using this approach.[30]

To illustrate this procedure, the OkCupid data was used once again. A single-layer, feed-forward neural network was used to model the probability of being in the STEM field using the same predictors as the previous two models. This model is an extremely flexible nonlinear classification system with many tuning parameters. See Goodfellow et al. (2016) for an excellent primer on neural networks and deep learning models.

---

[28]This number of neighbors is abnormally large compared to other data sets. This model tends to prefer values of K in, at most, the dozens (or less).

[29]This is generally not the case in other data sets; visualizing the individual profiles can often show an excessive amount of noise from resample-to-resample that is averaged out in the end.

[30]However, a regular grid may be more efficient. For some models, there are optimizations that can be used to compute the results for a candidate parameter set that can be determined *without refitting* that model. The nature of random search cancels this benefit.

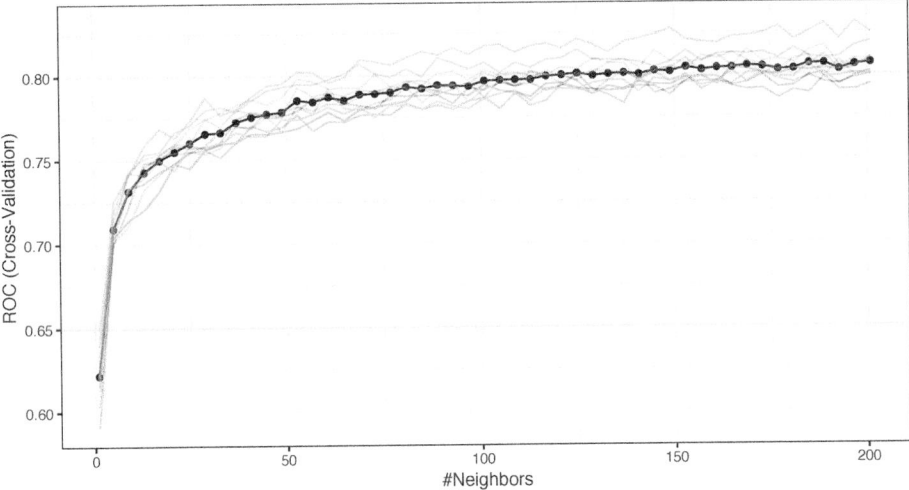

Figure 3.11: The resampling profile generated by a simple grid search on the number of neighbors in a K-NN classification model for the OkC data. The black line represents the averaged performance across 10 resamples while the other lines represent the performance across each individual resample.

The main tuning parameters for the model are:

- The number of hidden units. This parameter controls the complexity of the neural network. Larger values enable higher performance but also increase the risk of overfitting. For these data, the number of units in the hidden layers were randomly selected to be between 2 and 20.
- The activation function. The nonlinear function set by this parameter links different parts of the network. Three different functions were used: traditional sigmoidal curves, `tanh`, or rectified linear unit functions (ReLU).
- The dropout rate. This is the rate at which coefficients are randomly set to zero during training iterations and can attenuate overfitting (Nitish et al., 2014). Rates between 0 and 80% were considered.

The fitting procedure for neural network coefficients can be very numerically challenging. There are usually a large number of coefficients to estimate and there is a significant risk of finding a local optima. Here, we use a gradient-based optimization method called RMSProp[31] to fit the model. This is a modern algorithm for finding coefficient values and there are several model tuning parameters for this procedure:[32]

- The batch size controls how many of the training set data points are randomly exposed to the optimization process at each epoch (i.e., optimization itera-

---

[31]RMSprop is an general optimization method that uses gradients. The details are beyond the scope of this book but more information can be found in Goodfellow et al. (2016) and at `https://en.wikipedia.org/wiki/Stochastic_gradient_descent#RMSProp`.

[32]Many of these parameters are fairly arcane for those not well acquainted with modern derivative-based optimization. Chapter 8 of Goodfellow et al. (2016) has a substantial discussion of these methods.

Table 3.3: The settings and results for a random search of the neural network parameter space.

Units	Dropout	Batch Size	Learn Rate	Grad. Scaling	Decay	Act. Fun.	ROC
9	*0.592*	*26231*	*0.004*	*0.457*	*0.04458*	*relu*	*0.837*
16	0.572	25342	0.631	0.298	0.51018	relu	0.836
13	0.600	10196	0.333	0.007	0.17589	relu	0.835
10	0.286	19893	0.947	0.495	0.46124	relu	0.832
8	0.534	36267	0.387	0.505	0.14937	relu	0.832
14	0.617	21154	0.252	0.297	0.00654	sigmoid	0.829
17	0.250	11638	0.227	0.721	0.01380	tanh	0.825
20	0.392	17235	0.017	0.363	0.00029	relu	0.821
16	0.194	8477	0.204	0.155	0.01819	tanh	0.820
10	0.282	18989	0.878	0.260	0.06387	tanh	0.820
2	0.586	19404	0.721	0.159	0.18514	sigmoid	0.818
18	0.305	5563	0.836	0.457	0.01051	sigmoid	0.805
10	0.513	17877	0.496	0.770	0.00004	tanh	0.796
4	0.536	32689	0.719	0.483	0.00011	sigmoid	0.755
12	0.150	4015	0.558	0.213	0.00001	relu	0.732
14	0.368	7582	0.938	0.126	0.00001	relu	0.718
3	0.688	12905	0.633	0.698	0.00001	relu	0.598
3	0.645	5327	0.952	0.717	0.00001	sigmoid	0.560
2	0.688	14660	0.686	0.445	0.00003	relu	0.546
15	0.697	4800	0.806	0.139	0.00003	sigmoid	0.538

tion). This has the effect of reducing potential overfitting by providing some randomness to the optimization process. Batch sizes between 10 and 40K were considered.

- The learning rate parameter controls the rate of descent during the parameter estimation iterations and these values were contrasted to be between zero and one.

- A decay rate that decreases the learning rate over time (ranging between zero and one).

- The root mean square gradient scaling factor (ρ) controls how much the gradient is normalized by recent values of the squared gradient values. Smaller values of this parameter give more emphasis to recent gradients. The range of this parameter was set to be [0.0, 1.0].

For this model, 20 different seven-dimensional tuning parameter combinations were created randomly using uniform distributions to sample within the ranges above. Each of these settings was evaluated using the same 10-fold cross-validation splits used previously.

The resampled ROC values significantly varied between the candidate parameter values. The best setting is italicized in Table 3.3 which had a corresponding area under the ROC curve of 0.837.

As previously mentioned, grid search and random search methods have the tuning parameters specified in advance and the search does not adapt to look for novel values. There are other approaches that can be taken which do. For example, there are many nonlinear search methods such as the Nelder-Mead simplex search procedure, simulated annealing, and genetic algorithms that can be employed (Chong and Zak, 2008). These methods conduct very thorough searches of the grid space but tend to be computationally expensive. One reason for this is that each evaluation of a new parameter requires a good estimate of performance to guide the search. Resampling is One of the best methods for doing this. Another approach to searching the space is called Bayesian optimization (Mockus, 1994). Here, an initial pool of samples is evaluated using grid or random search. The optimization procedure creates a separate model to predict performance as a function of the tuning parameters and can then make a recommendation as to the next candidate set to evaluate. Once this new point is assessed, the model is updated and the process continues for a set number of iterations (Jones et al., 1998).[33]

One final point about the interpretation of resampling results is that, by choosing the best settings based on the results and representing the model's performance using these values, there is the risk of *optimization bias*. Depending on the problem, this bias might overestimate the model's true performance. There are nested resampling procedures that can be used to mitigate these biases. See Boulesteix and Strobl (2009) for more information.

3.7 | Comparing Models Using the Training Set

When multiple models are in contention, there is often the need to have formal evaluations between them to understand if any differences in performance are above and beyond what one would expect at random. In our proposed workflows, resampling is heavily relied on to estimate model performance. It is good practice to use the same resamples across any models that are evaluated. That enables apples-to-apples comparisons between models. It also allows for formal comparisons to be made between models prior to the involvement of the test set. Consider logistic regression and neural network models created for the OkCupid data. How do they compare? Since the two models used the same resamples to fit and assess the models, this leads to a set of 10 paired comparisons. Table 3.4 shows the specific ROC results per resample and their paired differences. The correlation between these two sets of values is 0.96, indicating that there is likely to be a resample-to-resample effect in the results.

Given this set of paired differences, formal statistical inference can be done to compare models. A simple approach would be to consider a paired t-test between the two models or an ordinary one-sample t-test on the differences. The estimated difference in the ROC values is -0.003 with 95% confidence interval (-0.004, -0.001). *There does appear to be evidence of a real (but negligible) performance difference between*

---

[33]The GitHub repository http://bit.ly/2yjzB5V has example code for nonlinear search methods and Bayesian optimization of tuning parameters.

Table 3.4: Matched resampling results for two models for predicting the OkCupid data. The ROC metric was used to tune each model. Because each model uses the same resampling sets, we can formally compare the performance between the models.

	Logistic Regression	Neural Network	Difference
Fold 1	0.830	0.827	-0.003
Fold 2	0.854	0.853	-0.002
Fold 3	0.844	0.843	-0.001
Fold 4	0.836	0.834	-0.001
Fold 5	0.838	0.834	-0.004
Fold 6	0.840	0.833	-0.007
Fold 7	0.839	0.838	-0.001
Fold 8	0.837	0.837	0.000
Fold 9	0.835	0.832	-0.003
Fold 10	0.838	0.835	-0.003

these models. This approach was previously used in Section 2.3 when the potential variables for predicting stroke outcome were ranked by their improvement in the area under the ROC above and beyond the null model. This approach can be used to compare models or different approaches for the same model (e.g., preprocessing differences or feature sets).

The value in this technique is two-fold:

1. It prevents the test set from being used during the model development process and
2. Many evaluations (via assessment sets) are used to gauge the differences.

The second point is more important. By using multiple differences, the variability in the performance statistics can be measured. While a single static test set has its advantages, it is a single realization of performance for a model and we have no idea of the precision of this value.

More than two models can also be compared, although the analysis must account for the within-resample correlation using a Bayesian hierarchical model (McElreath, 2015) or a repeated measures model (West and Galecki, 2014). The idea for this methodology originated with Hothorn et al. (2005). Benavoli et al. (2016) also provides a Bayesian approach to the analysis of resampling results between models and data sets.

3.8   Feature Engineering without Overfitting

Section 3.5 discussed approaches to tuning models using resampling. The crux of this approach was to evaluate a parameter value on data that were not used to build the model. This is not a controversial notion. In a way, the assessment set data are used as a potentially dissenting opinion on the validity of a particular tuning value. It is fairly clear that simply re-predicting the training set does not include such an opportunity.

As will be stated many times in subsequent chapters, the same idea should be applied to any other feature-related activity, such as engineering new features/encodings or when deciding on whether to include a new term into the model. There should always be a chance to have an independent piece of data evaluate the appropriateness of the design choice.

For example, for the Chicago train data, some days had their ridership numbers drastically *over-predicted*. The reason for the overprediction can be determined by investigating the training data. The root of the problem is due to holidays and ancillary trends that need to be addressed in the predictors (e.g., "Christmas on a Friday" and so on).[34] For a support vector machine, the impact of these new features are shown in Figure 3.12. The panels show the *assessment set* predictions for the two years that were resampled. The dashed line shows the predictions prior to adding holiday predictors and the solid line corresponds to the model after these features have been added. Clearly these predictors helped improve performance during these time slices. This had an obvious commonsense justification but the assessment sets were evaluated to confirm the results.

However, suppose that exploratory data analysis clearly showed that there were decreases in ridership when the Chicago Bulls and Bears had *away* games. Even if this hypothetical pattern had a very clear signal in the training set, resampling would be the best method to determine if it held up when some of the training data were held back as an unbiased control.

As another example, in Section 5.6, the OkCupid data were used to derive keyword features from text essay data. The approach taken there was to use a small set of the training set (chosen at random) to discover the keyword features. This small subset was kept in the training set and resampled as normal when model fitting. Some of the discovery data set were in the training so we would expect the resampling estimates to be slightly optimistically biased. Despite this, there is still a fairly large amount of data that were not involved in the discovery of these features that are used to determine their validity.[35]

In any case, it is best to follow up trends discovered during exploratory data analysis with "stress testing" using resampling to get a more objective sense of whether a trend or pattern is likely to be real.

---

[34] These holiday-related indicators comprise Set 5 in Figure 1.7.

[35] As noted in the last paragraph of Section 3.4.7, a better approach would be to allocate a separate data set from the training and test sets and to use this for discovery. This is most useful when there is an abundance of data.

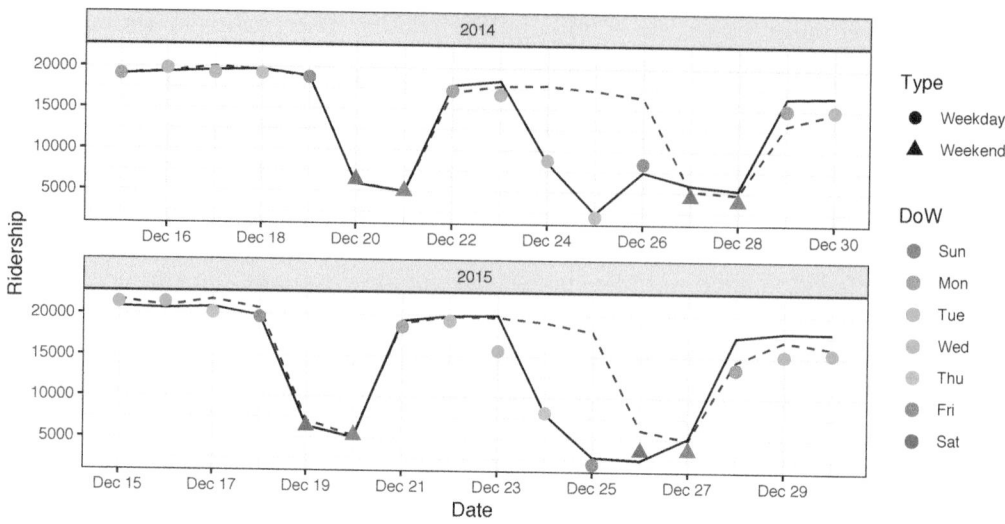

Figure 3.12: The effect of adding holiday predictors to an SVM model created for the Chicago train data. The dashed line corresponds to the model of the original predictors. The solid line corresponds to the model when an indicator of holiday was included as a predictor.

3.9 Summary

Before diving into methods of feature engineering and feature selection, we must first have a solid understanding of the predictive modeling process. A thorough understanding of this process helps the analyst to make a number of critical choices, most of which occur prior to building any models. Moreover, the more the analyst knows about the data, the mechanism that generated the data, and the question(s) to be answered, the better the initial choices the analyst can make. The initial choices include the model performance metric, approach to resampling, model(s) and corresponding tuning parameter(s). Without good choices, the overall modeling process often takes more time and leads to a sub-optimal model selection.

The next step in the process prior to modeling is to begin to understand characteristics of the available data. One of the best ways to do this is by visualizing the data, which is the focus of the next chapter.

3.10 Computing

The website `http://bit.ly/fes-review` contains R programs for reproducing these analyses.

4 | Exploratory Visualizations

The driving goal in everything that we do in the modeling process is to find reproducible ways to explain the variation in the response. As discussed in the previous chapter, discovering patterns among the predictors that are related to the response involves selecting a resampling scheme to protect against overfitting, choosing a performance metric, tuning and training multiple models, and comparing model performance to identify which models have the best performance. When presented with a new data set, it is tempting to jump directly into the predictive modeling process to see if we can quickly develop a model that meets the performance expectations. Or, in the case where there are many predictors, the initial goal may be to use the modeling results to identify the most important predictors related to the response. But as illustrated in Figure 1.4, a sufficient amount of time should be spent exploring the data. The focus of this chapter will be to present approaches for visually exploring data and to demonstrate how this approach can be used to help guide feature engineering.

One of the first steps of the exploratory data process when the ultimate purpose is to predict a response, is to create visualizations that help elucidate knowledge of the response and then to uncover relationships between the predictors and the response. Therefore, our visualizations should start with the response, understanding the characteristics of its distribution, and then build outward from that with the additional information provided in the predictors. Knowledge about the response can be gained by creating a histogram or box plot. This simple visualization will reveal the amount of variation in the response and if the response was generated by a process that has unusual characteristics that must be investigated further. Next, we can move on to exploring relationships among the predictors and between predictors and the response. Important characteristics can be identified by examining

- scatter plots of individual predictors and the response,
- a pairwise correlation plot among the predictors,
- a projection of high-dimensional predictors into a lower dimensional space,
- line plots for time-based predictors,
- the first few levels of a regression or classification tree,
- a heatmap across the samples and predictors, or
- mosaic plots for examining associations among categorical variables.

These visualizations provide insights that should be used to inform the initial models. It is important to note that some of the most useful visualizations for exploring the data are not necessarily complex or difficult to create. In fact, a simple scatter plot can elicit insights that a model may not be able to uncover, and can lead to the creation of a new predictor or to a transformation of a predictor or the response that improves model performance. The challenge here lies in developing intuition for knowing how to visually explore data to extract information for improvement. As illustrated in Figure 1.4, exploratory data analysis should not stop at this point, but should continue after initial models have been built. Post model building, visual tools can be used to assess model lack-of-fit and to evaluate the potential effectiveness of new predictors that were not in the original model.

In this chapter, we will delve into a variety of useful visualization tools for exploring data prior to constructing the initial model. Some of these tools can then be used after the model is built to identify features that can improve model performance. Following the outline of Figure 1.4, we will look at visualizations prior to modeling, then during the modeling process. We also refer the reader to Tufte (1990), Cleveland (1993), and Healy (2018) which are excellent resources for visualizing data.

To illustrate these tools, the Chicago Train Ridership data will be used for numeric visualizations and the OkCupid data for categorical visualizations.

4.1 | Introduction to the Chicago Train Ridership Data

To illustrate how exploratory visualizations are a critical part of understanding a data set and how visualizations can be used to identify and uncover representations of the predictors that aid in improving the predictive ability of a model, we will be using data collected on ridership on the Chicago Transit Authority (CTA) "L" train system[36] (Figure 4.1). In the short term, understanding ridership over the next one or two weeks would allow the CTA to ensure that an optimal number of cars were available to service the Chicagoland population. As a simple example of demand fluctuation, we would expect that ridership in a metropolitan area would be stronger on weekdays and weaker on weekends. Two common mistakes of misunderstanding demand could be made. At one extreme, having too few cars on a line to meet weekday demand would delay riders from reaching their destination and would lead to overcrowding and tension. At the other extreme, having too many cars on the weekend would be inefficient leading to higher operational costs and lower profitability. Good forecasts of demand would help the CTA to get closer to optimally meeting demand.

In the long term, forecasting could be used to project when line service may be necessary, to change the frequency of stops to optimally service ridership, or to add or eliminate stops as populations shift and demand strengthens or weakens.

---

[36]http://bit.ly/FES-Chicago

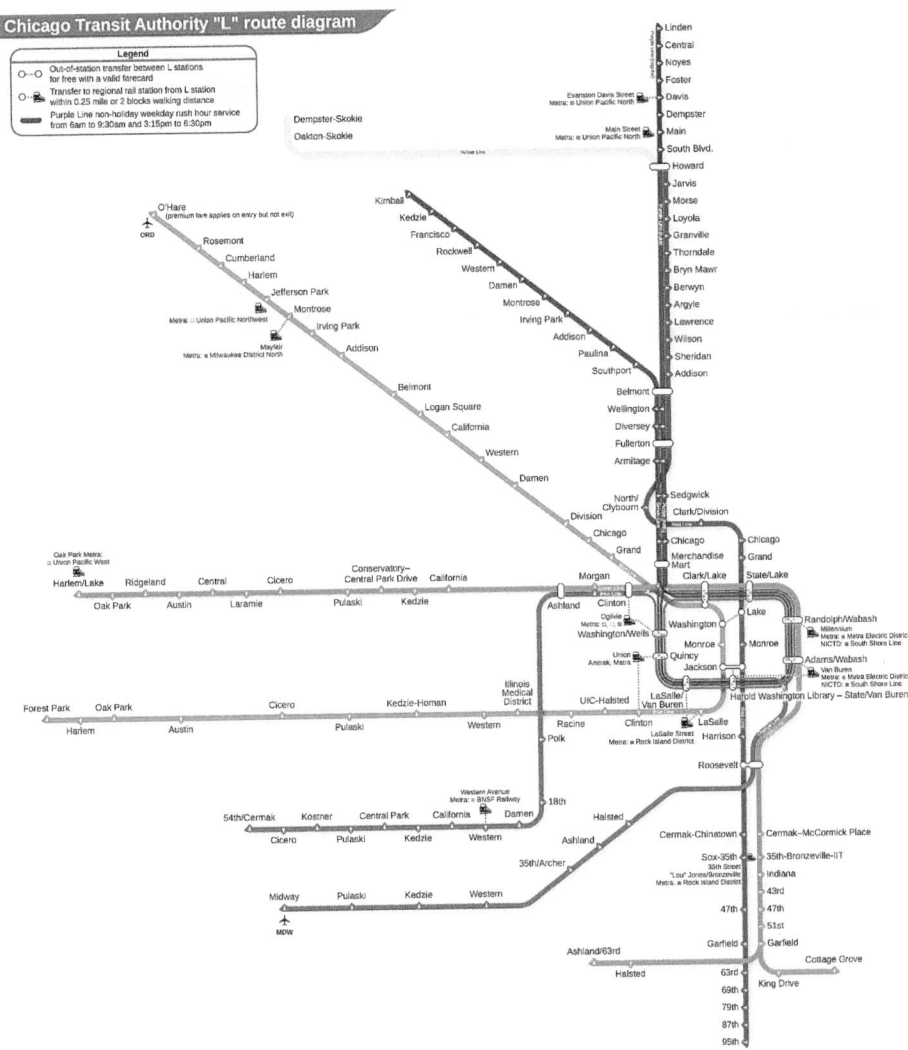

Figure 4.1: Chicago Transit Authority "L" map. For this illustration, we are interested in predicting the ridership at the Clark/Lake station in the Chicago Loop. (Source: Wikimedia Commons, Creative Commons license)

For illustrating the importance of exploratory visualizations, we will focus on short-term forecasting of daily ridership. Daily ridership data were obtained for 126 stations[37] between January 22, 2001 and September 11, 2016. Ridership is measured by the number of entries into a station across all turnstiles, and the number of daily riders across stations during this time period varied considerably, ranging between 0 and 36,323 per day. For ease of presentation, ridership will be shown and analyzed in units of thousands of riders.

Our illustration will narrow to predicting daily ridership at the Clark/Lake stop.

---

[37]The stations selected contained no missing values. See Section 8.5 for more details.

This station is an important stop to understand; it is located in the downtown loop, has one of the highest riderships across the train system, and services four different lines covering northern, western, and southern Chicago.

For these data, the ridership numbers for the other stations might be important to the models. However, since the goal is to predict *future ridership volume*, only historical data would be available at the time of prediction. For time series data, predictors are often formed by *lagging* the data. For this application, when predicting day D, predictors were created using the lag–14 data from every station (e.g., ridership at day $D - 14$). Other lags can also be added to the model, if necessary.

Other potential regional factors that may affect public transportation ridership are the weather, gasoline prices, and the employment rate. For example, it would be reasonable to expect that more people would use public transportation as the cost of using personal transportation (i.e., gas prices) increases for extended periods of time. Likewise, the use of public transportation would likely increase as the unemployment rate decreases. Weather data were obtained for the same time period as the ridership data. The weather information was recorded hourly and included many conditions such as overcast, freezing rain, snow, etc. A set of predictors was created that reflects conditions related to rain, snow/ice, clouds, storms, and clear skies. Each of these categories is determined by pooling more granular recorded conditions. For example, the rain predictor reflects recorded conditions that included "rain," "drizzle," and "mist". New predictors were encoded by summarizing the hourly data across a day by calculating the percentage of observations within a day where the conditions were observed. As an illustration, on December 20, 2012, conditions were recorded as snowy (12.8% of the day), cloudy (15.4%), as well as stormy (12.8%), and rainy (71.8%). Clearly, there is some overlap in the conditions that make up these categories, such as cloudy and stormy.

Other hourly weather data were also available related to the temperature, dew point, humidity, air pressure, precipitation and wind speeds. Temperature was summarized by the daily minimum, median, and maximum as well as the daily change. Pressure change was similarly calculated. In most other cases, the median daily value was used to summarize the hourly records. As with the ridership data, future weather data are not available so lagged versions of these data were used in the models.[38] In total, there were 18 weather-related predictors available. Summarizations of these types of data are discussed more in Chapter 9.

In addition to daily weather data, average weekly gasoline prices were obtained for the Chicago region (U.S. Energy Information Administration, 2017a) from 2001 through 2016. For the same time period, monthly unemployment rates were pulled from United States Census Bureau (2017). The potential usefulness of these predictors will be discussed below.

---

[38]Another alternative would be to simply compute the daily average values for these conditions from the entire training set.

4.2 Visualizations for Numeric Data: Exploring Train Ridership Data

4.2.1 Box Plots, Violin Plots, and Histograms

Univariate visualizations are used to understand the distribution of a single variable. A few common univariate visualizations are box-and-whisker plots (i.e., box plot), violin plots, or histograms. While these are simple graphical tools, they provide great value in comprehending characteristics of the quantity of interest.

Because the foremost goal of modeling is to understand variation in the response, the first step should be to understand the distribution of the response. For a continuous response such as the ridership at the Clark/Lake station, it is important to understand if the response has a symmetric distribution, if the distribution has a decreasing frequency of larger observations (i.e., the distribution is skewed), if the distribution appears to be made up of two or more individual distributions (i.e., the distribution has multiple peaks or modes), or if there appears to be unusually low or high observations (i.e outliers).

Understanding the distribution of the response as well as its variation provides a lower bound of the expectations of model performance. That is, if a model contains meaningful predictors, then the residuals from a model that contains these predictors should have less variation than the variation of the response. Furthermore, the distribution of the response may indicate that the response should be transformed prior to analysis. For example, responses that have a distribution where the frequency of response proportionally decreases with larger values may indicate that the response follows a log-normal distribution. In this case, log-transforming the response would induce a normal (bell-shaped, symmetric) distribution and often will enable a model to have better predictive performance. A third reason why we should work to understand the response is because the distribution may provide clues for including or creating features that help explain the response.

As a simple example of the importance of understanding the response distribution, consider Figure 4.2 which displays a box plot of the response for the ridership at the Clark/Lake station. The box plot was originally developed by John Tukey as a quick way to assess a variable's distribution (Tukey, 1977), and consists of the minimum, lower quartile, median, upper quartile and maximum of the data. Alternative versions of the box plot extend the whiskers to a value beyond which samples would be considered unusually high (or low) (Frigge et al., 1989). A variable that has a symmetric distribution has equal spacing across the quartiles making the box and whiskers also appear symmetric. Alternatively, a variable that has fewer values in a wider range of space will not appear symmetric.

A drawback of the box plot is that it is not effective at identifying distributions that have multiple peaks or modes. As an example, consider the distribution of ridership at the Clark/Lake station (Figure 4.3). Part (a) of this figure is a histogram of the data. To create a histogram, the data are binned into equal regions of the variable's

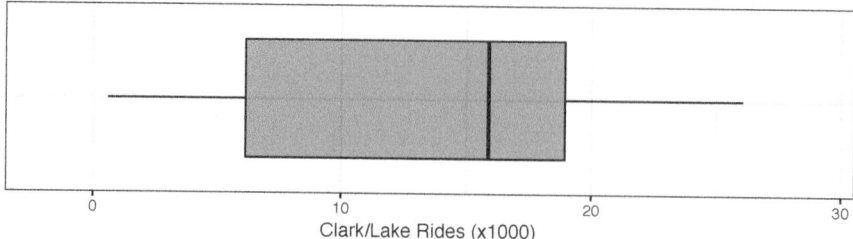

Figure 4.2: A box-and-whisker plot of the daily ridership at the Clark/Lake station from 2001 through 2016. The blue region is the box and represents the middle 50% of the data. The whiskers extend from the box and indicate the region of the lower and upper 25% of the data.

value. The number of samples are counted in each region, and a bar is created with the height of the frequency (or percentage) of samples in that region. Like box plots, histograms are simple to create, and these figures offer the ability to see additional distributional characteristics. In the ridership distribution, there are two peaks, which could represent two different mechanisms that affect ridership. The box plot (b) is unable to capture this important nuance. To achieve a compact visualization of the distribution that retains histogram-like characteristics, Hintze and Nelson (1998) developed the violin plot. This plot is created by generating a density or distribution of the data and its mirror image. Figure 4.3 (c) is the violin plot, where we can now see the two distinct peaks in ridership distribution. The lower quartile, median, and upper quartile can be added to a violin plot to also consider this information in the overall assessment of the distribution.

These data will be analyzed in several chapters. Given the range of the daily ridership numbers, there was some question as to whether the outcome should be modeled in the natural units or on the log scale. On one hand, the natural units make interpretation of the results easier since the RMSE would be in terms of riders. However, if the outcome were transformed prior to modeling, it would ensure that *negative* ridership could not be predicted. The bimodal nature of these data, as well as distributions of ridership for each year that have a longer tail on the right made this decision difficult. In the end, a handful of models were fit both ways to make the determination. The models computed in the natural units appeared to have slightly better performance and, for this reason, all models were analyzed in the natural units.

Examining the distribution of each predictor can help to guide our decisions about the need to engineer the features through transformation prior to analysis. When we have a moderate number of predictors ($< \sim 100$) and when the predictors are on the same order of magnitude, we can visualize the distributions simultaneously using side-by-side box or violin plots. Consider again the ridership data with the two-week lag in ridership as predictors. The distributions across selected stations for weekday ridership for 2016 are provided in Figure 4.4. Here, our focus is on the variability and

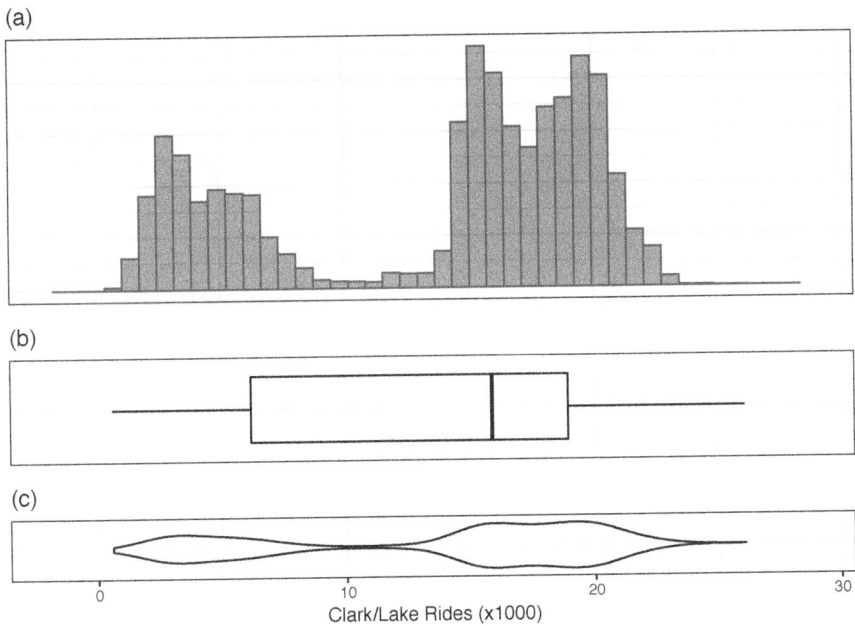

Figure 4.3: Distribution of daily ridership at the Clark/Lake stop from 2001 to 2016. The histogram (a) allows us to see that there are two peaks or modes in ridership distribution indicating that there may be two mechanisms affecting ridership. The box plot (b) does not have the ability to see multiple peaks in the data. However the violin plot (c) provides a compact visualization that identifies the distributional nuance.

range of ridership across stations. Despite the earlier warning, box plots are good at characterizing these aspects of data. To help see patterns more clearly, ridership is ordered from the largest median (left) to the smallest median (right). Several characteristics stand out: variability in ridership increases with the median ridership, there are a number of unusually low and unusually high values for each station, and a few stations have distinctly large variation. One station particularly stands out, which is about one-quarter of the way from the left. This happens to be the Addison station which is the nearest stop to Wrigley Field. The wider distribution is due to ridership associated with the weekday home games for the Chicago Cubs, with attendance at its peak reaching close to the most frequently traveled stations. If the goal was to predict ridership at the Addison station, then the Cubs' home game schedule would be important information for any model. The unusually low values for the majority of the stations will be discussed next.

As the number of predictors grows, the ability to visualize the individual distributions lessens and may be practically impossible. In this situation, a subset of predictors that are thought to be important can be examined using these techniques.

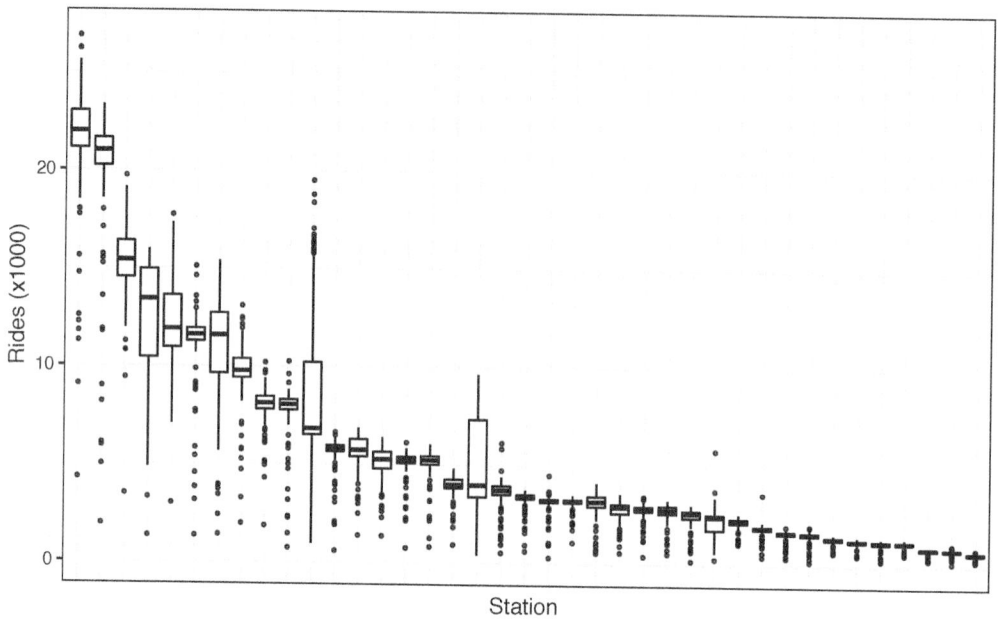

Figure 4.4: Distribution of weekday ridership at selected stations during 2016.

4.2.2 | Augmenting Visualizations through Faceting, Colors, and Shapes

Additional dimensions can be added to almost any figure by using faceting, colors, and shapes. *Faceting* refers to creating the same type of plot (e.g., a scatter plot) and splitting the plot into different panels based on some variable.[39] Figure 3.2 is a good example. While this is a simple approach, these types of augmentation can be powerful tools for seeing important patterns that can be used to direct the engineering of new features. The Clark/Lake station ridership distribution is a prime candidate for adding another dimension. As shown above, Figure 4.3 has two distinct peaks. A reasonable explanation for this would be that ridership is different for weekdays than for weekends. Figure 4.5 partitions the ridership distribution by part of the week through color and faceting (for ease of visualization). Part of the week was not a predictor in the original data set; by using intuition and carefully examining the distribution, we have found a feature that is important and necessary for explaining ridership and should be included in a model.

Figure 4.5 invites us to pursue understanding of these data further. Careful viewing of the weekday ridership distribution should draw our eye to a long tail on the left, which is a result of a number of days with lower ridership similar to the range of ridership on weekends. What would cause weekday ridership to be low? If the cause can be uncovered, then a feature can be engineered. A model that has the ability to explain these lower values will have better predictive performance than a model that does not.

---

[39]This visualization approach is also referred to as "trellising" or "conditioning" in different software.

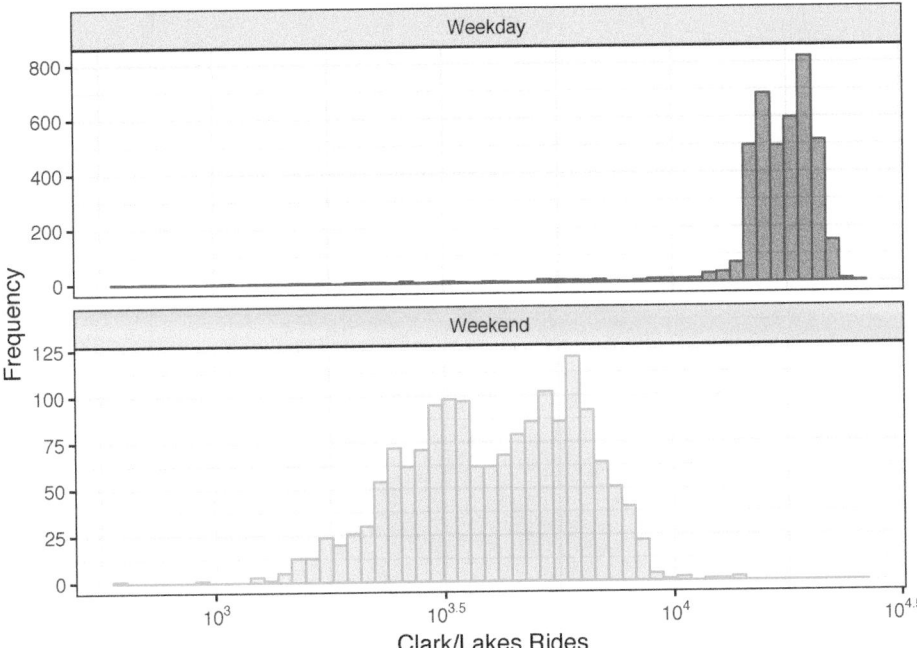

Figure 4.5: Distribution of daily ridership at the Clark/Lake stop from 2001 to 2016 colored and faceted by weekday and weekend. Note that the y-axis counts for each panel are on different scales and the long left tail in the weekday data.

The use of colors and shapes for elucidating predictive information will be illustrated several of the following sections.

4.2.3 | Scatter Plots

Augmenting visualizations through the use of faceting, color, or shapes is one way to incorporate an additional dimension in a figure. Another approach is to directly add another dimension to a graph. When working with two numeric variables, this type of graph is called a scatter plot. A scatter plot arranges one variable on the x-axis and another on the y-axis. Each sample is then plotted in this coordinate space. We can use this type of figure to assess the relationship between a predictor and the response, to uncover relationships between pairs of predictors, and to understand if a new predictor may be useful to include in a model. These simple relationships are the first to provide clues to assist in engineering characteristics that may not be directly available in the data.

If the goal is to predict ridership at the Clark/Lake station, then we could anticipate that recent past ridership information should be related to current ridership. That is to say, another potential predictor to consider would be the previous day's or previous week's ridership information. Because we know that weekday and weekend have different distributions, a one-day lag would be less useful for predicting ridership on Monday or Saturday. A week-based lag would not have this difficulty (although it

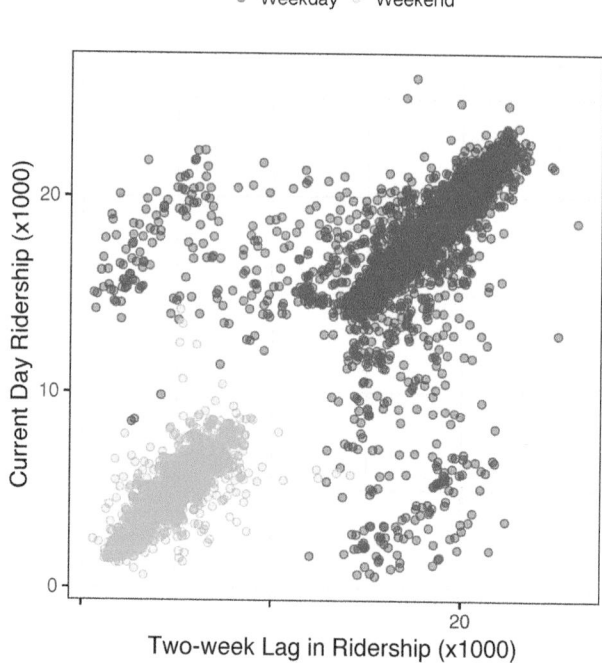

Figure 4.6: A scatter plot of the 14-day lag ridership at the Clark/Lake station and the current-day ridership at the same station.

would be further apart in time) since the information occurs on the same day of the week. Because the primary interest is in predicting ridership two weeks in advance, we will create the 14-day lag in ridership for the Clark/Lake station.

In this case, the relationship between these variables can be directly understood by creating a scatter plot (Figure 4.6). This figure highlights several characteristics that we need to know: there is a strong linear relationship between the 14-day lag and current-day ridership, there are two distinct groups of points (due to the part of the week), and there are many 14-day lag/current day pairs of days that lie far off from the overall scatter of points. These results indicate that the 14-day lag will be a crucial predictor for explaining current-day ridership. Moreover, uncovering the explanation of samples that are far off from the overall pattern visualized here will lead to a new feature that will be useful as a input to models.

4.2.4 | Heatmaps

Low weekday ridership as illustrated in Figure 4.5 might be due to annual occurrences; to investigate this hypothesis, the data will need to be augmented. The first step is to create an indicator variable for weekdays with ridership less than 10,000 or greater than or equal to 10,000. We then need a visualization that allows us to see when these unusual values occur. A visualization that would elucidate annual patterns in this context is a heatmap. A heatmap is a versatile plot that can be created utilizing almost any type of predictor and displays one predictor on the x-axis and another predictor on the y-axis. In this figure the x- and y-axis predictors must be able to be categorized. The categorized predictors then form a grid, and the grid is filled by another variable. The filling variable can be either continuous or categorical. If continuous, then the boxes in the grid are colored on a continuous scale from the lowest value of the filling predictor to the highest value. If the filling variable is categorical, then the boxes have distinct colors for each category.

For the ridership data, we will create a month and day predictor, a year predictor, and an indicator of weekday ridership less than 10,000 rides.

These new features are the inputs to the heatmap (Figure 4.7). In this figure, the x-axis represents the year and the y-axis represents the month and day. Red boxes indicate weekdays that have ridership less than 10,000 for the Clark/Lake station. The heatmap of the data in this form brings out some clear trends. Low ridership occurs on or around the beginning of the year, mid-January, mid-February until 2007, late-May, early-July, early-September, late-November, and late-December. Readers in the US would recognize these patterns as regularly observed holidays. Because holidays are known in advance, adding a feature for common weekday holidays will be beneficial for models to explain ridership.

Carefully observing the heatmap points to two days that do not follow the annual patterns: February 2, 2011 and January 6, 2014. These anomalies were due to extreme weather. On February 2, 2011, Chicago set a record low temperature of -16F. Then on January 6, 2014, there was a blizzard that dumped 21.2 inches of snow on the region. Extreme weather instances are infrequent, so adding this predictor will have limited usefulness in a model. If the frequency of extreme weather increases in the future, then using forecast data could become a valuable predictor for explaining ridership.

Now that we understand the effect of major US holidays, these values will be excluded from the scatter plot of 14-day lag versus current-day ridership (Figure 4.8). Most of the points that fell off the diagonal of Figure 4.6 are now gone. However, a couple of points remain. The day associated with these points was June 11, 2010, which was the city's celebration for the Chicago Blackhawks winning the Stanley Cup. While these types of celebrations are infrequent, engineering a feature to anticipate these unusual events will aid in reducing the prediction error for a model.[40]

---

[40]This does lead to an interesting dilemma for this analysis: should such an aberrant instance be allowed to potentially influence the model? We have left the value untouched but a strong case could be made for imputing what the value would be from previous data and using this value in the analysis.

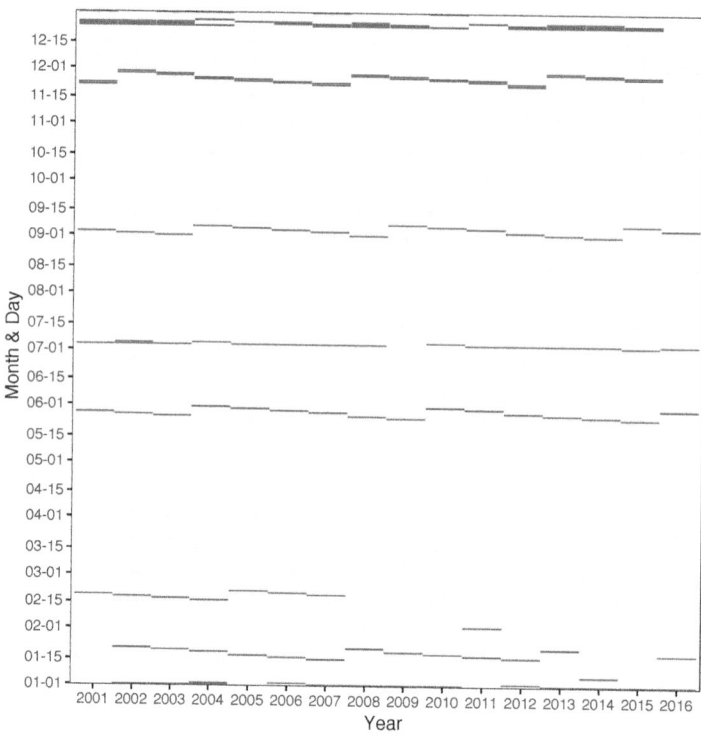

Figure 4.7: Heatmap of ridership from 2001 through 2016 for weekdays that have less than 10,000 rides at the Clark/Lake station. This visualization indicates that the distinct patterns of low ridership on weekdays occur on and around major US holidays.

4.2.5 | Correlation Matrix Plots

An extension of the scatter plot is the correlation matrix plot. In this plot, the correlation between each pair of variables is plotted in the form of a matrix. Every variable is represented on the outer x-axis and outer y-axis of the matrix, and the strength of the correlation is represented by the color in the respective location in the matrix. This visualization was first shown in Figure 2.3. Here we will construct a similar image for the 14-day lag in ridership across stations for non-holiday, weekdays in 2016 for the Chicago data. The correlation structure of these 14-day lagged predictors is almost the same as the original (unlagged) predictors; using the lagged versions ensures the correct number of rows are used.

The correlation matrix in Figure 4.9 leads to additional understandings. First, ridership across stations is positively correlated (red) for nearly all pairs of stations; this means that low ridership at one station corresponds to relatively low ridership at another station, and high ridership at one station corresponds to relatively high ridership at another station. Second, the correlation for a majority of pairs of stations is extremely high. In fact, more than 18.7% of the predictor pairs have a correlation

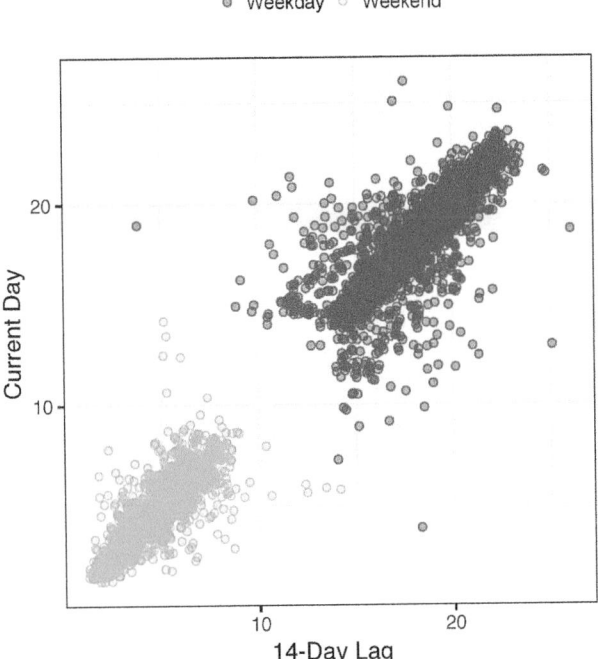

Figure 4.8: Two-week lag in daily ridership versus daily ridership for the Clark/Lake station, with common US holidays excluded, and colored by part of the week.

greater than 0.90 and 3.1% have a correlation greater than 0.95. The high degree of correlation is a clear indicator that the information present across the stations is redundant and could be eliminated or reduced. Filtering techniques as discussed in Chapter 3 could be used to eliminate predictors. Also, feature engineering through dimension reduction (Chapter 6) could be an effective alternative representation of the data in settings like these. We will address dimension reduction as an exploratory visualization technique in Section 4.2.7.

This version of the correlation plot includes an organization structure of the rows and columns based on hierarchical cluster analysis (Dillon and Goldstein, 1984). The overarching goal of cluster analysis is to arrange samples in a way that those that are 'close' in the measurement space are also nearby in their location on the axis. For these data, the distances between any two stations are based on the stations' vectors of correlation values. Therefore, stations that have similar correlation vectors will be nearby in their arrangement on each axis, whereas stations that have dissimilar correlation vectors will be located far away. The tree-like structure on the x- and y-axes is called a dendrogram and connects the samples based on their correlation vector proximity. This organizational structure helps to elucidate some visually distinct groupings of stations. These groupings, in turn,

Figure 4.9: Visualization of the correlation matrix of the 14-day lag ridership station predictors for non-holiday weekdays in 2016. Stations are organized using hierarchical cluster analysis, and the organizational structure is visualized using a dendrogram (top and right).

may point to features that could be important to include in models for explaining ridership.

As an example, consider the stations shown on the very left-hand side of the x-axis where there are some low and/or negative correlations. One station in this group has a median correlation of 0.23 with the others. This station services O'Hare airport, which is one of the two major airports in the area. One could imagine that ridership at this station has a different driver than for other stations. Ridership here is likely influenced by incoming and outgoing plane schedules, while the other stations are not. For example, this station has a negative correlation (-0.46) with the UIC-Halsted station on the same line. The second-most dissimilar station is the previously mentioned Addison station as it is driven by game attendance.

4.2.6 | Line Plots

A variable that is collected over time presents unique challenges in the modeling process. This type of variable is likely to have trends or patterns that are associated incrementally with time. This means that a variable's current value is more related to recent values than to values further apart in time. Therefore, knowing what a variable's value is today will be more predictive of tomorrow's value than last week,

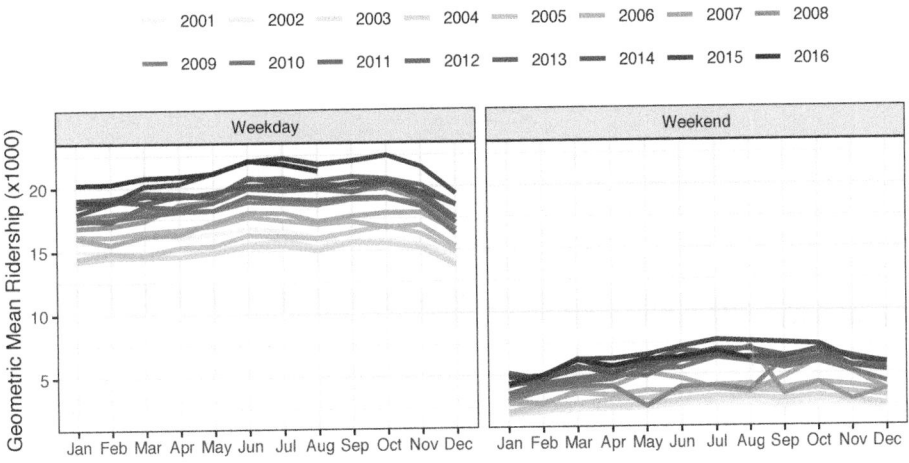

Figure 4.10: Monthly average ridership per year by weekday (excluding holidays) or weekend.

last month's, or last year's value. We can assess the relationship between time and the value of a variable by creating a line plot, which is an extension of a scatter plot. In a line plot, time is on the x-axis, and the value of the variable is on the y-axis. The value of the variable at adjacent time points is connected with a line. Identifying time trends can lead to other features or to engineering other features that are associated with the response. The Chicago data provide a good illustration of line plots.

The Chicago data are collected over time, so we should also look for potential trends due to time. To look for these trends, the mean of ridership per month during weekdays and separately during the weekend was computed (Figure 4.10). Here we see that since 2001, ridership has steadily increased during weekdays and weekends. This would make sense because the population of the Chicago metropolitan area has increased during this time period (United States Census Bureau, 2017). The line plot additionally reveals that within each year, ridership generally increases from January through October then decreases through December. These findings imply that the time proximity of ridership information should be useful to a model. That is, understanding ridership information within the last week or month will be more useful in predicting future ridership.

Weekend ridership also shows annual trends but exhibits more variation within the trends for some years. Uncovering predictors associated with this increased variation could help reduce forecasting error. Specifically, the weekend line plots have the highest variation during 2008, with much higher ridership in the summer. A potential driver for increased ridership on public transportation is gasoline prices. Weekly gas prices in the Chicago area were collected from the U.S. Energy Information Administration, and the monthly average price by year is shown in the line plot of Figure 4.11.

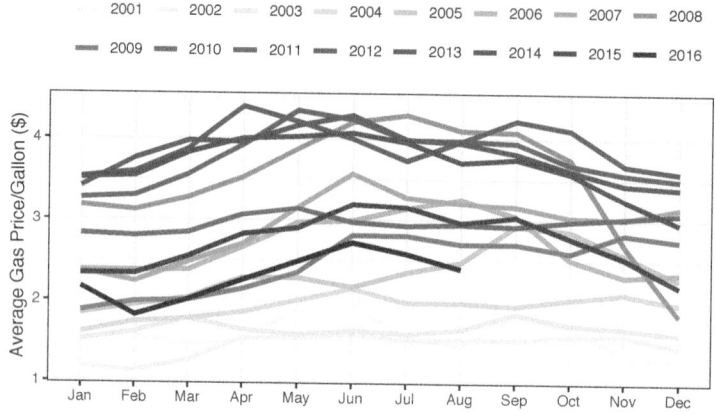

Figure 4.11: Monthly average gas price per gallon (USD) per year. Prices spike in the summer of 2008, which is at the same time that weekend ridership spikes.

Next, let's see if a relationship can be established between gas prices and ridership. To do this, we will calculate the monthly average 2-week lag in gas price and the geometric mean of ridership at the Clark/Lake station. Figure 4.12 illustrates this relationship; there is a positive association between the 2-week lag in gas price and the geometric mean of ridership. From 2001 through 2014, the higher the gas price, the higher the ridership, with 2008 data appearing at the far right of both the Weekday and Weekend scatter plots. This trend is slightly different for 2015 and 2016 when oil prices dropped due to a marked increase in supply (U.S. Energy Information Administration, 2017b). Here we see that by digging into characteristics of the original line plot, another feature can be found that will be useful for explaining some of the variation in ridership.

4.2.7 | Principal Components Analysis

It is possible to visualize five or six dimensions of data in a two-dimensional figure by using colors, shapes, and faceting. But almost any data set today contains many more than just a handful of variables. Being able to visualize many dimensions in the physical space that we can actually see is crucial to understanding the data and to understanding if there are characteristics of the data that point to the need for feature engineering. One way to condense many dimensions into just two or three is to use projection techniques such as principal components analysis (PCA), partial least squares (PLS), or multidimensional scaling (MDS). Dimension reduction techniques for the purpose of feature engineering will be more fully addressed in Section 6.3. Here we will highlight PCA, and how this method can be used to engineer features that effectively condense the original predictors' information.

Predictors that are highly correlated, like the station ridership illustrated in Figure 4.9, can be thought of as existing in a lower dimensional space than the original data. That is, the data represented here could be approximately represented by

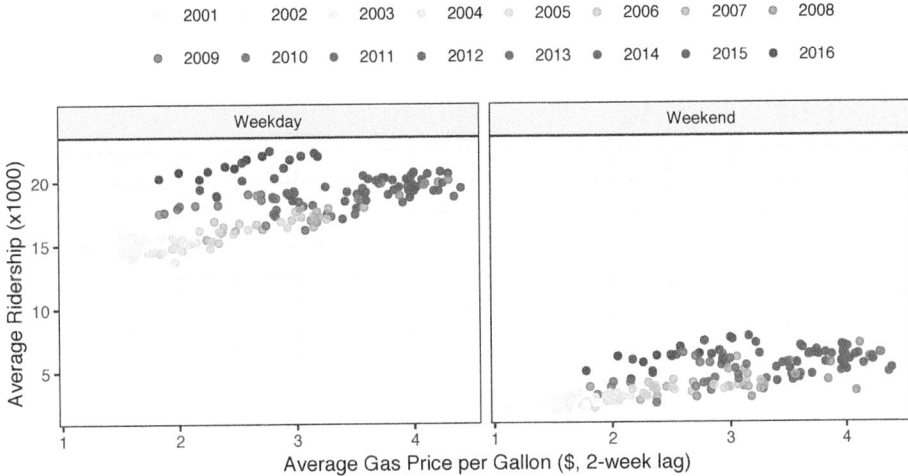

Figure 4.12: Monthly average two-week lag in gas price per gallon (USD) versus average monthly ridership. There is a positive relationship between gas price and ridership indicating that gas prices could help to explain some of the ridership variation.

combinations of similar stations. Principal components analysis finds combinations of the variables that best summarize the variability in the original data (Dillon and Goldstein, 1984). The combinations are a simpler representation of the data and often identify underlying characteristics within the data that will help guide the feature engineering process.

PCA will now be applied to the 14-day lagged station ridership data. Because the objective of PCA is to optimally summarize variability in the data, the cumulative percentage of variation is summarized by the top components. For these data, the first component captures 76.7% of the overall variability, while the first two components capture 83.1%. This is a large percentage of variation given that there are 125 total stations, indicating that station ridership information is redundant and can likely be summarized in a more condensed fashion.

Figure 4.13 provides a summary of the analysis. Part (a) displays the cumulative amount of variation summarized across the first 50 components. This type of plot is used to visually determine how many components are required to summarize a sufficient amount of variation in the data. Examining a scatter plot of the first two components (b) reveals that PCA is focusing on variation due to the part of the week with the weekday samples with lower component 1 scores and weekend samples with higher component 1 scores. The second component focuses on variation due to change over time with earlier samples receiving lower component 2 scores and later samples receiving higher component 2 scores. These patterns are revealed more clearly in parts (c) and (d) where the first and second components are plotted against the underlying variables that appear to affect them the most.

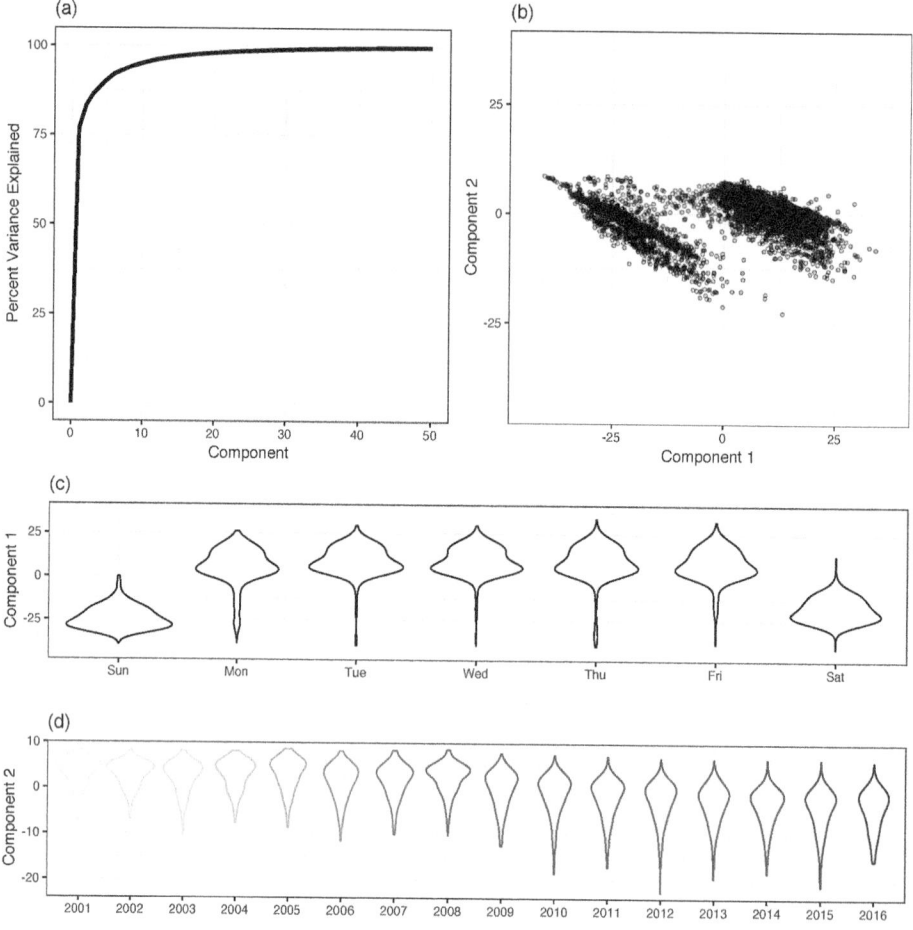

Figure 4.13: Principal component analysis of the 14-day station lag ridership. (a) The cumulative variability summarized across the first 10 components. (b) A scatter plot of the first two principal components. The first component focuses on variation due to part of the week while the second component focuses on variation due to time (year). (c) The relationship between the first principal component and ridership for each day of the week at the Clark/Lake station. (d) The relationship between second principal component and ridership for each year at the Clark/Lake station.

Other visualizations earlier in this chapter already alerted us to the importance of part of the week and year relative to the response. PCA now helps to confirm these findings, and will enable the creation of new features that simplify our data while retaining crucial predictive information. We will delve into this techniques and other similar techniques later in Section 6.3.

4.3 | Visualizations for Categorical Data: Exploring the OkCupid Data

To illustrate different visualization techniques for qualitative data, the OkCupid data are used. These data were first introduced in Section 3.1. Recall that the training set consisted of 38,809 profiles and the goal was to predict whether the profile's author worked in a STEM field. The event rate is 18.5% and most predictors were categorical in nature. These data are discussed more in the next chapter.

4.3.1 | Visualizing Relationships between Outcomes and Predictors

Traditionally, bar charts are used to represent counts of categorical values. For example, Figure 4.14(a) shows the frequency of the stated religion, partitioned and colored by outcome category. The virtue of this plot is that it is easy to see the most and least frequent categories. However, it is otherwise problematic for several reasons:

1. To understand if any religions are associated with the outcome, the reader's task is to visually judge the ratio of each dark blue bar to the corresponding light blue bar across all religions and to then determine if any of the ratios are different from random chance. This figure is ordered from greatest ratio (left) to least ratio (right) which, in this form, may be difficult for the reader see.

2. The plot is *indirectly* illustrating the characteristic of data that we are interested in, specifically, the ratio of frequencies between the STEM and non-STEM profiles. We don't care how many Hindus are in STEM fields; instead, the ratio of fields within the Hindu religion is the focus. In other words, the plot obscures the statistical hypothesis that we are interested in: is the rate of Hindu STEM profiles different than what we would expect by chance?

3. If the rate of STEM profiles within a religion is the focus, bar charts give no sense of *uncertainty* in that quantity. In this case, the uncertainty comes from two sources. First, the number of profiles in each religion can obviously affect the variation in the proportion of STEM profiles. This is illustrated by the height of the bars but this isn't a precise way to illustrate the noise. Second, since the statistic of interest is a proportion, the variability in the statistic becomes larger as the rate of STEM profiles approaches 50% (all other things being equal).

To solve the first two issues, we might show the within-religion percentages of the STEM profiles. Figure 4.14(b) shows this alternative version of the bar chart, which is an improvement since the proportion of STEM profiles is now the focus. It is more obvious how the religions were ordered and we can see how much each deviates from the baseline rate of 18.4%. However, we still haven't illustrated the uncertainty. Therefore, we cannot assess if the rates for the profiles with no stated religion are truly different from agnostics. Also, while Figure 4.14(b) directly compares proportions across religions, it does not give any sense of the frequency of each religion. For these data there are very few Islamic profiles; this important information cannot be seen in this display.

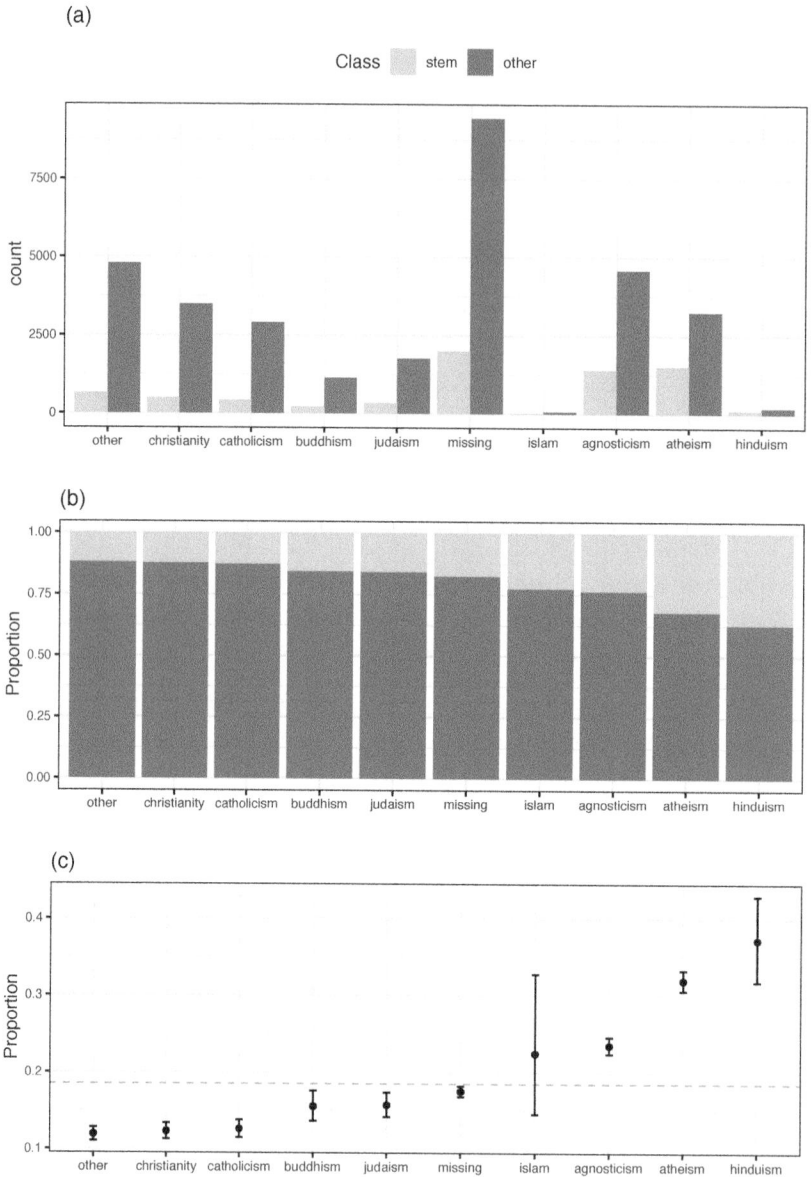

Figure 4.14: Three different visualizations of the relationship between religion and the outcome.

Figure 4.14(c) solves all three of the problems listed above. For each religion, the proportion of STEM profiles is calculated and a 95% confidence interval is shown to help understand what the noise is around this value. We can clearly see which religions deviate from randomness and the width of the error bars helps the reader understand how much each number should be trusted. This image is the best of the three since it directly shows the *magnitude* of the difference as well as the *uncertainty*. In situations where the number of categories on the x-axis is large, a volcano plot can be used to show the results (see Figures 2.6 and 5.4).

The point of this discussion is not that summary statistics with confidence intervals are always the solution to a visualization problem. The takeaway message is that each graph should have a clearly defined hypothesis and that this hypothesis is shown concisely in a way that allows the reader to make quick and informative judgments based on the data.

Finally, does religion appear to be related to the outcome? Since there is a gradation of rates of STEM professions between the groups, it would appear so. If there were no relationship, all of the rates would be approximately the same.

How would one visualize the relationship between a categorical outcome and a numeric predictor? As an example, the total length of all the profile essays will be used to illustrate a possible solution. There were 1181 profiles in the training set where the user did not fill out any of the open text fields. In this analysis, all of the nine answers were concatenated into a single text string. The distribution of the text length was very left-skewed with a median value of 1,862 characters. The maximum length was approximately 59K characters, although 10% of the profiles contained fewer than 444 characters. To investigate this, the distribution of the total number of essay characters is shown in Figure 4.15(a). The x-axis is the log of the character count and profiles without essay text are shown here as a zero. The distributions appear to be extremely similar between classes, so this predictor (in isolation) is unlikely to be important by itself. However, as discussed above, it would be better to try to directly answer the question.

To do this, another *smoother* is used to model the data. In this case, a regression spline smoother (Wood, 2006) is used to model the probability of a STEM profile as a function of the (log) essay length. This involves fitting a logistic regression model using a *basis expansion*. This means that our original factor, log essay length, is used to create a set of artificial features that will go into a logistic regression model. The nature of these predictors will allow a flexible, local representation of the class probability across values of essay length and are discussed more in Section 6.2.1.[41]

The results are shown in Figure 4.15(b). The black line represents the class probability of the logistic regression model and the bands denote 95% confidence intervals around the fit. The horizontal red line indicates the baseline probability of STEM profiles from the training set. Prior to a length of about $10^{1.5}$, the profile is slightly less likely than chance to be STEM. Larger profiles show an increase in the probability. This might seem like a worthwhile predictor but consider the scale of the y-axis. If this were put in the full probability scale of [0, 1], the trend would appear virtually flat. At most, the increase in the likelihood of being a STEM profile is almost 3.5%.

---

[41]There are other types of smoothers that can be used to discover potential nonlinear patterns in the data. One, called `loess`, is very effective and uses a series of moving regression lines across the predictor values to make predictions at a particular point (Cleveland, 1979).

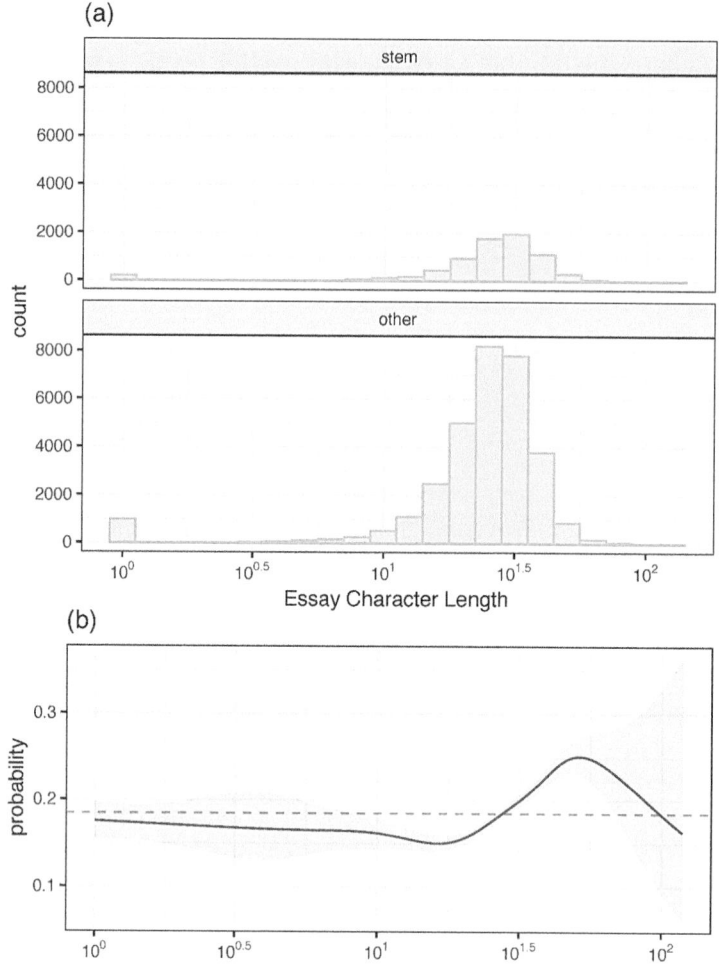

Figure 4.15: The effect of essay length on the outcome.

Also, note that the confidence bands rapidly expand around $10^{1.75}$ mostly due to a decreasing number of data points in this range so the potential increase in probability has a high degree of uncertainty. This predictor might be worth including in a model but is unlikely to show a strong effect on its own.

4.3.2   Exploring Relationships between Categorical Predictors

Before deciding how to use predictors that contain non-numeric data, it is critical to understand their characteristics and relationships with other predictors. The unfortunate practice that is often seen relies on a large number of basic statistical analyses of two-way tables that boil relationships down to numerical summaries such as the Chi-squared (χ^2) test of association. Often the best approach is to visualize the data. When considering relationships between categorical data, there are several options. Once a cross-tabulation between variables is created, mosaic plots can once again be used to understand the relationship between variables. For the OkCupid data, it is conceivable that the questionnaires for drug and alcohol use might be related and, for these variables, Figure 4.16 shows the mosaic plot. For alcohol, the majority of the data indicate social drinking while the vast majority of the drug responses were "never" or missing. Is there any relationship between these variables? Do any of the responses "cluster" with others?

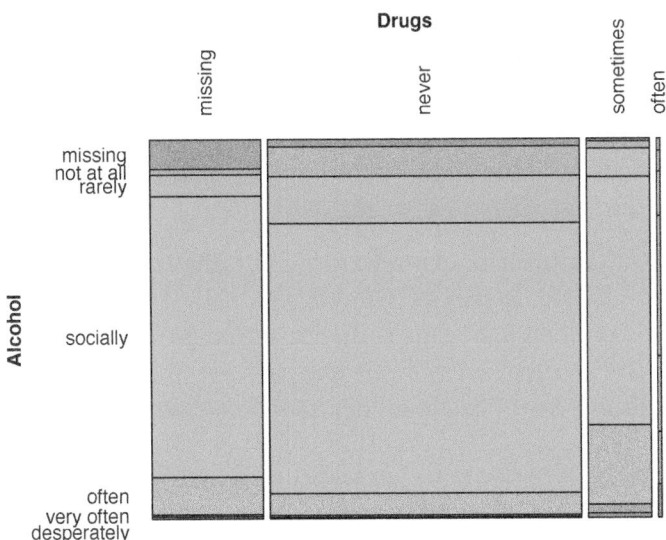

Figure 4.16: A mosaic plot of the drug and alcohol data in the OkCupid data.

These questions can be answered using *correspondence analysis* (Greenacre, 2017) where the cross-tabulation is analyzed. In a contingency table, the frequency distributions of the variables can be used to determine the *expected cell counts* which mimic what would occur if the two variables had no relationship. The traditional χ^2 test uses the deviations from these expected values to assess the association between the variables by adding up functions of this type of cell residuals. If the two variables in the table are strongly associated, the overall χ^2 statistic is large. Instead of summing up these residual functions, correspondence analysis analyzes them to determine new variables that account for the largest fraction of these statistics.[42] These new variables, called the *principal coordinates*, can be computed for both variables in the table and shown in the same plot. These plots can contain several features:

- The data on each axis would be evaluated in terms of how much of the information in the original table that the principal coordinate accounts for. If a coordinate only captures a small percentage of the overall χ^2 statistic, the patterns shown in that direction should not be over-interpreted.

- Categories that fall near the origin represent the "average" value of the data

---

[42]The mechanisms are very similar to principal component analysis (PCA), discussed in Chapter 6. For example, both use a singular value decomposition to compute the new variables.

From the mosaic plot, it is clear that there are categories for each variable that have the largest cell frequencies (e.g., "never" drugs and "social" alcohol consumption). More unusual categories are located on the outskirts of the principal coordinate scatter plot.

- Categories for a single variable whose principal coordinates are close to one another are indicative of redundancy, meaning that there may be the possibility of pooling these groups.

- Categories in different variables that fall near each other in the principal coordinate space indicate an association between these categories.

In the cross-tabulation between alcohol and drug use, the χ^2 statistic is very large (4114.2) for its degrees of freedom (18) and is associated with a very small p-value (0). This indicates that there is a strong association between these two variables.

Figure 4.17 shows the principal coordinates from the correspondence analysis. The component on the x-axis accounts for more than half of the χ^2 statistics. The values around zero in this dimension tends to indicate that no choice was made or sporadic substance use. To the right, away from zero, are less frequently occurring values. A small cluster indicates that occasional drug use and frequent drinking tend to have a specific association in the data. Another, more extreme, cluster shows that using alcohol very often and drugs have an association. The y-axis of the plot is mostly driven by the missing data (understandably since 25.8% of the table has at least one missing response) and accounts for another third of the χ^2 statistics. These results indicate that these two variables have similar results and might be measuring the same underlying characteristic.

4.4   Postmodeling Exploratory Visualizations

As previously mentioned in Section 3.2.2, predictions on the assessment data produced during resampling can be used to understand performance of a model. It can also guide the modeler to understand the next set of improvements that could be made using visualization and analysis. Here, this process is illustrated using the train ridership data.

Multiple linear regression has a rich set of diagnostics based on model residuals that aid in understanding the model fit and in identifying relationships that may be useful to include in the model. While multiple linear regression lags in predictive performance to other modeling techniques, the available diagnostics are magnificent tools that should not be underestimated in uncovering predictors and predictor relationships that can benefit more complex modeling techniques.

One tool from regression diagnosis that is helpful for identifying useful predictors is the *partial regression plot* (Neter et al., 1996). This plot utilizes residuals from two distinct linear regression models to unearth the potential usefulness of a predictor in a model. To begin the process, we start by fitting the following model and computing the residuals (ϵ_i):

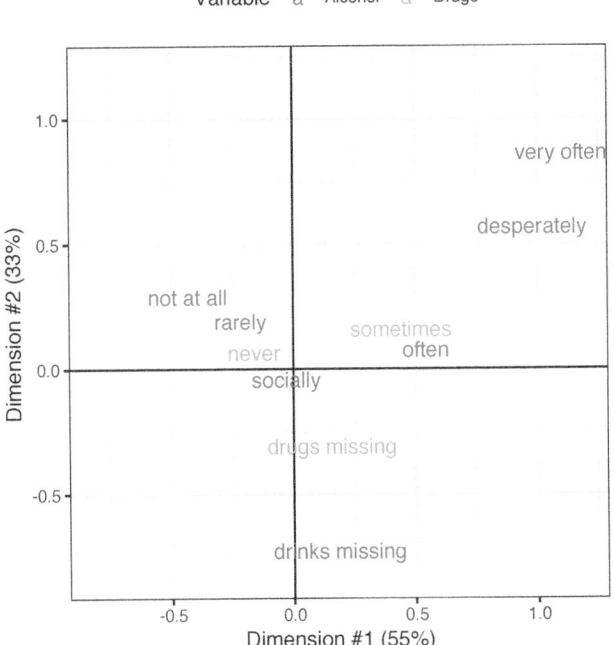

Figure 4.17: The correspondence analysis principal coordinates for the drug and alcohol data in the OkCupid data.

$$y_i = \beta_1 x_1 + \beta_2 x_2 + \cdots + \beta_p x_p + \epsilon_i.$$

Next, we select a predictor that was not in the model, but may contain additional predictive information with respect to the response. For the potential predictor, we then fit:

$$x_{new_i} = \beta_1 x_1 + \beta_2 x_2 + \cdots + \beta_p x_p + \eta_i.$$

The residuals (ϵ_i and η_i) from the models are then plotted against each other in a simple scatter plot. A linear or curvi-linear pattern between these sets of residuals are an indication that the new predictor as a linear or quadratic (or some other non-linear variant) term would be a useful addition in the model.

As previously mentioned, it is problematic to examine the model fit via residuals when those values are determined by simply pre-predicting the training set data. A better strategy is to use the residuals from the various assessment sets created during resampling.

For the Chicago data, a rolling forecast origin scheme (Section 3.4.4) was used for resampling. There are 5698 data points in the training set, each representing a single day. The resampling used here contains a *base set* of samples before September 01, 2014 and the analysis/assessment split begins at this date. In doing this, the analysis set grows cumulatively; once an assessment set is evaluated, it is put into

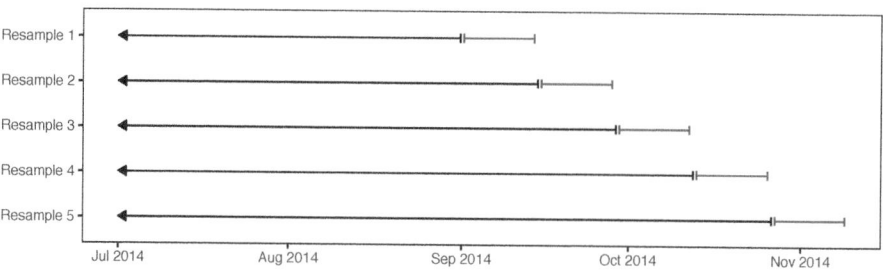

Figure 4.18: A diagram of the resampling scheme used for the Chicago L data. The arrow on the left side indicates that the analysis set begins on January 22, 2001. The red bars on the right indicate the rolling two-week assessment sets.

the analysis set on the next iteration of resampling. Each assessment set contains the 14 days immediately after the last value in the analysis set. As a result, there are 52 resamples and each assessment set is a mutually exclusive collection of the latest dates. This scheme is meant to mimic how the data would be repeatedly analyzed; once a new set of data is captured, the previous set is used to train the model and the new data are used as a test set. Figure 4.18 shows an illustration of the first few resamples. The arrows on the left-hand side indicate that the full analysis set starts on January 22, 2001.

For any model fit to these data using such a scheme, the collection of 14 sets of residuals can be used to understand the strengths and weaknesses of the model with minimal risk of overfitting. Also, since the assessment sets move over blocks of time, it also allows the analyst to understand if there are any specific times of year that the model does poorly.

The response for the regression model is the ridership at the Clark/Lake station, and our initial model will contain the predictors of week, month and year. The distribution of the hold-out residuals from this model are provided in Figure 4.19(a). As we saw earlier in this chapter, the distribution has two peaks, which we found were due to the part of the week (weekday versus weekend). To investigate the importance of part of the week, we then regress the base predictors on part of the week and compute the hold-out residuals from this model. The relationship between sets of hold-out residuals is provided in (b) which demonstrates the non-random relationship, indicating that part of the week contains additional predictive information for the Clark/Lake station ridership. We can see that including part of the week in the model further reduces the residual distribution as illustrated in the histogram labeled Base + Part of Week.

Next, let's explore the importance of the 14-day lag of ridership at the Clark/Lake station. Part (c) of the figure demonstrates the importance of including the 14-day lag of ridership at the Clark/Lake station. In this part of the figure we see a strong linear relationship between the model residuals in the mainstream of the data, with a handful of days lying outside of the overall pattern. These days happen to be holidays

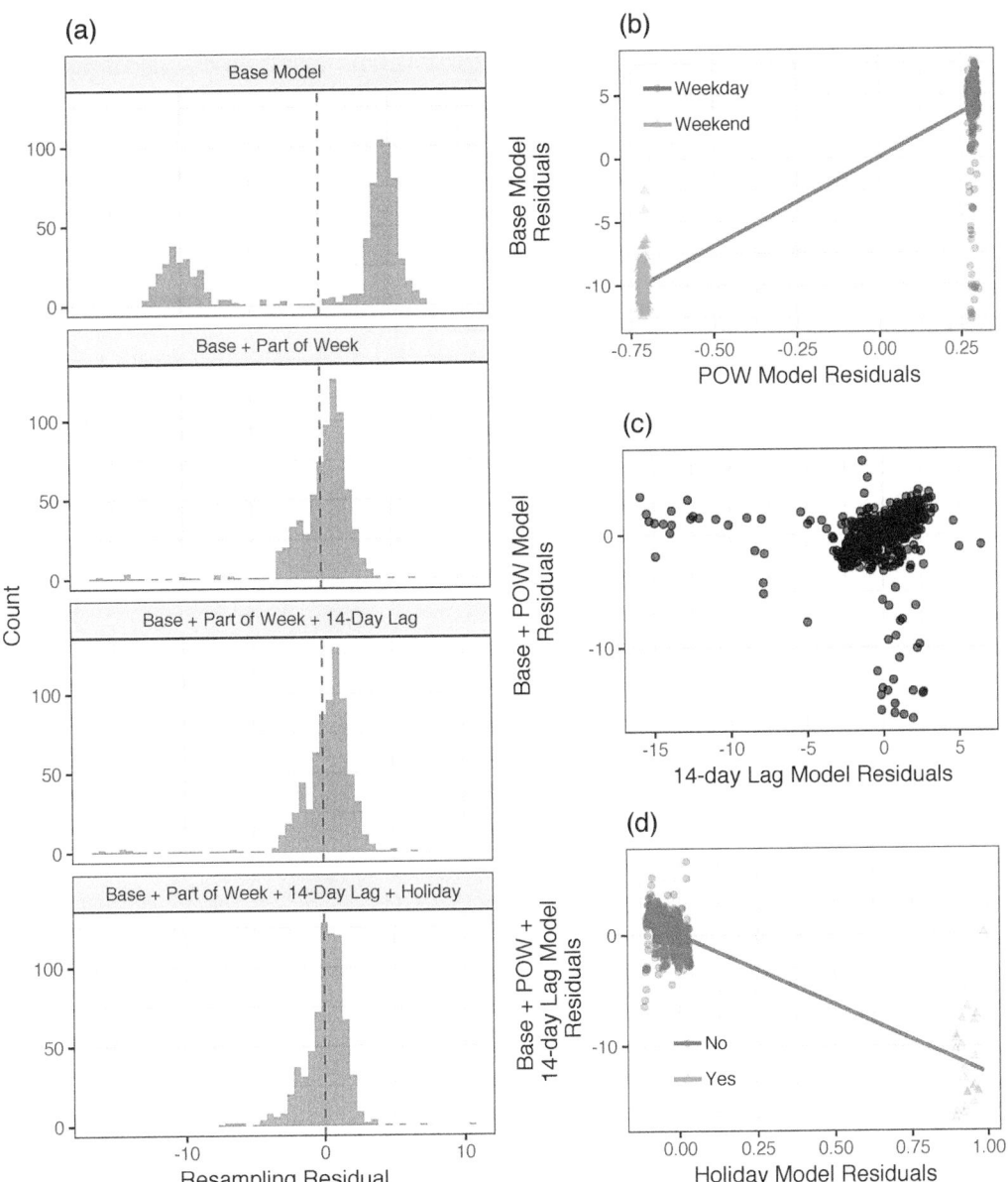

Figure 4.19: (a) The distribution of residuals from the model resampling process for the base model and the base model plus other potentially useful predictors for explaining ridership at the Clark/Lake station. (b) The partial regression plot for the effect of part of the week. (c) The partial regression plot for the 14-day lag predictor of the Clark/Lake station. (d) The partial regression plot for holiday classification.

for this time period. The potential predictive importance of holidays are reflected in part (d). In this figure, the reader's eye may be drawn to one holiday that lies far away from the rest of the holiday samples. It turns out that this sample is July 4, 2015, and is the only day in the training data that is both a holiday and weekend day. Because the model already accounted for part of the week, the additional information that this day is a holiday has virtually no effect on the predicted value for this sample.

4.5   Summary

Advanced predictive modeling and machine learning techniques offer the allure of being able to extract complex relationships between predictors and the response with little effort by the analyst. This hands-off approach to modeling will only put the analyst at a disadvantage. Spending time visualizing the response, predictors, relationships among the predictors, and relationships between predictors and the response can only lead to better understandings of the data. Moreover, this knowledge may provide crucial insights as to what features may be missing in the data and may need to be included to improve a model's predictive performance.

Data visualization is a foundational tool of feature engineering. The next chapter uses this base and begins the development of feature engineering for categorical predictors.

4.6   Computing

The website `http://bit.ly/fes-eda` contains R programs for reproducing these analyses.

5 | Encoding Categorical Predictors

Categorical or nominal predictors are those that contain qualitative data. For the OkCupid data, examples include education (e.g., high school, two-year college, college, etc.) and diet (e.g., anything, vegan, vegetarian, etc.). In the Ames data, the type of house and neighborhood are predictors that have no numeric scale. However, while numeric, the ZIP Code also qualifies as a qualitative predictor because the numeric values have no continuous meaning. Simply put, there is no reason that a ZIP Code of 21212 is 14,827 "more" than a ZIP Code of 06385.

Categorical predictors also can be derived from unstructured or open text. The OkCupid data contains sets of optional essays that describe individuals' interests and other personal information. Predictive models that depend on numeric inputs cannot directly handle open text fields. Instead, these information-rich data need to be processed prior to presenting the information to a model. Text information can be processed in different ways. For example, if keywords are extracted from text to use as predictors, should these include single words or strings of words? Should words be *stemmed* so that a root word is used (e.g., "comput" being short for "computer", "computers", "computing", "computed" etc.) or should a regular expression pattern-matching approach be used (e.g., `^comput`)?

Simple categorical variables can also be classified as *ordered* or *unordered*. A variable with values "Bad", "Good", and "Better" shows a clear progression of values. While the difference between these categories may not be precisely numerically quantifiable, there is a meaningful ordering. To contrast, consider another variable that takes values of "French", "Indian", or "Peruvian". These categories have no meaningful ordering. Ordered and unordered factors might require different approaches for including the embedded information in a model.

As with other preprocessing steps, the approach to including the predictors depends on the type of model. A large majority of models require that all predictors be numeric. There are, however, some exceptions. Algorithms for tree-based models can naturally handle splitting numeric or categorical predictors. These algorithms employ a series `if/then` statements that sequentially split the data into groups. For example, in the Chicago data, the day of the week is a strong predictor and a tree-based model would likely include a model component such as

```
if day in {Sun, Sat} then ridership = 4.4K
  else ridership = 17.3K
```

As another example, a naive Bayes model (Section 12.1) can create a cross-tabulation between a categorical predictor and the outcome class and this frequency distribution is factored into the model's probability calculations. In this situation, the categories can be processed in their natural format. The final section of this chapter investigates how categorical predictors interact with tree-based models.

Tree-based and naive Bayes models are exceptions; most models require that the predictors take numeric form. This chapter focuses primarily on methods that encode categorical data to numeric values.

Many of the analyses in this chapter use the OkCupid data that were introduced in Section 3.1 and discussed in the previous chapter. In this chapter, we will explore specific variables in more detail. One issue in doing this is related to overfitting. Compared to other data sets discussed in this book, the OkCupid data set contains a large number of potential categorical predictors, many of which have a low prevalence. It would be problematic to take the entire training set, examine specific trends and variables, then add features to the model based on these analyses. Even though we cross-validate the models on these data, we may overfit and find predictive relationships that are not generalizable to other data. The data set is not small; the training set consists of 38,809 data points. However, for the analyses here, a subsample of 5,000 STEM profiles and 5,000 non-STEM profiles were selected from the training set for the purpose of discovering new predictors for the models (consistent with Section 3.8).

5.1 | Creating Dummy Variables for Unordered Categories

The most basic approach to representing categorical values as numeric data is to create *dummy* or *indicator variables*. These are artificial numeric variables that capture some aspect of one (or more) of the categorical values. There are many methods for doing this and, to illustrate, consider a simple example for the day of the week. If we take the seven possible values and convert them into *binary* dummy variables, the mathematical function required to make the translation is often referred to as a *contrast* or *parameterization* function. An example of a contrast function is called the "reference cell" or "treatment" contrast, where one of the values of the predictor is left unaccounted for in the resulting dummy variables. Using Sunday as the reference cell, the contrast function would create *six* dummy variables (as shown in Table 5.1). These six numeric predictors would take the place of the original categorical variable.

Why only six? There are two related reasons. First, if the values of the six dummy variables are known, then the seventh can be directly inferred. The second reason is

Table 5.1: A set of dummy variables.

Original Value	Dummy Variables					
	Mon	Tues	Wed	Thurs	Fri	Sat
Sun	0	0	0	0	0	0
Mon	1	0	0	0	0	0
Tues	0	1	0	0	0	0
Wed	0	0	1	0	0	0
Thurs	0	0	0	1	0	0
Fri	0	0	0	0	1	0
Sat	0	0	0	0	0	1

more technical. When fitting linear models, the *design matrix* X is created. When the model has an intercept, an additional initial column of ones for all rows is included. Estimating the parameters for a linear model (as well as other similar models) involves inverting the matrix $(X'X)$. If the model includes an intercept and contains dummy variables for all seven days, then the seven day columns would add up (row-wise) to the intercept and this linear combination would prevent the matrix inverse from being computed (as it is singular). When this occurs, the design matrix said to be *less than full rank* or *overdetermined*. When there are C possible values of the predictor and only $C - 1$ dummy variables are used, the matrix inverse can be computed and the contrast method is said to be a *full rank* parameterization (Timm and Carlson, 1975; Haase, 2011). Less than full rank encodings are sometimes called "one-hot" encodings. Generating the full set of indicator variables may be advantageous for some models that are insensitive to linear dependencies (such as the glmnet model described in Section 7.3.2). Also, the splits created by trees and rule-based models might be more interpretable when all dummy variables are used.

What is the interpretation of the dummy variables? That depends on what type of model is being used. Consider a linear model for the Chicago transit data that only uses the day of the week in the model with the reference cell parameterization above. Using the training set to fit the model, the intercept value estimates the mean of the reference cell, which is the average number of Sunday riders in the training set, and was estimated to be 3.84K people. The second model parameter, for Monday, is estimated to be 12.61K. In the reference cell model, the dummy variables represent the mean value *above and beyond* the reference cell mean. In this case, the estimate indicates that there were 12.61K *more* riders on Monday than Sunday. The overall estimate of Monday ridership adds the estimates from the intercept and dummy variable (16.45K rides).

When there is more than one categorical predictor, the reference cell becomes multidimensional. Suppose there was a predictor for the weather that has only a few values: "clear", "cloudy", "rain", and "snow". Let's consider "clear" to be the reference cell. If this variable was included in the model, the intercept would correspond to the mean of the Sundays with a clear sky. But, the interpretation of each set of dummy variables does not change. The average ridership for a cloudy Monday would augment the average clear Sunday ridership with the average incremental effect of cloudy and the average incremental effect of Monday.

There are other contrast functions for dummy variables. The "cell means" parameterization (Timm and Carlson, 1975) would create a dummy variable for each day of the week and would not include an intercept to avoid the problem of singularity of the design matrix. In this case, the estimates for each dummy variable would correspond to the average value for that category. For the Chicago data, the parameter estimate for the Monday dummy variable would simply be 16.45K.

There are several other contrast methods for constructing dummy variables. Another, based on polynomial functions, is discussed below for ordinal data.

5.2 │ Encoding Predictors with Many Categories

If there are C categories, what happens when C becomes very large? For example, ZIP Code in the United States may be an important predictor for outcomes that are affected by a geographic component. There are more than 40K possible ZIP Codes and, depending on how the data are collected, this might produce an overabundance of dummy variables (relative to the number of data points). As mentioned in the previous section, this can cause the data matrix to be overdetermined and restrict the use of certain models. Also, ZIP Codes in highly populated areas may have a higher rate of occurrence in the data, leading to a "long tail" of locations that are infrequently observed.

One potential issue is that resampling might exclude some of the rarer categories from the analysis set. This would lead to dummy variable columns in the data that contain all zeros and, for many models, this would become a numerical issue that will cause an error. Moreover, the model will not be able to provide a relevant prediction for new samples that contain this predictor. When a predictor contains a single value, we call this a *zero-variance predictor* because there truly is no variation displayed by the predictor.

The first way to handle this issue is to create the full set of dummy variables and simply remove the zero-variance predictors. This is a simple and effective approach but it may be difficult to know *a priori* what terms will be in the model. In other words, during resampling, there may be a different number of model parameters across resamples. This can be a good side-effect since it captures the variance caused by omitting rarely occurring values and propagates this noise into the resampling estimates of performance.

Prior to producing dummy variables, one might consider determining which (if any) of these variables are *near-zero variance predictors* or have the potential to have near zero variance during the resampling process. These are predictors that have few unique values (such as two values for binary dummy variables) *and* occur infrequently

Table 5.2: A demonstration of feature hashing for five of the locations in the OkC data.

| | | Hash Feature Number | |
	Hash Integer Value	16 cols	256 cols
belvedere tiburon	582753783	8	248
berkeley	1166288024	9	153
martinez	-157684639	2	98
mountain view	-1267876914	15	207
san leandro	1219729949	14	30
san mateo	986716290	3	131
south san francisco	-373608504	9	201

in the data (Kuhn and Johnson, 2013). For the training set, we would consider the ratio of the frequencies of the most commonly occurring value to the second-most commonly occurring. For dummy variables, this is simply the ratio of the numbers of ones and zeros. Suppose that a dummy variable has 990 values of zero and 10 ones. The ratio of these frequencies, 99, indicates that it could easily be converted into all zeros during resampling. We suggest a rough cutoff of 19 to declare such a variable "too rare," although the context may raise or lower this bar for different problems.

Although near-zero variance predictors likely contain little valuable predictive information, we may not desire to filter these out. One way to avoid filtering these predictors is to redefine the predictor's categories prior to creating dummy variables. Instead, an "other" category can be created that pools the rarely occurring categories, assuming that such a pooling is sensible. A cutoff based on the training set frequency can be specified that indicates which categories should be combined. Again, this should occur during resampling so that the final estimates reflect the variability in which predictor values are included in the "other" group.

Another way to combine categories is to use a *hashing function* (or *hashes*). Hashes are used to map one set of values to another set of values and are typically used in databases and cryptography (Preneel, 2010). The original values are called the *keys* and are typically mapped to a smaller set of artificial *hash values*. In our context, a potentially large number of predictor categories are the keys and we would like to represent them using a smaller number of categories (i.e., the hashes). The number of possible hashes is set by the user and, for numerical purposes, is a power of 2. Some computationally interesting aspects to hash functions are as follows:

1. The only data required is the value being hashed and the resulting number of hashes. This is different from the previously described contrast function using a reference cell.
2. The translation process is completely deterministic.
3. In traditional applications, it is important that the number of keys that are mapped to hashes is relatively *uniform*. This would be important if the

hashes are used in cryptography since it increases the difficulty of guessing the result. As discussed below, this might not be a good characteristic for data analysis.

4. Hash functions are unidirectional; once the hash values are created, there is no way of knowing the original values. If there are a known and finite set of original values, a table can be created to do the translation but, otherwise, the keys are indeterminable when only the hash value is known.

5. There is no free lunch when using this procedure; some of the original categories will be mapped to the same hash value (called a "collision"). The number of collisions will be largely determined by the number of features that are produced.

When using hashes to create dummy variables, the procedure is called "feature hashing" or the "hash trick" (Weinberger et al., 2009). Since there are many different hashing functions, there are different methods for producing the table that maps the original set to the reduced set of hashes.

As a simple example, the locations from the OkCupid data were hashed.[43] Table 5.2 shows 7 of these cities to illustrate the details. The hash value is an integer value derived solely from each city's text value. In order to convert this to a new feature in the data set, *modular arithmetic* is used. Suppose that 16 features are desired. To create a mapping from the integer to one of the 16 feature columns, integer division is used on the hash value using the formula *column* = (*integer* mod 16) + 1. For example, for the string "mountain view", we have (−1267876914 mod 16) + 1 = 15, so this city is mapped to the fifteenth dummy variable feature. The last column above shows the assignments for the cities when 256 features are requested; mod 256 has more possible remainders than mod 16 so more features are possible. With 16 features, the dummy variable assignments for a selection of locations in the data are shown in Table 5.3.

There are several characteristics of this table to notice. First there are 4 hashes (columns) that are not shown in this table since they contain all zeros. This occurs because none of the hash integer values had those specific mod 16 remainders. This could be a problem for predictors that have the chance to take different string values than the training data and would be placed in one of these columns. In this case, it would be wise to create an 'other' category to ensure that new strings have a hashed representation. Next, notice that there were 6 hashes with no collisions (i.e., have a single 1 in their column). But several columns exhibit collisions. An example of a collision is hash number 14 which encodes for both Menlo Park and San Leandro. In statistical terms, these two categories are said to be *aliased* or *confounded*, meaning that a parameter estimated from this hash cannot isolate the effect of either city. Alias structures have long been studied in the field of statistical experimental design, although usually in the context of aliasing between variables rather than within

---

[43]These computations use the "MurmurHash3" hash found at https://github.com/aappleby/smhasher and implemented in the FeatureHashing R package.

Table 5.3: Binary hashed features that encode a categorical predictor.

	1	2	3	4	5	6	9	12	13	14	15	16
alameda	1	0	0	0	1	0	0	0	0	0	0	0
belmont	1	0	0	0	0	0	0	0	0	0	1	0
benicia	1	0	0	1	0	0	0	0	0	0	0	0
berkeley	1	0	0	0	0	0	1	0	0	0	0	0
castro valley	1	0	0	0	0	0	0	1	0	0	0	0
daly city	1	0	0	0	0	0	0	0	0	0	1	0
emeryville	1	0	0	0	0	0	0	0	1	0	0	0
fairfax	1	1	0	0	0	0	0	0	0	0	0	0
martinez	1	1	0	0	0	0	0	0	0	0	0	0
menlo park	1	0	0	0	0	0	0	0	0	1	0	0
mountain view	1	0	0	0	0	0	0	0	0	0	1	0
oakland	1	0	0	0	0	0	0	1	0	0	0	0
other	1	0	0	0	0	0	1	0	0	0	0	0
palo alto	1	1	0	0	0	0	0	0	0	0	0	0
san francisco	1	0	0	0	0	1	0	0	0	0	0	0
san leandro	1	0	0	0	0	0	0	0	0	1	0	0
san mateo	1	0	1	0	0	0	0	0	0	0	0	0
san rafael	1	0	0	0	0	0	0	0	0	0	0	1
south san francisco	1	0	0	0	0	0	1	0	0	0	0	0
walnut creek	1	1	0	0	0	0	0	0	0	0	0	0

variables (Box et al., 2005). Regardless, from a statistical perspective, minimizing the amount and degree of aliasing is an important aspect of a contrast scheme.

Some hash functions can also be *signed*, meaning that instead of producing a binary indicator for the new feature, possible values could be -1, 0, or +1. A zero would indicate that the category is not associated with that specific feature and the ± 1 values can indicate different values of the original data. Suppose that two categories map to the same hash value. A binary hash would result in a collision but a signed hash could avoid this by encoding one category as +1 and the other as -1. Table 5.4 shows the same data as Table 5.3 with a signed hash of the same size.

Notice that the collision in hash 14 is now gone since the value of 1 corresponds to Menlo Park and San Leandro is encoded as -1. For a linear regression model, the coefficient for the 14th hash now represents the positive or negative shift in the response due to Menlo Park or San Leandro, respectively.

What level of aliasing should we expect? To investigate, data were simulated so that there were 10^4 predictor categories and each category consisted of 20 unique, randomly generated character values. Binary and signed hashes of various sizes were generated and the collision rate was determined. This simulation was done 10

Table 5.4: Signed integer hashed features that encode a categorical predictor.

	1	2	3	4	5	6	9	12	13	14	15	16
alameda	1	0	0	0	1	0	0	0	0	0	0	0
belmont	1	0	0	0	0	0	0	0	0	0	1	0
benicia	1	0	0	1	0	0	0	0	0	0	0	0
berkeley	1	0	0	0	0	0	-1	0	0	0	0	0
castro valley	1	0	0	0	0	0	0	-1	0	0	0	0
daly city	1	0	0	0	0	0	0	0	0	0	1	0
emeryville	1	0	0	0	0	0	0	0	1	0	0	0
fairfax	1	-1	0	0	0	0	0	0	0	0	0	0
martinez	1	1	0	0	0	0	0	0	0	0	0	0
menlo park	1	0	0	0	0	0	0	0	0	1	0	0
mountain view	1	0	0	0	0	0	0	0	0	0	-1	0
oakland	1	0	0	0	0	0	0	1	0	0	0	0
other	1	0	0	0	0	0	1	0	0	0	0	0
palo alto	1	1	0	0	0	0	0	0	0	0	0	0
san francisco	1	0	0	0	0	-1	0	0	0	0	0	0
san leandro	1	0	0	0	0	0	0	0	0	-1	0	0
san mateo	1	0	1	0	0	0	0	0	0	0	0	0
san rafael	1	0	0	0	0	0	0	0	0	0	0	-1
south san francisco	1	0	0	0	0	0	-1	0	0	0	0	0
walnut creek	1	1	0	0	0	0	0	0	0	0	0	0

different times and the average percent collision was calculated. Figure 5.1 shows the results where the y-axis is the rate of *any* level of collision in the features. As expected, the signed values are associated with fewer collisions than the binary encoding. The green line corresponds to the full set of dummy variables (without using hashes). For these data, using signed features would give the best aliasing results but the percentage of collisions is fairly high.

Because of how the new features are created, feature hashing is oblivious to the concept of predictor aliasing. Uniformity of the hash function is an important characteristic when hashes are used in their typical applications. However, it may be a liability in data analysis. The reduction of aliasing is an important concept in statistics; we would like to have the most specific representations of categories as often as possible for a few different reasons:

- Less aliasing allows for better interpretation of the results. If a predictor has a significant impact on the model, it is probably a good idea to understand why. When multiple categories are aliased due to a collision, untangling the true effect may be very difficult, although this may help narrow down which

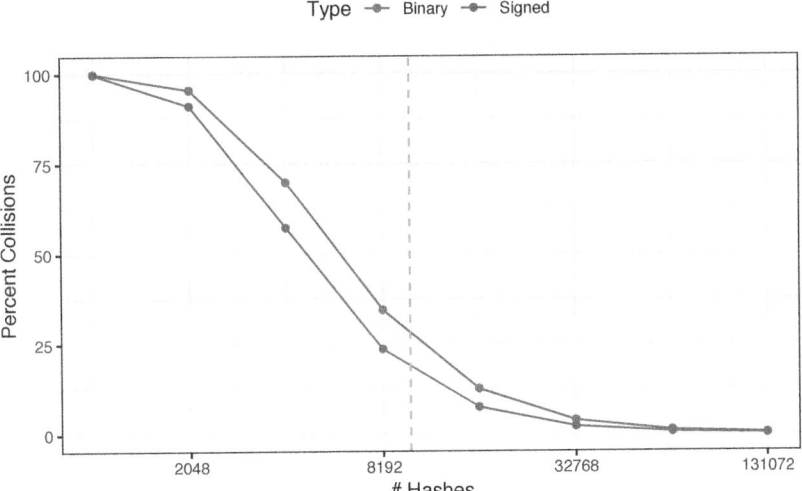

Figure 5.1: The average rate of collisions in a simulation study on a predictor with 10K categories.

categories are affecting the response.

- Categories involved in collisions are not related in any meaningful way. For example, San Leandro and Menlo Park were not aliased because they share a particular similarity or geographic closeness. Because of the arbitrary nature of the collisions, it is possible to have different categories whose true underlying effect are counter to one another. This might have the effect of negating the impact of the hashed feature.

- Hashing functions have no notion of the probability that each key will occur. As such, it is conceivable that a category that occurs with great frequency is aliased with one that is rare. In this case, the more abundant value will have a much larger influence on the effect of that hashing feature.

Although we are not aware of any statistically conscious competitor to feature hashing, there are some characteristics that a contrast function should have when used with a predictor containing a large number of categories.[44] For example, if there is the capacity to have some categories unaliased with any others, the most frequently occurring values should be chosen for these roles. The reason for this suggestion is that it would generate a high degree of specificity for the categories that are most likely to be exposed to the contrast function. One side effect of this strategy would be to alias values that occur less frequently to one another. This might be beneficial since it reduces the *sparsity* of those dummy variables. For example, if two categories that each account for 5% of the training set values are combined, the

---

[44]Traditional analysis of alias structures could be applied to this problem. However, more research would be needed since those methods are usually applied to *balanced designs* where the values of the predictors have the same prevalence. Also, as previously mentioned, they tend to focus on aliasing between multiple variables. In any case, there is much room for improvement on this front.

dummy variable has a two-fold increase in the number of ones. This would reduce the chances of being filtered out during resampling and might also lessen the potential effect that such a near-zero variance predictor might have on the analysis.

5.3 Approaches for Novel Categories

Suppose that a model is built to predict the probability that an individual works in a STEM profession and that this model depends on geographic location (e.g., city). The model will be able to predict the probability of a STEM profession if a new individual lives in one of the 20 cities. But what happens to the model prediction when a new individual lives in a city that is not represented in the original data? If the models are solely based on dummy variables, then the models will not have seen this information and will not be able to generate a prediction.

If there is a possibility of encountering a new category in the future, one strategy would be to use the previously mentioned "other" category to capture new values. While this approach may not be the most effective at extracting predictive information relative to the response for this specific category, it does enable the original model to be applied to new data without completely refitting it. This approach can also be used with feature hashing since the new categories are converted into a unique "other" category that is hashed in the same manner as the rest of the predictor values; however, we do need to ensure that the "other" category is present in the training/testing data. Alternatively, we could ensure that the "other" category collides with another hashed category so that the model can be used to predict a new sample. More approaches to novel categories are given in the next section.

Note that the concept of a predictor with novel categories is not an issue in the training/testing phase of a model. In this phase the categories for all predictors in the existing data are known at the time of modeling. If a specific predictor category is present in only the training set (or vice versa), a dummy variable can still be created for both data sets, although it will be an zero-variance predictor in one of them.

5.4 Supervised Encoding Methods

There are several methods of encoding categorical predictors to numeric columns using the outcome data as a guide (so that they are *supervised* methods). These techniques are well suited to cases where the predictor has many possible values or when new levels appear after model training.

The first method is a simple translation that is sometimes called *effect* or *likelihood encoding* (Micci-Barreca, 2001; Zumel and Mount, 2016). In essence, the effect of the factor level on the outcome is measured and this effect is used as the numeric encoding. For example, for the Ames housing data, we might calculate the mean or median sale price of a house for each neighborhood from the training data and use this statistic to represent the factor level in the model. There are a variety of ways to estimate the effects and these will be discussed below.

For classification problems, a simple logistic regression model can be used to measure the effect between the categorical outcome and the categorical predictor. If the outcome event occurs with rate p, the *odds* of that event are defined as $p/(1-p)$. As an example, with the OkC data, the rate of STEM profiles in Mountain View California is 0.53 so that the odds would be 1.125. Logistic regression models the log-odds of the outcome as a function of the predictors. If a single categorical predictor is included in the model, then the log-odds can be calculated for each predictor value and can be used as the encoding. Using the same location previously shown in Table 5.2, the simple effect encodings for the OkC data are shown in Table 5.5 under the heading of "Raw".

Table 5.5: Supervised encoding examples for several towns in the OkC Data.

Location	Data		Log-Odds		Word Embeddings		
	Rate	n	Raw	Shrunk	Feat 1	Feat 2	Feat 3
belvedere tiburon	0.086	35	-2.367	-2.033	0.050	-0.003	0.003
berkeley	0.163	2676	-1.637	-1.635	0.059	0.077	0.033
martinez	0.091	197	-2.297	-2.210	0.008	0.047	0.041
mountain view	0.529	255	0.118	0.011	0.029	-0.232	-0.353
san leandro	0.128	431	-1.922	-1.911	-0.050	0.040	0.083
san mateo	0.277	880	-0.958	-0.974	0.030	-0.195	-0.150
south san francisco	0.178	258	-1.528	-1.554	0.026	-0.014	-0.007
<new location>				-1.787	0.008	0.007	-0.004

As previously mentioned, there are different methods for estimating the effects. A single generalized linear model (e.g., linear or logistic regression) can be used. While very fast, it has drawbacks. For example, what happens when a factor level has a single value? Theoretically, the log-odds should be infinite in the appropriate direction but, numerically, it is usually capped at a large (and inaccurate) value.

One way around this issue is to use some type of *shrinkage* method. For example, the overall log-odds can be determined and, if the *quality* of the data within a factor level is poor, then this level's effect estimate can be biased towards an overall estimate that disregards the levels of the predictor. "Poor quality" could be due to a small sample size or, for numeric outcomes, a large variance within the data for that level. Shrinkage methods can also move extreme estimates towards the middle of the distribution. For example, a well-replicated factor level might have an extreme mean or log-odds (as will be seen for Mountain View data below).

A common method for shrinking parameter estimates is Bayesian analysis (McElreath, 2015). In this case, expert judgement can be used prior to seeing the data to specify a *prior distribution* for the estimates (e.g., the means or log-odds). The prior would

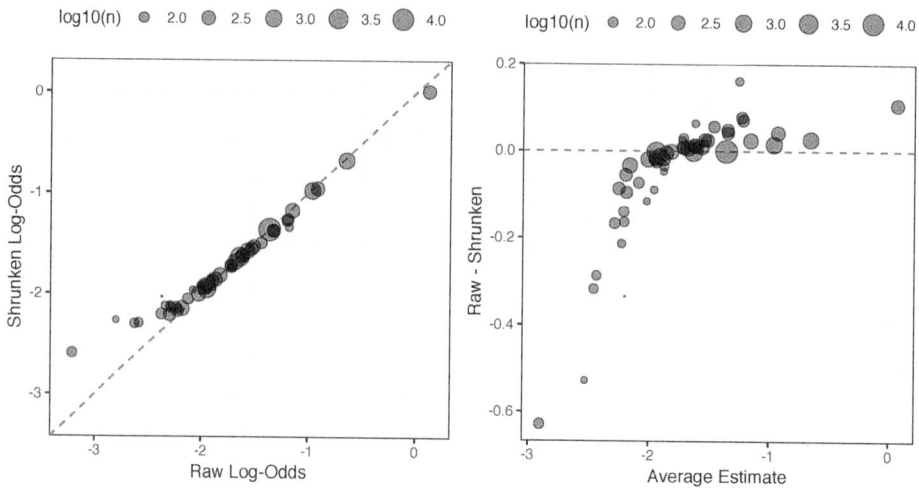

Figure 5.2: Left: Raw and shrunken estimates of the effects for the OkC data. Right: The same data shown as a function of the magnitude of the log-odds (on the x axis).

be a theoretical distribution that represents the overall distribution of effects. Almost any distribution can be used. If there is not a strong belief as to what this distribution should be, then it may be enough to focus on the shape of the distribution. If a bell-shaped distribution is reasonable, then a very diffuse or wide prior can be used so that it is not overly opinionated. Bayesian methods take the observed data and blend it with the prior distribution to come up with a *posterior distribution* that is a combination of the two. For categorical predictor values with poor data, the posterior estimate is shrunken closer to the center of the prior distribution. This can also occur when their raw estimates are relatively extreme.[45]

For the OkC data, a normal prior distribution was used for the log-odds that has a large standard deviation ($\sigma = 10$). Table 5.5 shows the results. For most locations, the raw and shrunken estimates were very similar. For Belvedere/Tiburon, there was a relatively small sample size (and a relatively small STEM rate) and the estimate was shrunken towards the center of the prior. Note that there is a row for "<new location>". When a new location is given to the Bayesian model, the procedure used here estimates its effect as the most likely value using the mean of the posterior distribution (a value of -1.79 for these data). A plot of the raw and shrunken effect estimates is also shown in Figure 5.2(a). Panel (b) shows the magnitude of the estimates (x-axis) versus the difference between the two. As the raw effect estimates dip lower than -2, the shrinkage pulled the values towards the center of the estimates. For example, the location with the largest log-odds value (Mountain View) was reduced from 0.118 to 0.011.

---

[45]In this situation, the shrinkage of the values towards the mean is called *partial pooling* of the estimates.

Empirical Bayes methods can also be used, in the form of linear (and generalized linear) mixed models (West and Galecki, 2014). These are non-Bayesian estimation methods that also incorporate shrinkage in their estimation procedures. While they tend to be faster than the types of Bayesian models used here, they offer less flexibility than a fully Bayesian approach.

One issue with effect encoding, independent of the estimation method, is that it increases the possibility of overfitting. This is because the estimated effects are taken from one model and put into another model (as variables). If these two models are based on the same data, it is somewhat of a self-fulfilling prophecy. If these encodings are not consistent with future data, this will result in overfitting that the model cannot detect since it is not exposed to any other data that might contradict the findings. Also, the use of summary statistics as predictors can drastically underestimate the variation in the data and might give a falsely optimistic opinion of the utility of the new encoding column.[46] As such, it is strongly recommended that either different data sets be used to estimate the encodings and the predictive model or that their derivation is conducted inside resampling so that the assessment set can measure the overfitting (if it exists).

Another supervised approach comes from the deep learning literature on the analysis of textual data. In this case, large amounts of text can be cut up into individual words. Rather than making each of these words into its own indicator variable, *word embedding* or *entity embedding* approaches have been developed. Similar to the dimension reduction methods described in the next chapter, the idea is to estimate a smaller set of numeric features that can be used to adequately represent the categorical predictors. Guo and Berkhahn (2016) and Chollet and Allaire (2018) describe this technique in more detail. In addition to the dimension reduction, there is the possibility that these methods can estimate semantic relationships between words so that words with similar themes (e.g., "dog", "pet", etc.) have similar values in the new encodings. This technique is not limited to text data and can be used to encode any type of qualitative variable.

Once the number of new features is specified, the model takes the traditional indicator variables and randomly assigns them to one of the new features. The model then tries to optimize both the allocation of the indicators to features as well as the parameter coefficients for the features themselves. The outcome in the model can be the same as the predictive model (e.g., sale price or the probability of a STEM profile). Any type of loss function can be optimized, but it is common to use the root mean squared error for numeric outcomes and cross-entropy for categorical outcomes (which are the loss functions for linear and logistic regression, respectively). Once the model is fit, the values of the embedding features are saved for each observed value of the quantitative factor. These values serve as a look-up table that is used for prediction. Additionally, an extra level can be allocated to the original predictor to serve as a place-holder for any new values for the predictor that are encountered after model training.

---

[46]This is discussed more, with an example, in Section 6.2.2.

A more typical neural network structure can be used that places one or more sets of nonlinear variables positioned between the predictors and outcomes (i.e., the *hidden layers*). This allows the model to make more complex representations of the underlying patterns. While a complete description of neural network models is beyond the scope of this book, Chollet and Allaire (2018) provide a very accessible guide to fitting these models and specialized software exists to create the encodings.

For the OkCupid data, there are 52 locations in the training set. We can try to represent these locations, plus a slot for potential new locations, using a set *embedding features*. A schematic for the model is:

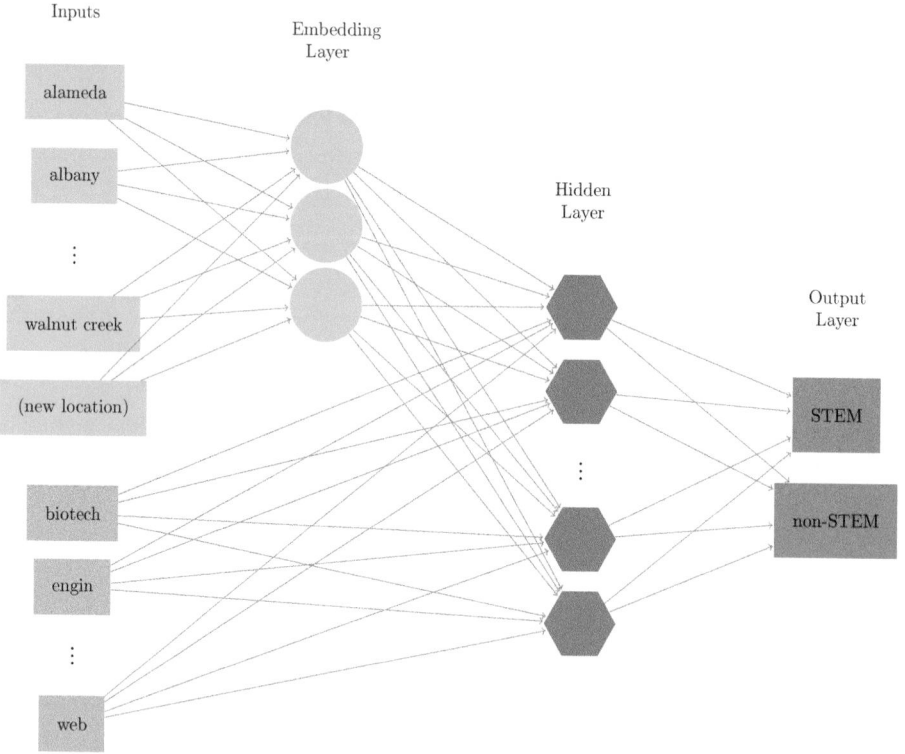

Here, each green node represents the part of the network that is dedicated to reducing the dimensionality of the location data to a smaller set. In this diagram, three embedding features are used. The model can also include a set of other predictors unrelated to the embedding components. In this diagram an additional set of indicator variables, derived later in Section 5.6, is shown in orange. This allows the embeddings to be estimated in the presence of other potentially important predictors so that they can be adjusted accordingly. Each line represents a slope coefficient for the model. The connections between the indicator variable layer on the left and the embedding layer, represented by the green circular nodes, are the values that will be used to represent the original locations. The activation function connecting the embeddings and other predictors to the hidden layer used rectified linear unit (ReLU) connections (introduced in Section 6.2.1). This allows the model additional complexity using nonlinear relationships. Here, ten hidden nodes were used during model estimation. In total, the network model estimated 678 parameters in the course of deriving the embedding features. Cross-entropy was used to fit the model and it was trained for 30 iterations, where it was found to converge. These data are not especially complex since there are few unique values that tend to have very little textual information since each data point only belongs to a single location (unlike the case where text is cut into individual words).

Table 5.5 shows the results for several locations and Figure 5.3 contains the relationship between each of the resulting encodings (i.e., features) and the raw log-odds. Each of the features has a relationship with the raw log-odds, with rank correlations of -0.01, -0.58, and -0.69. Some of the new features are correlated with the others; absolute correlation values range between 0.01 and 0.69. This raises the possibility that, for these data, fewer features might be required. Finally, as a reminder, while an entire supervised neural network was used here, the embedding features can be used in other models for the purpose of preprocessing the location values.

5.5   Encodings for Ordered Data

In the previous sections, an unordered predictor with C categories was represented by $C - 1$ binary dummy variables or a hashed version of binary dummy variables. These methods effectively present the categorical information to the models. But now suppose that the C categories have a relative ordering. For example, consider a predictor that has the categories of "low", "medium", and "high." This information could be converted to 2 binary predictors: $X_{low} = 1$ for samples categorized as "low" and $= 0$ otherwise, and $X_{medium} = 1$ for samples categorized as "medium" and $= 0$ otherwise. These new predictors would accurately identify low, medium, and high samples, but the two new predictors would miss the information contained in the relative ordering, which could be very important relative to the response.

Ordered categorical predictors would need a different way to be presented to a model to uncover the relationship of the order to the response. Ordered categories may have a linear relationship with the response. For instance, we may see an increase of approximately 5 units in the response as we move from low to medium and an increase of approximately 5 units in the response as we move from medium to high. To allow a model to uncover this relationship, the model must be presented with a numeric encoding of the ordered categories that represents a linear ordering. In the field of Statistics, this type of encoding is referred to as a *polynomial contrast*. A contrast has the characteristic that it is a single comparison (i.e., one degree of freedom) and its coefficients sum to zero. For the "low", "medium", "high" example above, the contrast to uncover a linear trend would be -0.71, 0, 0.71, where low samples encode to -0.71, medium samples to 0, and high samples to 0.71. Polynomial contrasts can extend to nonlinear shapes, too. If the relationship between the predictor and the response was best described by a quadratic trend, then the contrast would be 0.41, -0.82, and 0.41 (Table 5.6). Conveniently, these types of contrasts can be generated

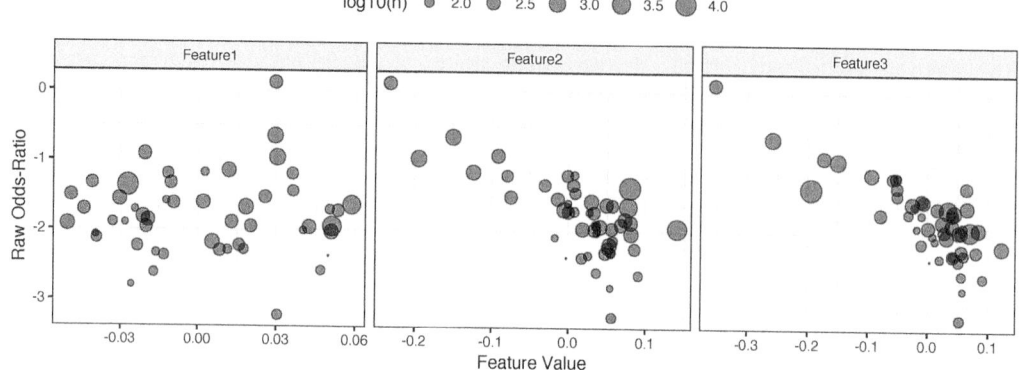

Figure 5.3: Three word embedding features for each unique location in the OkC data and their relationship to the raw odds-ratios.

for predictors with any number of ordered factors, but the complexity of the contrast is constrained to one less than the number of categories in the original predictor. For instance, we could not explore a cubic relationship between a predictor and the response with a predictor that has only 3 categories.

By employing polynomial contrasts, we can investigate multiple relationships (linear, quadratic, etc.) simultaneously by including these in the same model. When we do this, the form of the numeric representation is important. Specifically, the new predictors would contain unique information. To do this the numeric representations are required to be orthogonal. This means that the dot product of the contrast vectors is 0.

Table 5.6: An example of linear and quadratic polynomial contrasts for an ordered categorical predictor with three levels.

Original Value	Dummy Variables	
	Linear	Quadratic
low	-0.71	0.41
medium	0.00	-0.82
high	0.71	0.41

It is important to recognize that patterns described by polynomial contrasts may not effectively relate a predictor to the response. For example, in some cases, one might expect a trend where "low" and "middle" samples have a roughly equivalent response but "high" samples have a much different response. In this case, polynomial contrasts are unlikely to be effective at modeling this trend. Another downside to polynomial contrasts for ordered categories occurs when there are moderate to high number of categories. If an ordered predictor has C levels, the encoding into dummy variables uses polynomials up to degree $C - 1$. It is very unlikely that these higher-level polynomials are modeling important trends (e.g., octic patterns) and it might make sense to place a limit on the polynomial degree. In practice, we rarely explore the effectiveness of anything more than a quadratic polynomial.

As an alternative to polynomial contrasts, one could:

- Treat the predictors as unordered factors. This would allow for patterns that are not covered by the polynomial feature set. Of course, if the true underlying pattern is linear or quadratic, unordered dummy variables may not effectively uncover this trend.

- Translate the ordered categories into a single set of numeric *scores* based on context-specific information. For example, when discussing failure modes of a piece of computer hardware, experts would be able to rank the severity of a type of failure on an integer scale. A minor failure might be scored as a "1" while a catastrophic failure mode could be given a score of "10" and so on.

Simple visualizations and context-specific expertise can be used to understand whether either of these approaches is a good ideas.

5.6 | Creating Features from Text Data

Often, data contain textual fields that are gathered from questionnaires, articles, reviews, tweets, and other sources. For example, the OkCupid data contains the responses to nine open text questions, such as "my self-summary" and "six things I could never do without". The open responses are likely to contain important information relative to the outcome. Individuals in a STEM field are likely to have a different vocabulary than, say, someone in the humanities. Therefore, the words or phrases used in these open text fields could be very important for predicting the outcome. This kind of data is qualitative and requires more effort to put in a form that models can consume. How, then, can these data be explored and represented for a model?

For example, some profile text answers contained links to external websites. A reasonable hypothesis is that the presence of a hyperlink could be related to the person's profession. Table 5.7 shows the results for the random subset of profiles. The rate of hyperlinks in the STEM profiles was 21%, while this rate was 12.4% in the non-STEM profiles. One way to evaluate a difference in two proportions is called the *odds-ratio* (Agresti, 2012). First, the odds of an event that occurs with rate p is defined as $p/(1 - p)$. For the STEM profiles, the odds of containing a hyperlink are relatively small with a value of $0.21/0.79 = 0.27$. For the non-STEM profiles, it is even smaller (0.142). However, the ratio of these two quantities can be used to understand what the effect of having a hyperlink would be between the two professions. In this case, this indicates that the odds of a STEM profile is nearly 1.9-times higher when the profile contains a link. Basic statistics can be used to assign a lower 95% confidence interval on this quantity. For these data the lower bound is 1.7, which indicates that this increase in the odds is unlikely to be due

Table 5.7: A cross-tabulation between the existence of at least one hyperlink in the OkCupid essay text and the profession.

	stem	other
Link	1063	620
No Link	3937	4380

to random noise since it does not include a value of 1.0. Given these results, the indicator of a hyperlink will likely benefit a model and should be included.

Are there words or phrases that would make good predictors of the outcome? To determine this, the text data must first be processed and cleaned. At this point, there were 63,440 distinct words in the subsample of 10,000 profiles. A variety of features were computed on the data such as the number of commas, hashtags, mentions, exclamation points, and so on. This resulted in a set of 13 new "text-related" features. In these data, there is an abundance of HTML markup tags in the text, such as `<br />` and `<a href= \link\>`. These were removed from the data, as well as punctuation, line breaks, and other symbols. Additionally, there were some words that were nonsensical repeated characters, such as `***` or `aaaaaaaaa`, that were removed. Additional types of preprocessing of text are described below and, for more information, Christopher et al. (2008) is a good introduction to the analysis of text data while Silge and Robinson (2017) is an excellent reference on the computational aspects of analyzing such data.

Given this set of 63,440 words and their associated outcome classes, odds-ratios can be computed. First, the words were again filtered so that terms with at least 50 occurrences in the 10,000 profiles would be analyzed. This cutoff was determined so that there was a sufficient frequency for modeling. With this constraint the potential number of keywords was reduced to 4,918 terms. For each of these, the odds-ratio and associated p-value were computed. The p-value tests the hypothesis that the odds of the keyword occurring in either professional group are equal (i.e., 1.0). However, the p-values can easily provide misleading results for two reasons:

- In isolation, the p-value only relates the question "Is there a difference in the odds between the two groups?" The more appropriate question is, "How much of a difference is there between the groups?" The p-value does not measure the magnitude of the differences.

- When testing a single (predefined) hypothesis, the false positive rate for a hypothesis test using $\alpha = 0.05$ is 5%. However, there are 4,918 tests on the same data set, which raises the false positive rate exponentially. The false-discovery rate (FDR) p-value correction (Efron and Hastie, 2016) was developed for just this type of scenario. This procedure uses the entire distribution of p-values and attenuates the false positive rate. If a particular keyword has an FDR value of 0.30, this implies that the collection of keywords with FDR values

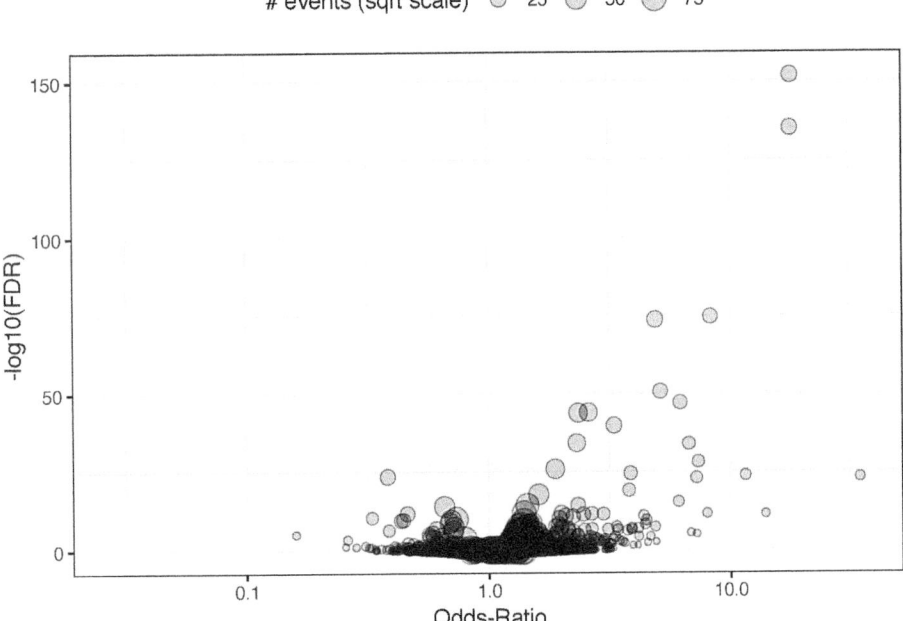

Figure 5.4: A volcano plot of the keyword analysis. Each dot represents a word from the OkCupid essays, and the size of the dot represents the frequency of the occurrence. Words in the upper right are strongly associated with STEM profiles.

less than 0.30 has a collective 30% false discovery rate. For this reason, the focus here will be on the FDR values generated using the Benjamini-Hochberg correction (Benjamini and Hochberg, 1995).

To characterize these results, a *volcano* plot is shown in Figure 5.4 where the estimated odds-ratio is shown on the x-axis and the minus log of the FDR values are shown on the y-axis (where larger values indicate higher statistical significance). The size of the points is associated with the number of events found in the sampled data set. Keywords falling in the upper left- and right-rand side would indicate a strong difference between the classes that is unlikely to be random result. The plot shows far more keywords that have a higher likelihood to be found in the STEM profiles than in the non-STEM professions as more points fall on the upper right-hand side of one on the x-axis. Several of these have extremely high levels of statistical significance with FDR values that are vanishingly small.

As a rough criteria for "importance", keywords with an odds-ratio of at least 2 (in either direction) *and* an FDR value less than 10^{-5} will be considered for modeling. This results in 52 keywords:

```
alot, apps, biotech, code, coding, computer,
computers, data, developer, electronic, electronics, engineer,
engineering, firefly, fixing, futurama, geek, geeky,
im, internet, lab, law, lawyer, lol,
math, matrix, mechanical, mobile, neal, nerd,
pratchett, problems, programmer, programming, science, scientist,
silicon, software, solve, solving, startup, stephenson,
student, systems, teacher, tech, technical,
technology, valley, web, websites, wikipedia
```

Of these keywords, only 7 were enriched in the non-STEM profiles: `im`, `lol`, `teacher`, `student`, `law`, `alot`, `lawyer`. Of the STEM-enriched keywords, many of these make sense (and play to stereotypes). The majority are related to occupation (e.g., `engin`, `startup`, and `scienc`), while others are clearly related to popular geek culture, such as `firefli`,[47] `neal`, and `stephenson`,[48] and `scifi`.

One final set of 9 features were computed related to the sentiment and language of the essays. There are curated collections of words with assigned sentiment values. For example, "horrible" has a fairly negative connotation while "wonderful" is associated with positivity. Words can be assigned qualitative assessments (e.g., "positive", "neutral", etc.) or numeric scores where neutrality is given a value of zero. In some problems, sentiment might be a good predictor of the outcome. This feature set included sentiment-related measures, as well as measures of the point-of-view (i.e., first-, second-, or third-person text) and other language elements.

Do these features have an impact on a model? A series of logistic regression models were computed using different feature sets:

1. A basic set of profile characteristics unrelated to the essays (e.g., age, religion, etc.) consisting of 160 predictors (after generating dummy variables). This resulted in a baseline area under the ROC curve of 0.77. The individual resamples for this model are shown in Figure 5.5.

2. The addition of the simple text features increased the number of predictors to 173. Performance was not appreciably affected as the AUC value was 0.776. These features were discarded from further use.

3. The keywords were added to the basic profile features. In this case, performance jumped to 0.839. Recall that these were derived from a smaller portion of these data and that this estimate may be slightly optimistic. However, the entire training set was cross-validated and, if overfitting was severe, the resampled performance estimates would not be very good.

4. The sentiment and language features were added to this model, producing an AUC of 0.841. This indicates that these aspects of the essays did not provide

---

[47]https://en.wikipedia.org/wiki/Firefly_(TV_series)
[48]https://en.wikipedia.org/wiki/Neal_Stephenson

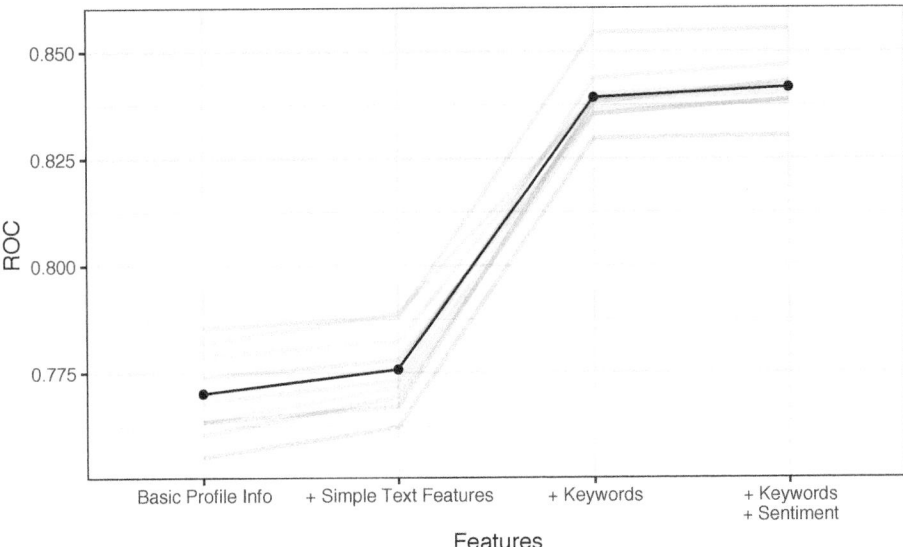

Figure 5.5: Resampling results for a series of logistic regression models computed with different feature sets.

any predictive value above and beyond what was already captured by the previous features.

Overall, the keyword and basic feature model would be the best version found here.[49]

The strategy shown here for computing and evaluating features from text is fairly simplistic and is not the only approach that could be taken. Other methods for preprocessing text data include:

- removing commonly used *stop words*, such as "is", "the", "and", etc.
- *stemming* the words so that similar words, such as the singular and plural versions, are represented as a single entity.

The SMART lexicon of stop words (Lewis et al., 2004) was filtered out of the current set of words, leaving 62,928 unique results. The words were then "stemmed" using the Porter algorithm (Willett, 2006) to convert similar words to a common root. For example, these 7 words are fairly similar: `teach, teacher, teachers, teaches, teachable, teaching, teachings`. Stemming would reduce these to 3 unique values: `teach, teacher, teachabl`. Once these values were stemmed, there were 45,486 unique words remaining. While this does reduce the potential number of terms to analyze, there are potential drawbacks related to a reduced specificity of the text. Schofield and Mimno (2016) and Schofield et al. (2017) demonstrate that there can be harm done using these preprocessors.

One other method for determining relevant features uses the term frequency-inverse

---

[49]This was the feature set that was used in Chapter 3 when resampling and model comparisons were discussed in Table 3.4.

document frequency (`tf-idf`) statistic (Amati and Van R, 2002). Here the goal is to find words or terms that are important to individual documents in the collection of documents at hand. For example, if the words in this book were processed, there are some that would have high frequency (e.g., `predictor`, `encode`, or `resample`) but are unusual in most other contexts.

For a word `W` in document `D`, the term frequency, `tf`, is the number of times that `W` is contained in `D` (usually adjusted for the length of `D`). The inverse document frequency, `idf`, is a weight that will normalize by how often the word occurs in the current collection of documents. As an example, suppose the term frequency of the word `feynman` in a specific profile was 2. In the sample of profiles used to derive the features, this word occurs 55 times out of 10,000. The inverse document frequency is usually represented as the log of the ratio, or $log_2(10000/55)$ or 7.5. The `tf-idf` value for feynman in this profile would be $2 \times log_2(10000/55)$ or 15. However, suppose that in the same profile, the word `internet` also occurs twice. This word is more prevalent across profiles and is contained at least once 1529 times. The `tf-idf` value here is 5.4; in this way the same raw count is down-weighted according to its abundance in the overall data set. The word `feynman` occurs the same number of times as `internet` in this hypothetical profile, but `feynman` is more distinctive or important (as measured by `tf-idf`) because it is rarer overall, and thus possibly more effective as a feature for prediction.

One potential issue with `td-idf` in predictive models is the notion of the "current collection of documents". This can be computed for the training set but what should be done when a single new sample is being predicted (i.e., there is no collection)? One approach is to use the overall `idf` values from the training set to weight the term frequency found in new samples.

Additionally, all of the previous approaches have considered a single word at a time. Sequences of consecutive words can also be considered and might provide additional information. *n*-grams are terms that are sequences of *n* consecutive words. One might imagine that the sequence "I hate computers" would be scored differently than the simple term "computer" when predicting whether a profile corresponds to a STEM profession.

5.7 | Factors versus Dummy Variables in Tree-Based Models

As previously mentioned, certain types of models have the ability to use categorical data in their natural form (i.e., without conversions to dummy variables). A simple regression tree (Breiman et al., 1984) using the day of the week predictor for the Chicago data resulted in this simple split for prediction:

```
if day in {Sun, Sat} then ridership = 4.4K
 else ridership = 17.3K
```

Suppose the day of the week *had* been converted to dummy variables. What would

Table 5.8: A summary of the data sets used to evaluate encodings of categorical predictors in tree- and rule-based models.

	Attrition	Cars	Churn	German Credit	HPC
n	1470	1728	5000	1000	4331
p	30	6	19	20	7
Classes	2	4	2	2	4
Numeric Predictors	16	0	15	7	5
Factor Predictors	14	6	4	13	2
Ordered Factors	7	0	0	1	0
Factor with 3+ Levels	12	7	2	11	3
Factor with 5+ Levels	3	0	1	5	2

have occurred? In this case, the model is slightly more complex since it can only create rules as a function of a single dummy variable at a time:

```
if day = Sun then ridership = 3.84K
  else if day = Sat then ridership = 4.96K
    else ridership = 17.30K
```

The non-dummy variable model could have resulted in the same structure if it had found that the results for Saturday and Sunday were sufficiently different to merit an extra set of splits. This leads to the question related to using categorical predictors in tree-based models: *Does it matter how the predictions are encoded?*

To answer this question, a series of experiments was conducted.[50] Several public classification data sets were used to make the comparison between the different encodings. These are summarized in Table 5.8. Two data sets contained both ordered and unordered factors. As described below, the ordered factors were treated in different ways.

For each iteration of the simulation, 75% of the data was used for the training set and 10-fold cross-validation was used to tune the models. When the outcome variable had two levels, the area under the ROC curve was maximized. For the other datasets, the multinomial log-likelihood was optimized. The same number of tuning parameter values were evaluated although, in some cases, the values of these parameters were different due to the number of predictors in the data before and after dummy variable generation. The same resamples and random numbers were used for the models with and without dummy variables. When the data contained a significant class imbalance, the data were downsampled to compensate.

For each data set, models were fit with the data in its original form as well as (unordered) dummy variables. For the two data sets with ordinal data, an additional model was created using ordinal dummy variables (i.e., polynomial contrasts).

---

[50]In Chapter 12, a similar but smaller-scale analysis is conducted with the naive Bayes model.

Several models were fit to the data sets. Unless otherwise stated, the default software parameters were used for each model.

- Single CART trees (Breiman et al., 1984). The complexity parameter was chosen using the "one-standard error rule" that is internal to the recursive partitioning algorithm.
- Bagged CART trees (Breiman, 1996). Each model contained 50 constituent trees.
- Single C5.0 trees and single C5.0 rulesets (Quinlan, 1993; Kuhn and Johnson, 2013)
- Single conditional inference trees (Hothorn et al., 2006). A grid of 10 values for the p-value threshold for splitting were evaluated.
- Boosted CART trees (a.k.a. stochastic gradient boosting, Friedman (2002)). The models were optimized over the number of trees, the learning rate, the tree depth, and the number of samples required to make additional splits. Twenty-five parameter combinations were evaluated using random search.
- Boosted C5.0 trees. These were tuned for the number of iterations.
- Boosted C5.0 rules. These were also tuned for the number of iterations.
- Random forests using CART trees (Breiman, 2001). Each forest contained 1,500 trees and the number of variables selected for a split was tuned over a grid of 10 values.
- Random forests using conditional (unbiased) inference trees (Strobl et al., 2007). Each forest contained 100 trees and the number of variables selected for a split was tuned over a grid of 10 values.

For each model, a variety of different performance metrics were estimated using resampling as well as the total time to train and tune the model.[51]

For the three data sets with two classes, Figure 5.6 shows a summary of the results for the area under the ROC curve. In the plot, the percent difference in performance is calculated using

$$\%Difference = \frac{Factor - Dummy}{Factor} \times 100$$

In this way, positive values indicate that the factor encodings have better performance. The image shows the median in addition to the lower and upper 5% percentiles of the distribution. The gray dotted line indicates no difference between the encodings.

The results of these simulations show that, for these data sets, there is no real difference in the area under the ROC curve between the encoding methods. This also appears to be true when simple factor encodings are compared to dummy variables generated from polynomial contrasts for ordered predictors.[52] Of the ten models,

---

[51]The programs used in the simulation are contained in the GitHub repository `https://github.com/topepo/dummies-vs-factors`.

[52]Note that many of the non-ensemble methods, such as C5.0 trees/rules, CART, and conditional inference trees, show a significant amount of variation. This is due to the fact that they are *unstable* models with high variance.

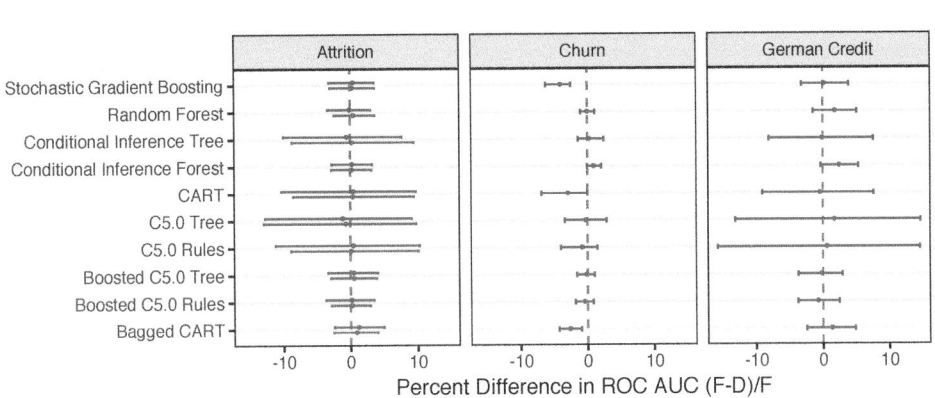

Figure 5.6: A comparison of encodings for the area under the ROC curve.

there are only two cases out of 40 scenarios where the mainstream of the distributions did not cover zero. Stochastic gradient boosting and bagged CART trees are both ensemble methods and these showed a 2%-4% drop in the ROC curve when using factors instead of dummy variables for a single data set.

Another metric, overall accuracy, can also be assessed. These results are shown in Figure 5.7 where all of the models can be taken into account. In this case, the results are mixed. When comparing factors to unordered dummy variables, two of the models show differences in encodings. The churn data shows results similar to the ROC curve metrics. The car evaluation data demonstrates a nearly uniform effect where factor encodings do better than dummy variables. Recall that in the car data, which has four classes, all of the predictors are categorical. For this reason, it is likely to show the most effect of all of the data sets.

In the case of dummy variables generated using polynomial contrasts, neither of the data sets show a difference between the two encodings. However, the car evaluation data shows a pattern where the factor encodings had no difference compared to polynomial contrasts but when compared to unordered dummy variables, the factor encoding is superior. This indicates that the underlying trend in the data follows a polynomial pattern.

In terms of performance, it appears that differences between the two encodings are rare (but can occur). One might infer that, since the car data contains all categorical variables, this situation would be a good indicator for when to use factors instead of dummy variables. However, two of the data sets (Attrition and German Credit), have a high proportion of categorical predictors and show no difference. In some data sets, the effect of the encoding would depend on whether the categorical predictor(s) are important to the outcome and in what way.

In summary, while few differences were seen, it is very difficult to predict when a difference will occur.

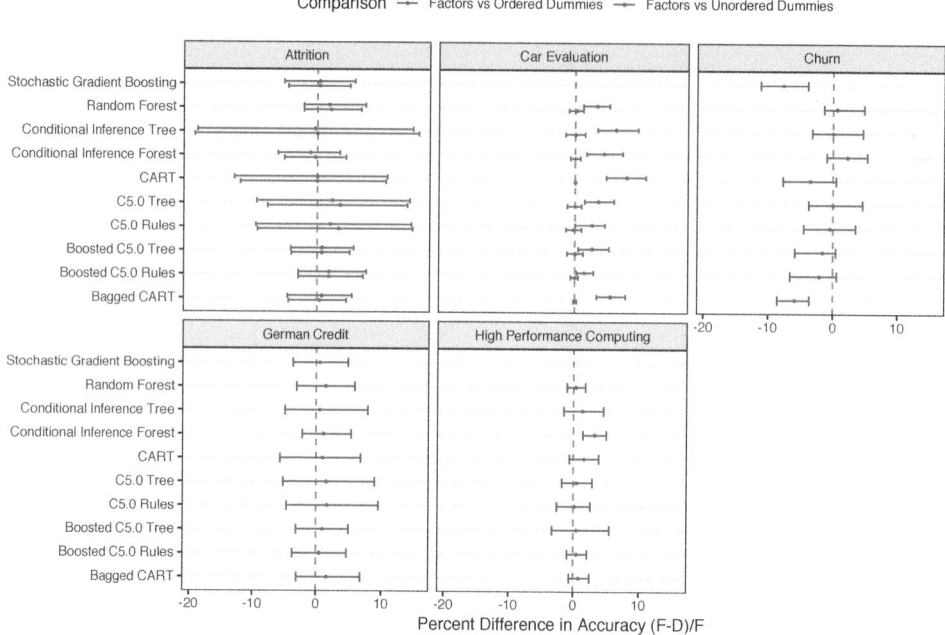

Figure 5.7: A comparison of encodings for the accuracy.

However, one other statistic was computed for each of the simulations: the time to train the models. Figure 5.8 shows the *speed-up* of using factors above and beyond dummy variables (i.e., a value of 2.5 indicates that dummy variable models are two and a half times slower than factor encoding models). Here, there is very strong trend that factor-based models are more efficiently trained than their dummy variable counterparts. The reason for this is likely to be that the expanded number of predictors (caused by generating dummy variables) requires more computational time than the method for determining the optimal split of factor levels. The exceptions to this trend are the models using conditional inference trees.

One other effect of how qualitative predictors are encoded is related to summary measures. Many of these techniques, especially tree-based models, calculate *variable importance scores* that are relative measures for how much a predictor affected the outcome. For example, trees measure the effect of a specific split on the improvement in model performance (e.g., impurity, residual error, etc.). As predictors are used in splits, these improvements are aggregated; these can be used as the importance scores. If a split involves all of the predictor's values (e.g., Saturday versus the other six days), the importance score for the *entire variable* is likely to be much larger than a similar importance score for an individual level (e.g., Saturday or not-Saturday). In the latter case, these fragmented scores for each level may not be ranked as highly as the analogous score that reflects all of the levels. A similar issue comes up during feature selection (Chapters 10 through 7) and interaction detection (Chapter 12). The choice of predictor encoding methods is discussed further there.

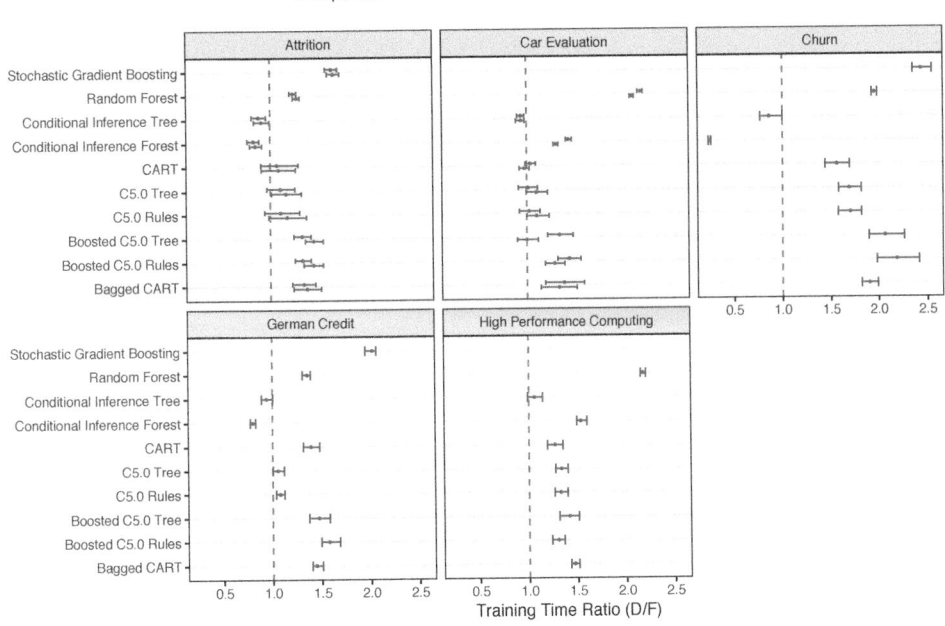

Figure 5.8: A comparison of encodings for time to train the model. Large values indicate that the factor encoding took less time to train than the model with dummy variables.

For a guideline, we suggest using the predictors without converting to dummy variables and, if the model appears promising, to also try refitting using dummy variables.

5.8 | Summary

Categorical predictors can take a variety of forms in the data that is to be modeled. With the exception of tree-based models, categorical predictors must first be converted to numeric representations to enable other models to use the information. The most simplistic feature engineering technique for a categorical predictor is to convert each category to a separate binary dummy predictor. But even this basic conversion comes with a primary caution that some models require one fewer dummy predictors than the number of categories. Creating dummy predictors may not be the most effective way of extracting predictive information from a categorical predictor. If, for instance, the predictor has ordered categories, then other techniques such as linear or polynomial contrasts may be better related to the outcome.

Text fields, too, can be viewed as an agglomeration of categorical predictors and must be converted to numeric values. There have been a recent host of approaches for converting text to numeric. Often these approaches must include a filtering step to remove highly frequent, non-descriptive words.

Done well, feature engineering for categorical predictors can unlock important predictive information related to the outcome. The next chapter will focus on using feature engineering techniques to uncover additional predictive information within continuous predictors.

5.9 | Computing

The website `http://bit.ly/fes-cat` contains R programs for reproducing these analyses.

6 | Engineering Numeric Predictors

The previous chapter provided methods for skillfully modifying qualitative predictors. Often, other predictors have continuous, real number values. The objective of this chapter is to develop tools for converting these types of predictors into a form that a model can better utilize.

Predictors that are on a continuous scale are subject to a host of potential issues that we may have to confront. Some of the problems that are prevalent with continuous predictors can be mitigated through the type of model that we choose. For example, models that construct relationships between the predictors and the response that are based on the rank of the predictor values rather than the actual value, like trees, are immune to predictor distributions that are skewed or to individual samples that have unusual values (i.e., outliers). Other models such as K-nearest neighbors and support vector machines are much more sensitive to predictors with skewed distributions or outliers. Continuous predictors that are highly correlated with each other is another regularly occurring scenario that presents a problem for some models but not for others. Partial least squares, for instance, is specifically built to directly handle highly correlated predictors. But models like multiple linear regression or neural networks are adversely affected in this situation.

If we desire to utilize and explore the predictive ability of more types of models, the issues presented by the predictors need to be addressed through engineering them in a useful way.

In this chapter we will provide such approaches for and illustrate how to handle continuous predictors with commonly occurring issues. The predictors may:

- be on vastly different scales.
- follow a skewed distribution where a small proportion of samples are orders of magnitude larger than the majority of the data (i.e., skewness).
- contain a small number of extreme values.
- be censored on the low and/or high end of the range.
- have a complex relationship with the response and be truly predictive but cannot be adequately represented with a simple function or extracted by sophisticated models.

- contain relevant and overly redundant information. That is, the information collected could be more effectively and efficiently represented with a smaller, consolidated number of new predictors while still preserving or enhancing the new predictors' relationships with the response.

The techniques in this chapter have been organized into three general categories. The first category of engineering techniques includes those that address problematic characteristics of individual predictors (Section 6.1). Section 6.2 illustrates methods for expanding individual predictors into many predictors in order to better represent more complex predictor-response relationships and to enable the extraction of predictive information. Last, Section 6.3 provides a methodology for consolidating redundant information across many predictors. In the end, the goal of all of these approaches is to convert the existing continuous predictors into a form that can be utilized by any model and presents the most useful information to the model.

As with many techniques discussed in this text, the need for these can be very data- and model-dependent. For example, transformations to resolve skewness or outliers would not be needed for some models but would be critical for others to perform well. In this chapter, a guide to which models would benefit from specific preprocessing methods will be provided.

6.1 │ 1:1 Transformations

There are a variety of modifications that can be made to an individual predictor that might improve its utility in a model. The first type of transformations to a single predictor discussed here are those that change the *scale* of the data. A good example is the transformation described in Figure 1.3 of the first chapter. In that case, the two predictors had very skewed distributions and it was shown that using the inverse of the values improved model performance. Figure 6.1(a) shows the test set distribution for one of the predictors from Figure 1.3.

A Box-Cox transformation (Box and Cox, 1964) was used to *estimate* this transformation. The Box-Cox procedure, originally intended as a transformation of a model's *outcome*, uses maximum likelihood estimation to estimate a transformation parameter λ in the equation

$$x^* = \begin{cases} \frac{x^\lambda - 1}{\lambda \, \tilde{x}^{\lambda-1}}, & \lambda \neq 0 \\ \tilde{x} \, \log x, & \lambda = 0 \end{cases}$$

where $\tilde{x}$ is the geometric mean of the predictor data. In this procedure, λ is estimated from the data. Because the parameter of interest is in the exponent, this type of transformation is called a *power transformation*. Some values of λ map to common transformations, such as $\lambda = 1$ (no transformation), $\lambda = 0$ (log), $\lambda = 0.5$ (square root), and $\lambda = -1$ (inverse). As you can see, the Box-Cox transformation is quite flexible in its ability to address many different data distributions. For the data in

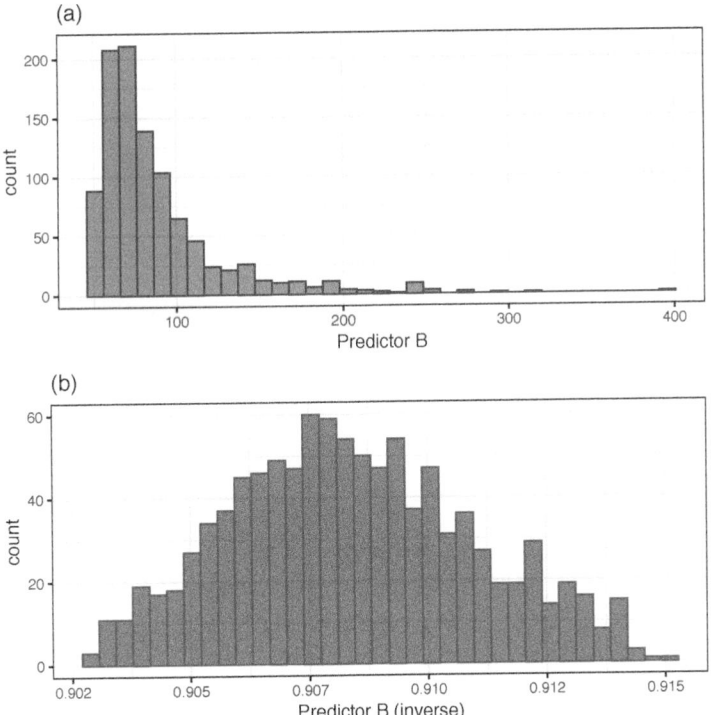

Figure 6.1: The distribution of a skewed predictor before (a) and after (b) applying the Box-Cox transformation.

Figure 6.1, the parameter was estimated from the training set data to be $\widehat{\lambda} = -1.09$. This is effectively the inverse transformation. Figure 6.1(b) shows the results when the transformation is applied to the test set, and yields a transformed distribution that is approximately symmetric. It is important to note that the Box-Cox procedure can only be applied to data that is strictly positive. To address this problem, Yeo and Johnson (2000) devised an analogous procedure that can be used on any numeric data.

Also, note that both transformations are *unsupervised* since, in this application, the outcome is not used in the computations. While the transformation might improve the predictor distribution, it has no guarantee of improving the model. However, there are a variety of parametric models that utilize polynomial calculations on the predictor data, such as most linear models, neural networks, and support vector machines. In these situations, a skewed predictor distribution can have a harmful effect on these models since the tails of the distribution can dominate the underlying calculations.

It should be noted that the Box-Cox transformation was originally used as a *supervised* transformation of the outcome. A simple linear model would be fit to the data and the transformation would be estimated from the model *residuals*. The outcome variable would be transformed using the results of the Box-Cox method. Here, the method has been appropriated to be independently applied to each predictor and uses their data, instead of the residuals, to determine an appropriate transformation.

Another important transformation to an individual variable is for a variable that has values bounded between zero and one, such as proportions. The problem with modeling this type of outcome is that model predictions might may not be guaranteed to be within the same boundaries. For data between zero and one, the *logit* transformation could be used. If π is the variable, the logit transformations is

$$logit(\pi) = log\left(\frac{\pi}{1 - \pi}\right).$$

This transformation changes the scale from values between zero and one to values between negative and positive infinity. On the extremes, when the data are absolute zero or one, a small constant can be added or subtracted to avoid division by zero. Once model predictions are created, the inverse logit transformation can be used to place the values back on their original scale. An alternative to the logit transformation is the *arcsine transformation*. This is primarily used on the square root of the proportions (e.g., $y^* = arcsine(\sqrt{\pi})$).

Another common technique for modifying the scale of a predictor is to standardize its value in order to have specific properties. *Centering* a predictor is a common technique. The predictor's training set average is subtracted from the predictor's individual values. When this is applied separately to each variable, the collection of variables would have a common mean value (i.e., zero). Similarly, *scaling* is the process of dividing a variable by the corresponding training set's standard deviation. This ensures that that variables have a standard deviation of one. Alternatively, *range* scaling uses the training set minimum and maximum values to translate the data to be within an arbitrary range (usually zero and one). Again, it is emphasized that the statistics required for the transformation (e.g., the mean) are estimated from the training set and are applied to all data sets (e.g., the test set or new samples).

These transformations are mostly innocuous and are typically needed when the model requires the predictors to be in common units. For example, when the distance or dot products between predictors are used (such as K-nearest neighbors or support vector machines) or when the variables are required to be a common scale in order to apply a penalty (e.g., the lasso or ridge regression described in Section 7.3), a standardization procedure is essential.

Another helpful preprocessing methodology that can be used on data containing a *time or sequence effect* is simple data smoothing. For example, a running mean can be used to reduce excess noise in the predictor or outcome data prior to modeling. For example, a running 5-point mean would replace each data point with the average of itself and the two data points before and after its position.[53] As one might expect,

---

[53]There are various approaches for the beginning and end of the data, such as leaving the data as is.

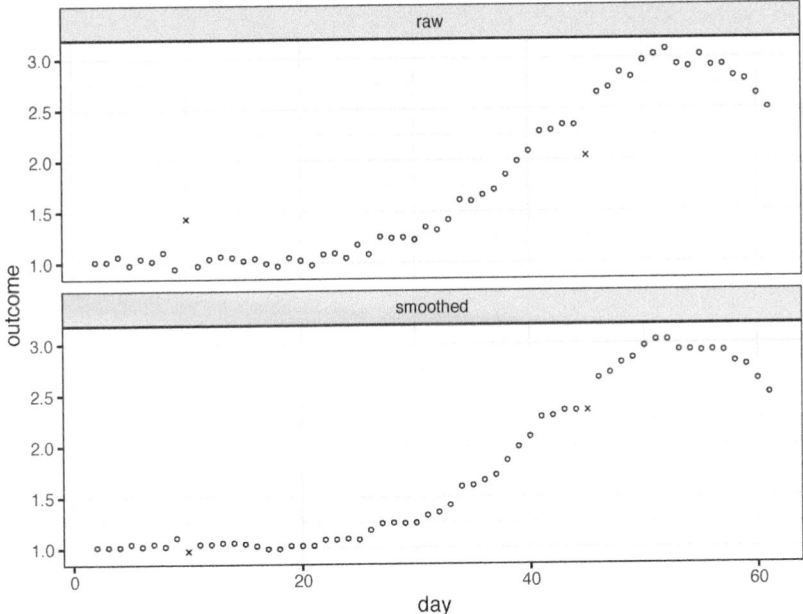

Figure 6.2: A sequence of outcome values over time. The raw data contain outliers on days 10 and 45. The smoothed values are the result of a 3-point running median.

the size of the moving window is important; too large and the smoothing effect can eliminate important trends, such as nonlinear patterns.

A short running median can also be helpful, especially if there are significant outliers. When an outlier falls into a moving window, the mean value is pulled towards the outlier. The median would be very insensitive to an aberrant value and is a better choice. It also has the effect of changing fewer of the original data points.

For example, Figure 6.2 shows a sequence of data over 61 days. The raw data are in the top panel and there are two aberrant data values on days 10 and 45. The lower panel shows the results of a 3-point running median smoother. It does not appear to blunt the nonlinear trend seen in the last 15 days of data but does seem to mitigate the outliers. Also, roughly 40% of the smoothed data have the same value as the original sequence.

Other smoothers can be used, such as *smoothing splines* (also described in this chapter). However, the simplicity and robustness of a short running median can be an attractive approach to this type of data. This smoothing operation can be applied to the outcome data (as in Figure 6.2) and/or to any sequential predictors. The latter case is mentioned in Section 3.4.7 in regard to information leakage. It is important to make sure that the test set predictor data are smoothed separately to avoid having the training set influence values in the test set (or new unknown samples). Smoothing predictor data to reduce noise is explored more in Section 9.4.

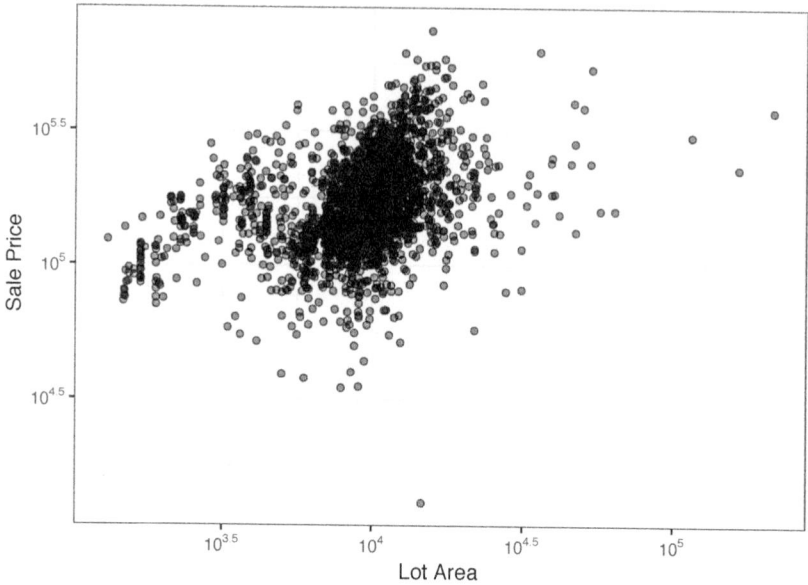

Figure 6.3: The sale price and lot area data for the training set.

6.2 | 1:Many Transformations

The previous chapter illustrated the process of creating multiple numeric indicator columns from a single qualitative predictor. Similarly, transformations can be made on a single numeric predictor to expand it to many predictors. These one-to-many transformations of the data can be used to improve model performance.

6.2.1 | Nonlinear Features via Basis Expansions and Splines

A *basis expansion* of a predictor x can be achieved by deriving a set of functions $f_i(x)$ that can be combined using a linear combination. For example, in the last chapter, polynomial contrast functions were used to encode ordered categories. For a continuous predictor x, a cubic basis expansion is

$$f(x) = \sum_{i=1}^{3} \beta_i f_i(x) = \beta_1 x + \beta_2 x^2 + \beta_3 x^3.$$

To use this basis expansion in the model, the original column is augmented by two new features with squared and cubed versions of the original. The β values could be estimated using basic linear regression. If the true trend were linear, the second two regression parameters would presumably be near zero (at least relative to their standard error). Also, the linear regression used to determine the coefficients for the basis expansion could be estimated in the presence of other regressors.

This type of basis expansion, where the pattern is applied *globally* to the predictor, can often be insufficient. For example, take the lot size variable in the Ames housing data. When plotted against the sale price[54], there is a linearly increasing trend in the mainstream of the data (between $10^{3.75}$ and $10^{4.25}$) but on either side of the bulk of the data, the patterns are negligible.

An alternative method to creating a global basis function that can be used in a regression model is a *polynomial spline* (Wood, 2006; Eilers and Marx, 2010). Here, the basis expansion creates different regions of the predictor space whose boundaries are called *knots*. The polynomial spline uses polynomial functions, usually cubic, within each of these regions to represent the data. We would like the cubic functions to be connected at the knots, so specialized functions can be created to ensure an overall continuous function.[55] Here, the number of knots controls the number of regions and also the potential complexity of the function. If the number of knots is low (perhaps three or less), the basis function can represent relatively simple trends. Functions with more knots are more adaptable but are also more likely to overfit the data.

The knots are typically chosen using percentiles of the data so that a relatively equal amount of data is contained within each region. For example, a spline with three regions typically places the knots at the 33.3% and 66.7% percentiles. Once the knots are determined, the basis functions can be created and used in a regression model. There are many types of polynomial splines and the approach described here is often called a *natural cubic spline*.

For the Ames lot area data, consider a spline with 6 regions. The procedure places the knots at the following percentiles: 16.7%, 33.3%, 50%, 66.7%, 83.3%. When creating the natural spline basis functions, the first function is typically taken as the intercept. The other functions of x generated by this procedure are shown in Figure 6.4(a). Here the blue lines indicate the knots. Note that the first three features contain regions where the basis function value is zero. This implies that those features do not affect that part of the predictor space. The other three features appear to put the most weight on areas outside the mainstream of the data. Panel (b) of this figure shows the final regression form. There is a strong linear trend in the middle of the cloud of points and weak relationships outside of the mainstream. The left-hand side does not appear to fit the data especially well. This might imply that more knots are required.

How many knots should be used? This is a tuning parameter that can be determined using grid search or through visual inspection of the smoothed results (e.g., Figure 6.4(b)). In a single dimension, there will be visual evidence if the model is overfitting. Also, some spline functions use a method called *generalized cross-validation* (GCV) to estimate the appropriate spline complexity using a computational shortcut for linear regression (Golub et al., 1979).

Note that the basis function was created *prior* to being exposed to the outcome data.

---

[54]Both variables are highly right skewed, so a log transformation was applied to both axes.

[55]These methods also ensure that the derivatives of the function, to a certain order, are also continuous. This is one reason that cubic functions are typically used.

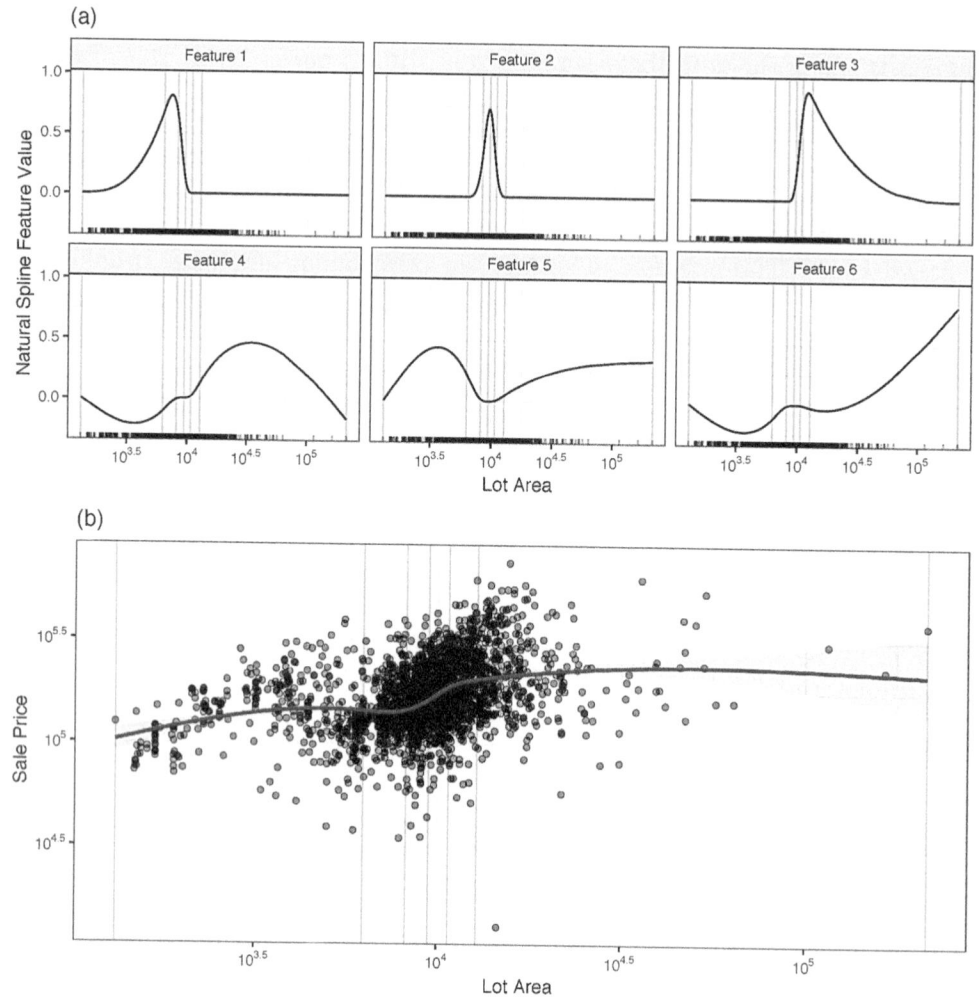

Figure 6.4: (a) Features created using a natural spline basis function for the lot area predictor. The blue lines correspond to the knots. (b) The fit after the regression model estimated the model coefficients for each feature.

This does not have to be the case. One method of determining the optimal complexity of the curve is to initially assign *every* training set point as a potential knot and uses regularized regression models (similar to weight decay or ridge regression) to determine which instances should be considered to be knots. This approach is used by the *smoothing spline* methodology (Yandell, 1993). Additionally, there is a rich class or models called *generalized additive models* (GAMs). These extend general linear models, which include linear and logistic regression, to have nonlinear terms for individual predictors (and cannot model interactions). GAM models can adaptively model separate basis functions for different variables and estimate the complexity for each. In other words, different predictors can be modeled with different levels of complexity. Additionally, there are many other types of supervised nonlinear smoothers that can be used. For example, the `loess` model fits a weighted moving regression line across a predictor to estimate the nonlinear pattern in the data. See Wood (2006) for more technical information about GAM models.

One other feature construction method related to splines and the multivariate adaptive regression spline (MARS) model (Friedman, 1991) is the single, fixed knot spline. The *hinge function* transformation used by that methodology is

$$h(x) = xI(x > 0)$$

where I is an indicator function that is x when x is greater than zero and zero otherwise. For example, if the log of the lot area were represented by x, $h(x - 3.75)$ generates a new feature that is zero when the log lot area is less than 3.75 and is equal to $x - 3.75$ otherwise. The opposite feature, $h(3.75 - x)$, is zero above a value of 3.75. The effect of the transformation can be seen in Figure 6.5(a) which illustrates how a pair of hinge functions isolate certain areas of the predictor space.

These features can be added to models to create *segmented regression models* that have distinct sections with different trends. As previously mentioned, the lot area data in Figure 6.4 exhibits a linear trend in the central region of the data and flat trends on either extreme. Suppose that two sets of hinge functions were added to a linear model with knots at 3.75 and 4.25. This would generate a separate linear trend in the region below a value of $10^{3.75}$, between $10^{3.75}$ and $10^{4.25}$, and above $10^{4.25}$. When added to a linear regression, the left-most region's slope would be driven by the "left-handed" hinge functions for both knots. The middle region would involve the slopes for both terms associated with a knot of $10^{3.75}$ as well as the left-handed hinge function with the knot at $10^{4.25}$ and so on.[56] The model fit associated with this strategy is shown in Figure 6.5(b).

This feature generation function is known in the neural network and deep learning fields as the rectified linear unit (ReLU) activation function. Nair and Hinton (2010) gives a summary of their use in these areas.

While basis functions can be effective at helping build better models by representing nonlinear patterns for a feature, they can be very effective for exploratory data analysis. Visualizations, such as Figures 6.4(b) and 6.5, can be used to inform the modeler about what the potential functional form of the prediction should be (e.g., log-linear, quadratic, segmented, etc.).

6.2.2 | Discretize Predictors as a Last Resort

Binning, also known as categorization or discretization, is the process of translating a quantitative variable into a set of two or more qualitative buckets (i.e., categories). For example, a variable might be translated into quantiles; the buckets would be for whether the numbers fell into the first 25% of the data, between the 25th and

---

[56]Note that the MARS model sequentially generates a set of knots adaptively and determines which should be retained in the model.

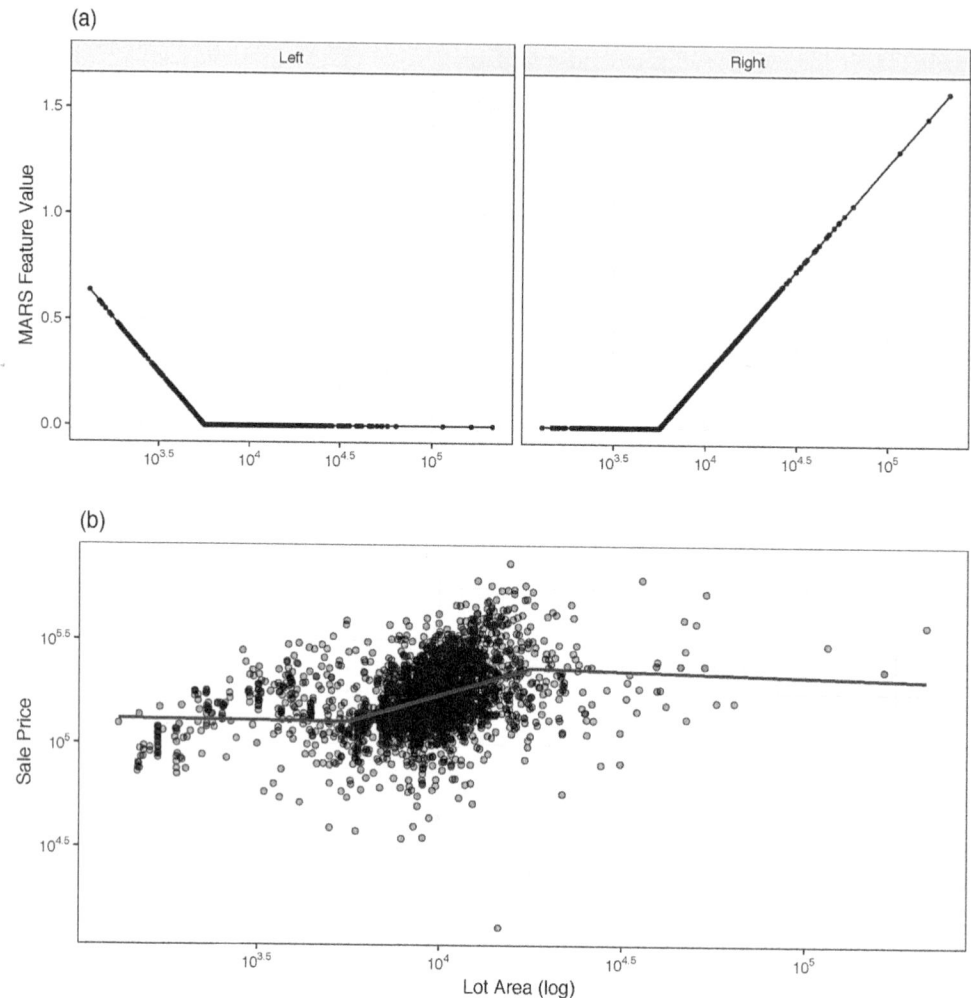

Figure 6.5: (a) Features created using a natural spline basis function for the lot area predictor. The blue lines correspond to the knots. (b) The fit after the regression model estimated the model coefficients for each feature.

median, etc. In this case, there would be four distinct values of the binned version of the data.

There are a few apparent reasons for subjecting the data to such a transformation:

- Some feel that it simplifies the analysis and/or interpretation of the results. Suppose that a person's age was a predictor and this was binned by whether someone was above 40 years old or not. One might be able to make a statement that there is a 25% increase in the probability of the event between younger and older people. There is no discussion of per-unit increases in the response.

- Binning *may* avoid the problem of having to specify the relationship between the predictor and outcome. A set of bins can be perceived as being able to model more patterns without having to visualize or think about the underlying pattern.

- Using qualitative versions of the predictors may give the perception that it reduces the variation in the data. This is discussed at length below.

There are a number of methods for binning data. Some are *unsupervised* and are based on user-driven cutoffs or estimated percentiles (e.g., the median). In other cases, the placement (and number) of cut-points are optimized to improve performance. For example, if a single split were required, an ROC curve (Section 3.2.2) could be used to find an appropriate exchange of sensitivity and specificity.

There are a number of problematic issues with turning continuous data categorical. First, it is extremely unlikely that the underlying trend is consistent with the new model. Secondly, when a real trend exists, discretizing the data is most likely making it *harder* for the model to do an effective job since all of the nuance in the data has been removed. Third, there is probably no objective rationale for a specific cut-point. Fourth, when there is no relationship between the outcome and the predictor, there is a substantial increase in the probability that an erroneous trend will be "discovered". This has been widely researched and verified. See Altman (1991), Altman et al. (1994), and the references therein.

Kenny and Montanari (2013) do an exceptional job of illustrating how this can occur and our example follows their research. Suppose a set of variables is being screened to find which ones have a relationship with a numeric outcome. Figure 6.6(a) shows an example of a simulated data set with a linear trend with a coefficient of 1.0 and normally distributed errors with a standard deviation of 4.0. A linear regression of these data (shown) would find an increasing trend and would have an estimated R^2 of 30.3%. The data were cut into 10 equally spaced bins and the average outcome was estimated per bin These data are shown in panel (b) along with the regression line. Since the average is being used, the estimated R^2 is much larger (64.2%) since the model *thinks* that the data are more precise (which, in reality, they are not).

Unfortunately, this artificial reduction in variation can lead to a high false positive rate when the *predictor is unrelated to the outcome*. To illustrate this, the same data were used but with a slope of zero (i.e., the predictor is uninformative). This simulation was run a large number of times with and without binning the data and Figure 6.6(c) shows the distribution of the estimated R^2 values across the simulations. For the raw data, the largest R^2 value seen in 1000 data sets was 18.3% although most data are less than 20%. When the data were discretized, the R^2 values tended to be much larger (with a maximum of 76.9%). In fact, 19% of the simulations had $R^2 \geq 20\%$.

There is a possibility that a discretization procedure might improve the model; the No Free Lunch Theorem discounts the idea that a methodology will *never* work. If a categorization strategy is to be used, especially if it is supervised, we suggest that:

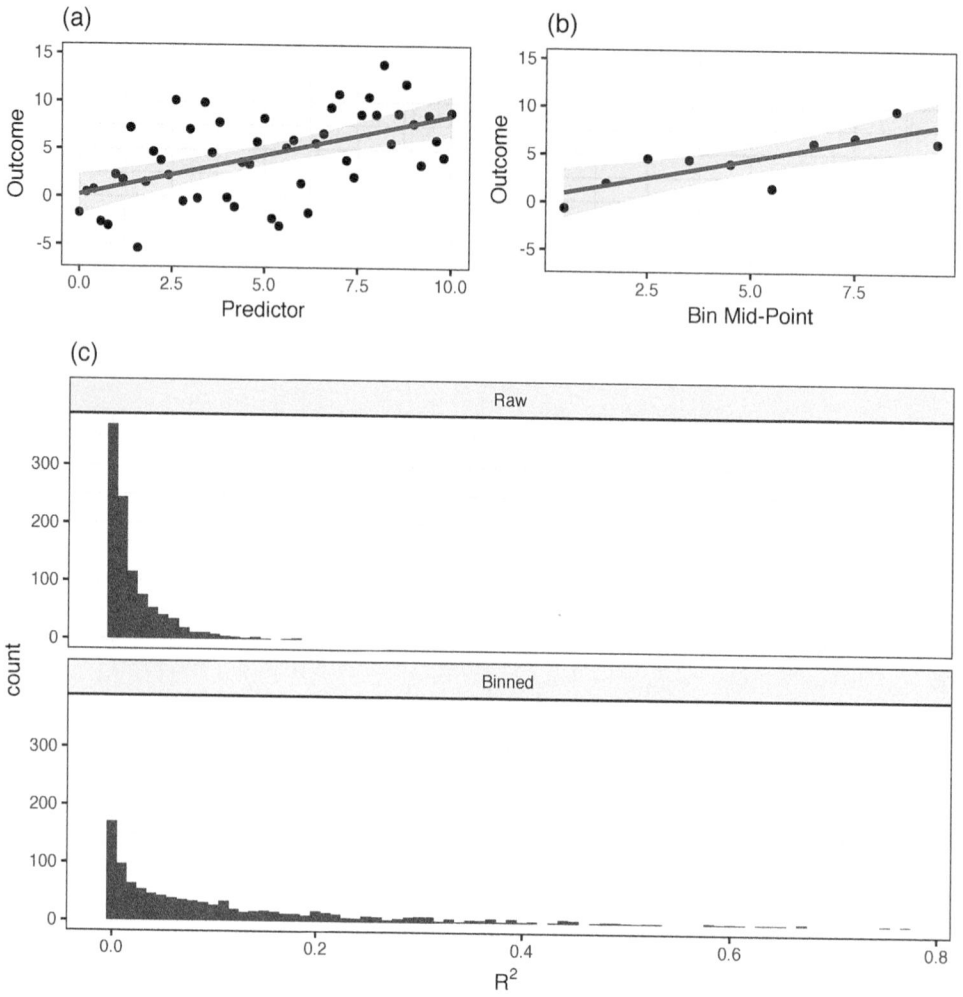

Figure 6.6: Raw data (a) and binned (b) simulation data. Also, the results of a simulation (c) where the predictor is noninformative.

1. The strategy should not be part of the normal operating procedure. Categorizing predictors should be a method of last resort.

2. The determination of the bins *must* be included inside of the resampling. process. This will help diagnose when nonexistent relationships are being induced by the procedure and will mitigate the overestimation of performance when an informative predictor is used. Also, since some sort of procedure would have to be defined to be included inside of resampling, it would prevent the typical approach of "eyeballing" the data to pick cut-points. We have found that a binning strategy based solely on visual inspection of the data has a higher propensity of overfitting.

6.3 | Many:Many Transformations

When creating new features from multiple predictors, there is a possibility of correcting a variety of issues such as outliers or collinearity. It can also help reduce the dimensionality of the predictor space in ways that might improve performance and reduce computational time for models.

6.3.1 | Linear Projection Methods

Observational or available data sets nowadays contain many predictors, and often more than the number of available samples. The predictors are usually any measurements that may have a remote chance of potentially being related the response. A conventional thought process is that the more predictors that are included, the better chance the model will have to find the signal. But this is not the case; including irrelevant predictors can have detrimental effects on the final model (see Chapter 10). The negative impacts include: increasing the computational time required for training a model, decreasing the predictive performance, and complicating the predictor importance calculation. By including any potential predictor, we end up with a sub-optimal model. This is illustrated in Section 10.3.

The tension in working with modern data is that we often do not fully know which predictors are relevant to the response. What may in fact be true is that some of the available predictors (or combinations) have a meaningful association with the response. Under these conditions, the task is to try to identify the simple or complex combination that is relevant to the response.

This section will cover projection methods that have been shown to effectively identify meaningful projections of the original predictors. The specific techniques discussed here are principal component analysis (PCA), kernel PCA, independent component analysis (ICA), non-negative matrix factorization (NNMF), and partial least squares (PLS). The first four techniques in this list are unsupervised techniques that are ignorant of the response. The final technique uses the response to direct the dimension reduction process so that the reduced predictor space is optimally associated with the response.

These methods are *linear* in the sense that they take a matrix of numeric predictor values (X) and create new components that are linear combinations of the original data. These components are usually called the *scores*. If there are n training set points and p predictors, the scores could be denoted as the $n \times p$ matrix X^* and are created using

$$X^* = XA$$

where the $p \times p$ matrix A is often called the *projection matrix*. As a result, the value of the first score for data point i would be:

$$x_{i1}^* = a_{11}x_{i1} + a_{21}x_{i2} + \ldots + a_{p1}x_{ip}.$$

One possibility is to only use a portion of A to reduce the number of scores down to $k < p$. For example, it may be that the last few columns of A show no benefit to describing the predictor set. This notion of *dimension reduction* should not be confused with *feature selection*. If the predictor space can be described with k score variables, there are less features in the data given to the model but each of these columns is a function of all of the original predictors.

The differences between the projection methods mentioned above[57] is in how the projection values are determined.

As will be shown below, each of the unsupervised techniques has its own objective for summarizing or projecting the original predictors into a lower dimensional subspace of size k by using only the first k columns of A. By condensing the predictor information, the predictor redundancy can be reduced and potential signals (as opposed to the noise) can be teased out within the subspace. A direct benefit of condensing information to a subspace of the predictors is a decrease in the computational time required to train a model. But employing an unsupervised technique does not guarantee an improvement in predictive performance. The primary underlying assumption when using an unsupervised projection method is that the identified subspace is predictive of the response. This may or may not be true since the objective of unsupervised techniques is completely disconnected with the response. At the conclusion of model training, we may find that unsupervised techniques are convenient tools to condense the original predictors, but do not help to effectively improve predictive performance.[58]

The supervised technique, PLS, improves on unsupervised techniques by focusing on the objective of finding a subspace of the original predictors that is directly associated with the response. Generally, this approach finds a smaller and more efficient subspace of the original predictors. The drawback of using a supervised method is that a larger data set is required to reserve samples for testing and validation to avoid overfitting.

6.3.1.1   Principal Component Analysis

Principal component analysis was introduced earlier as an exploratory visualization technique in Section 4.2.7. We will now explore PCA in more detail and investigate how and when it can be effectively used to create new features. Principal component analysis was originally developed by Karl Pearson more than a century ago and, despite its age, is still one of the most widely used dimension reduction tools. A good reference for this method is Abdi and Williams (2010). Its relevance in modern data stems from the underlying objective of the technique. Specifically, the objective of PCA is to find linear combinations of the original predictors such that the combinations summarize the maximal amount of variation in the original predictor space. From a statistical perspective, variation is synonymous with information. So, by finding combinations of the original predictors that capture variation, we find

---

[57]Except kernel PCA.

[58]This is a situation similar to the one described above for the Box-Cox transformation.

the subspace of the data that contains the information relevant to the predictors. Simultaneously, the new PCA scores are required to be orthogonal (i.e., uncorrelated) to each other. The property of orthogonality enables the predictor space variability to be neatly partitioned in a way that does not overlap. It turns out that the variability summarized across all of the principal components is exactly equal to the variability across the original predictors. In practice, enough new score variables are computed to account for some pre-specified amount of the variability in the original predictors (e.g., 95%).

An important side benefit of this technique is that the resulting PCA scores are uncorrelated. This property is very useful for modeling techniques (e.g., multiple linear regression, neural networks, support vector machines, and others) that need the predictors to be relatively uncorrelated. This is discussed again in the context of ICA below.

PCA is a particularly useful tool when the available data are composed of one or more clusters of predictors that contain redundant information (e.g., predictors that are highly correlated with one another). Predictors that are highly correlated can be thought of as existing in a lower dimensional space than the original data. As a trivial example, consider the scatter plot in Figure 6.7 (a), where two predictors are linearly related. While these data have been measured in two dimensions, only one dimension is needed (a line) to adequately summarize the position of the samples. That is, these data could be nearly equivalently represented by a combination of these two variables as displayed in part (b) of the figure. The red points in this figure are the orthogonal projections of the original data onto the first principal component score (imagine the diagonal line being rotated to horizontal). The new feature represents the original data by one numeric value which is the distance from the origin to the projected value. After creating the first PCA score, the "leftover" information is represented by the length of the blue lines in panel (b). These distances would be the final score (i.e., the second component).

For this simple example, a linear regression model was built using the two correlated predictors. A linear regression model was also built between the first principal component score and the response. Figure 6.8 provides the relationship between the observed and predicted values from each model. For this example, there is virtually no difference in performance between these models.

Principal component regression, which describes the process of first reducing dimension and then regressing the new component scores onto the response is a stepwise process (Massy, 1965). The two steps of this process are disjoint which means that the dimension reduction step is not directly connected with the response. Therefore, the principal component scores associated with the variability in the data may or may not be related to the response.

To provide a practical example of PCA, let's return to the Chicago ridership data that we first explored in Chapter 4. Recall that ridership across many stations was highly correlated (Figure 4.9). Of the 125 stations, 94 had at least one pairwise correlation

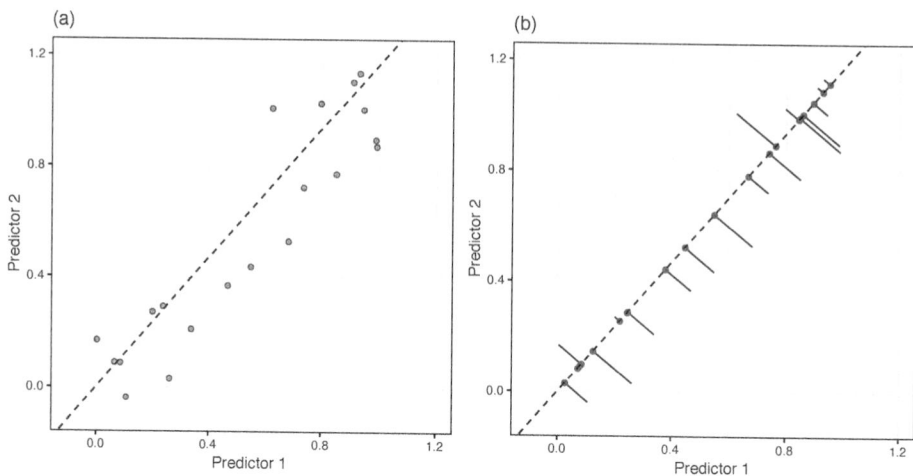

Figure 6.7: Principal component analysis of two highly correlated predictors. (a) A scatter plot of two highly correlated predictors with the first principal direction represented with a dashed line. (b) The orthogonal projection of the original data onto the first principal direction. The new feature is represented by the red points.

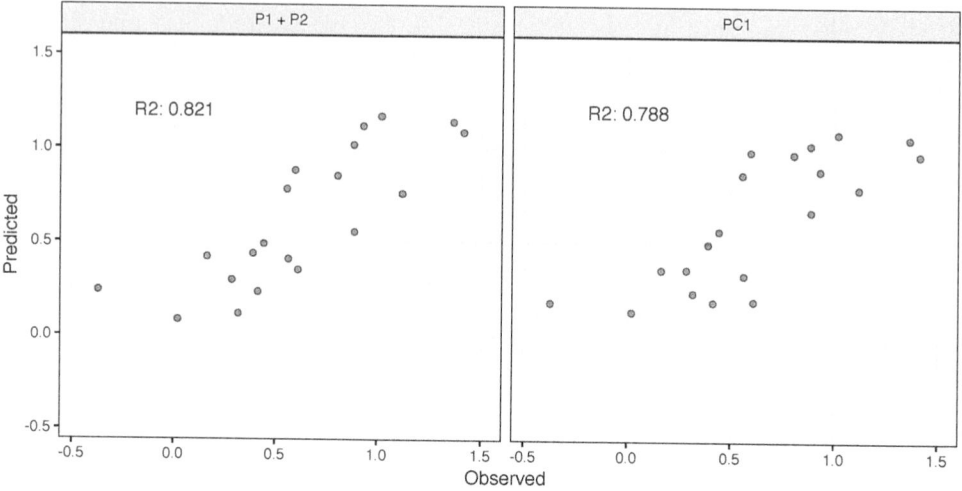

Figure 6.8: A linear combination of the two predictors chosen to summarize the maximum variation contains essentially the same information as both of the original predictors.

with another station greater than 0.9. What this means is that if the ridership is known at one station, the ridership at many other stations can be estimated with a high degree of accuracy. In other words, the information about ridership is more simply contained in a much lower dimension of these data.

The Chicago data can be used to demonstrate how projection methods work. Recall that the main driving force in the data are differences between weekend and weekday

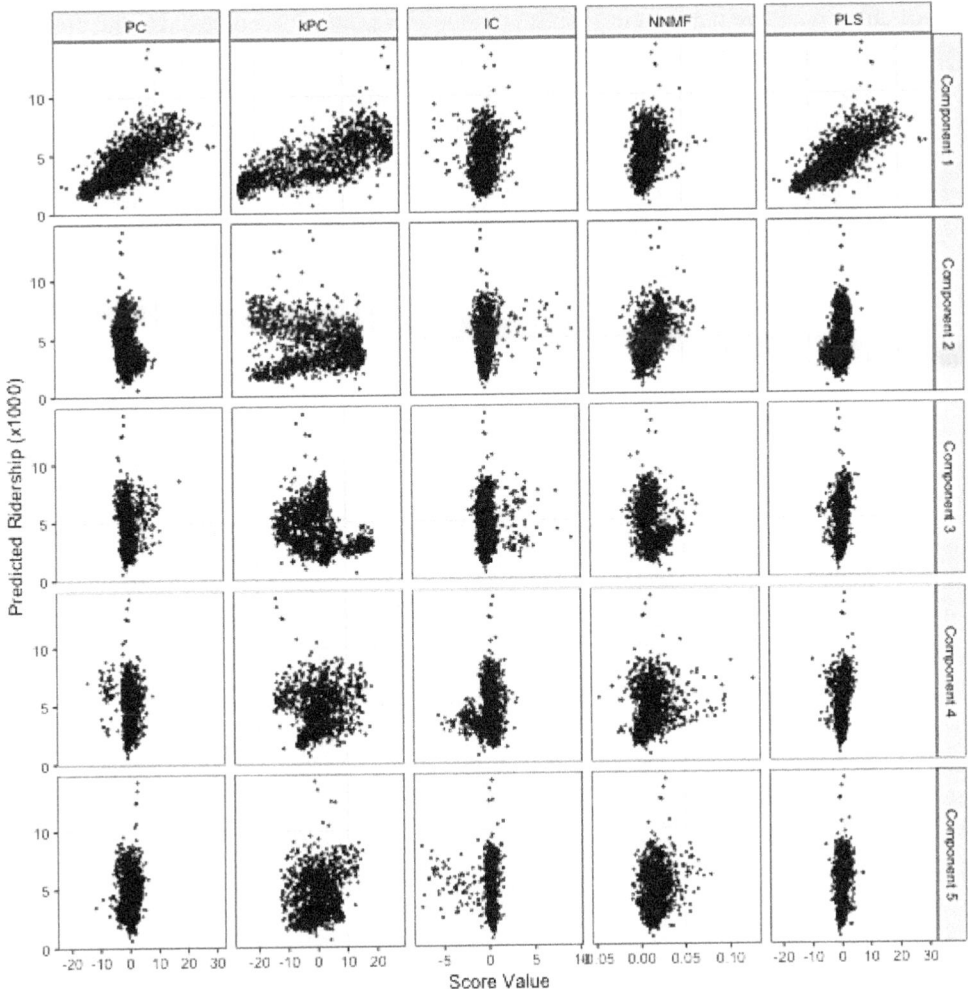

Figure 6.9: Score values from several linear projection methods for the weekend ridership at Clark and Lake. The x-axis values are the scores for each method and the y-axis is ridership (in thousands).

ridership. The gulf between the data for these two groups completely drives the principal component analysis since it accounts for most of what is going on in the data. Rather than showing this analysis, the weekend data were analyzed here.[59] There were 1628 days on either Saturday or Sunday in the training data. Using only the 14-day lag predictors, PCA was conducted on the weekend data after the columns were centered and scaled.

The left-most column of Figure 6.9 shows the first five components and their relationships to the outcome. The first component, which accounts for 60.4% of the variation in the lagged predictors, shows a strong linear relationship with ridership. For this

---

[59]We don't suggest this as a realistic analysis strategy for these data, only to make the illustration more effective.

component, all of the coefficients of the projection matrix A corresponding to this linear combination are positive and the five most influential stations were 35th/Archer, Ashland, Austin, Oak Park, and Western.[60] The five *least* important stations, to the first component, were Central, O'Hare, Garfield, Linden, and Washington/Wells.

The second component only accounts for 5% of the variability and has less correlation with the outcome. In fact, the remaining components capture smaller and smaller amounts of information in the lagged predictors. To cover 95% of the original information, a total of 36 principal components would be required.

One important question that PCA might help answer is, "Is there a line effect on the weekends?" To help answer this, the first five components are visualized in Figure 6.10 using a heatmap. In the heatmap, the color of the points represents the signs and magnitude of the elements of A. The rows have been reordered based on a clustering routine that uses the five components. The row labels show the line (or lines) for the corresponding station.

The first column shows that the relative weight for each station to the first component is reliably positive and small (in comparison to the other components); there is no line effect in the main underlying source of information. The other components appear to isolate specific lines (or subsets within lines). For example, the second component is strongly driven by stations along the Pink line (the dark red section), a section of the Blue line going towards the northwest, and a portion of the Green line that runs northbound of the city. Conversely, the fourth component has a strong *negative* cluster for the same Green line stations. In general, the labels along the y-axis show that the lines are fairly well clustered together, which indicates that there is a line effect after the first component's effects are removed.

Despite the small amount of variation that the second component captures, we'll use these coefficients in A to accentuate the geographic trends in the weekend data. The second PCA component's rotation values have both positive and negative values. Their sign and magnitude can help inform us as to which stations are driving this component. As seen in Figure 6.11, the stations on the Red and Purple lines[61] that head due north along the coast have substantial positive effects on the second component. Southwestern stations along the Pink line have large negative effects (as shown by the larger red markers) as did the O'Hare station (in the extreme northwest). Also, many of the stations clustered around Clark and Lake have small or moderately negative effects on the second station. It is interesting that two stations in close proximity to one another, Illinois Medical District and Polk, have substantial effects on the component but have opposing signs. These two stations are on different lines (Blue and Pink, respectively) and this may be another indication that there is a line effect in the data once the main piece of information in the data (i.e, the first component) is removed.

---

[60]In these cases, "influential" means the largest coefficients.

[61]Recall Figure 4.1 for the station lines.

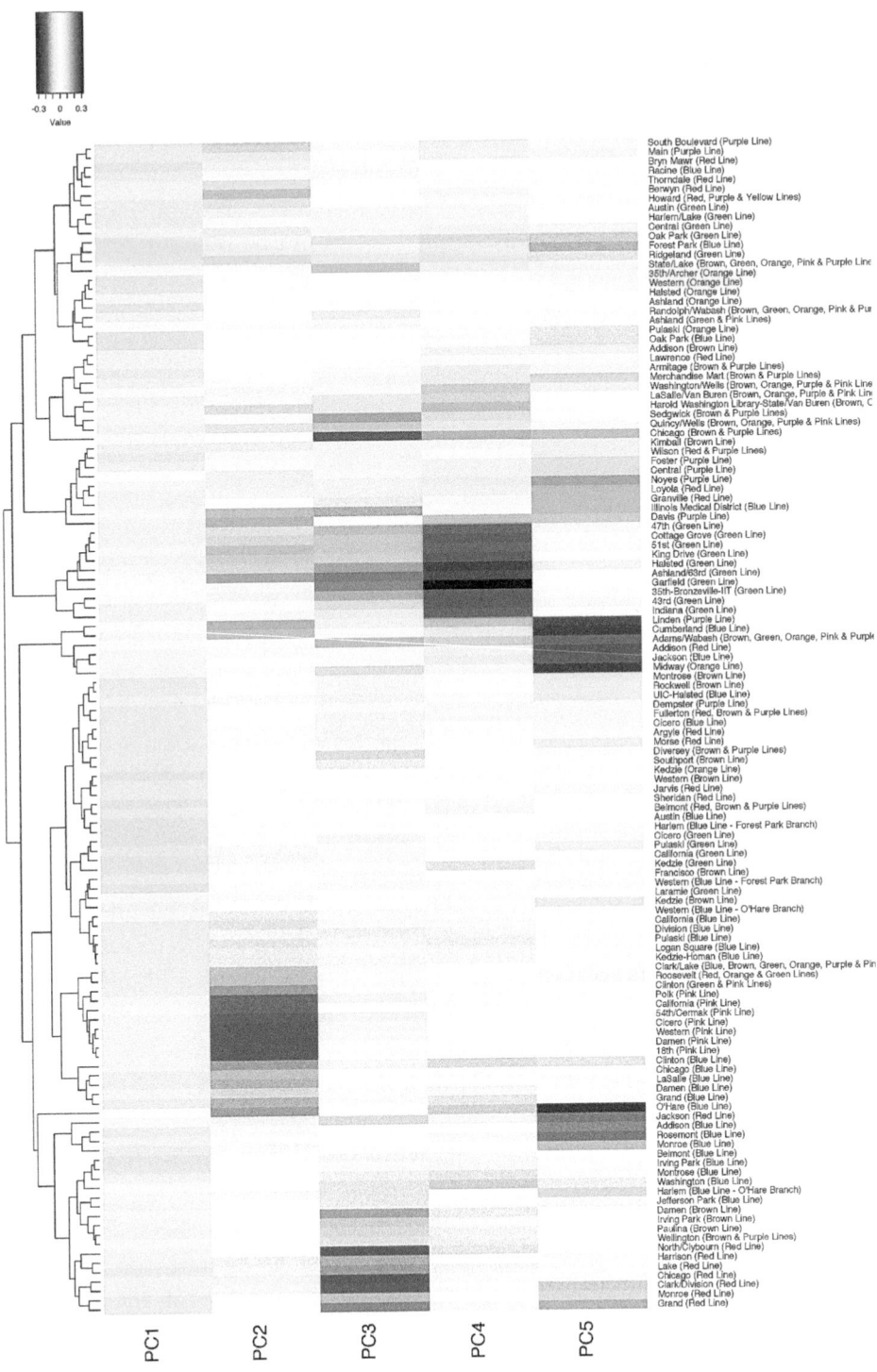

Figure 6.10: A heatmap of the first five principal component coefficients. The rows are clustered using all five components.

Figure 6.11: Coefficients for the second principal component. The size is relative magnitude of the coefficient while red indicates negativity, blue is positivity, and white denotes values near zero.

There are a variety of methods for visualizing the PCA results. It is common to plot the first few scores against one another to look for separation between classes, outliers, or other interesting patterns. Another visualization is the *biplot* (Greenacre, 2010), an informative graphic that overlays the scores along with the coefficients of A. In our data, there are 126 highly correlated stations involved in the PCA analysis and, in this case, the biplot is not terribly effective.

Finally, it is important to note that since each PCA component captures less and less variability, it is common for the *scale of the scores* to become smaller and smaller. For this reason, it is important to plot the component scores on a common axis scale that accounts for all of the data in the scores (i.e., the largest possible range).

In this way, small patterns in components that capture a small percentage of the information in the data will not be over-interpreted.

6.3.1.2 Kernel Principal Component Analysis

Principal component analysis is an effective dimension reduction technique when predictors are *linearly correlated* and when the resulting scores are associated with the response. However, the orthogonal partitioning of the predictor space may not provide a good predictive relationship with the response, especially if the true underlying relationship between the predictors and the response is non-linear. Consider, for example, a hypothetical relationship between a response (y) and two predictors (x_1 and x_2) as follows:

$$y = x_1 + x_1^2 + x_2 + x_2^2 + \epsilon.$$

In this equation ϵ represents random noise. Suppose also that the correlation between x_1 and x_2 is strong as in Figure 6.7. Applying traditional PCA to this example would summarize the relationship between x_1 and x_2 into one principal component. However this approach would ignore the important quadratic relationships that would be critically necessary for predicting the response. In this two-predictor example, a simple visual exploration of the predictors and the relationship between the predictors and the response would lead us to understanding that the quadratic versions of the predictors are required for predicting the relationship with the response. But the ability to visually explore important higher-order terms diminishes as the dimension of the data increases. Therefore traditional PCA will be a much less effective dimension reduction technique when the data should be augmented with additional features.

If we knew that one or more quadratic terms were necessary for explaining the response, then these terms could be added to the predictors to be evaluated in the modeling process. Another approach would be to use a nonlinear basis expansion as described earlier in this chapter. A third possible approach is kernel PCA (Schölkopf et al., 1998; Shawe-Taylor and Cristianini, 2004). The kernel PCA approach combines a specific mathematical view of PCA with *kernel functions* and the kernel 'trick' to enable PCA to expand the dimension of the predictor space in which dimension reduction is performed (not unlike basis expansions). To better understand what kernel PCA is doing, we first need to recognize that the variance summarization problem posed by PCA can be solved by finding the eigenvalues and eigenvectors of the covariance matrix of the predictors. It turns out that the form of the eigen decomposition of the covariance matrix can be equivalently written in terms of the inner product of the samples. This is known as the 'dual' representation of the problem. The inner-product is referred to as the linear kernel; the linear kernel between x_1 and x_2 is represented as:

$$k\left(\mathbf{x}_1, \mathbf{x}_2\right) = \langle \mathbf{x}_1, \mathbf{x}_2 \rangle = \mathbf{x}_1' \mathbf{x}_2.$$

For a set of $i = 1 \ldots n$ samples and two predictors, this corresponds to

$$k\left(\mathbf{x}_1, \mathbf{x}_2\right) = x_{11}x_{12} + \ldots + x_{n1}x_{n2}.$$

Kernel PCA then uses the kernel substitution trick to exchange the linear kernel with any other valid kernel. For a concise list of kernel properties for constructing valid kernels, see Bishop (2011). In addition to the linear kernel, the basic polynomial kernel has one parameter, d, and is defined as:

$$k\left(\mathbf{x}_1, \mathbf{x}_2\right) = \langle \mathbf{x}_1, \mathbf{x}_2 \rangle^d.$$

Again, with two predictors and $d = 2$:

$$k\left(\mathbf{x}_1, \mathbf{x}_2\right) = \left(x_{11}x_{12} + \ldots + x_{n1}x_{n2}\right)^2.$$

Using the polynomial kernel, we can directly expand a linear combination as a polynomial of degree d (similar to what was done for basis functions in Section 6.2.1).

The radial basis function (RBF) kernel has one parameter, σ and is defined as:

$$k\left(\mathbf{x}_1, \mathbf{x}_2\right) = \exp\left(-\frac{\|\mathbf{x}_1 - \mathbf{x}_2\|^2}{2\sigma^2}\right).$$

The RBF kernel is also known as the Gaussian kernel due to its mathematical form and its relationship to the normal distribution. The radial basis and polynomial kernels are popular starting points for kernel PCA.

Utilizing this view enables the kernel approach to expand the original dimension of the data into high dimensions that comprise potentially beneficial representations of the original predictors. What's even better is that the kernel trick side-steps the need to go through the hassle of creating these terms in the X matrix (as one would with basis expansions). Simply using a kernel function and solving the optimization problem directly explores the higher dimensional space.

To illustrate the utility of kernel PCA, we will simulate data based on the model earlier in this section. For this simulation, 200 samples were generated, with 150 selected for the training set. PCA and kernel PCA will be used separately as dimension reduction techniques followed by linear regression to compare how these methods perform. Figure 6.12(a) displays the relationship between x_1 and x_2 for the training data. The scatter plot shows the linear relationship between these two predictors. Panel (b) uncovers the quadratic relationship between x_1 and y, which could be easily be found and incorporated into a model if there were a small number of predictors. PCA found one dimension to summarize x_1 and x_2 and the relationship between the principal component and the response is presented in Figure 6.12(c). Often in practice, we use the number of components that summarize a threshold amount of variability as the predictors. If we blindly ran principal component regression, we would get a poor fit between PC_1 and y because these data are better described by a quadratic fit. The relationship between the response and the first component using a polynomial kernel of degree 2 is presented in panel (d). The expansion of the

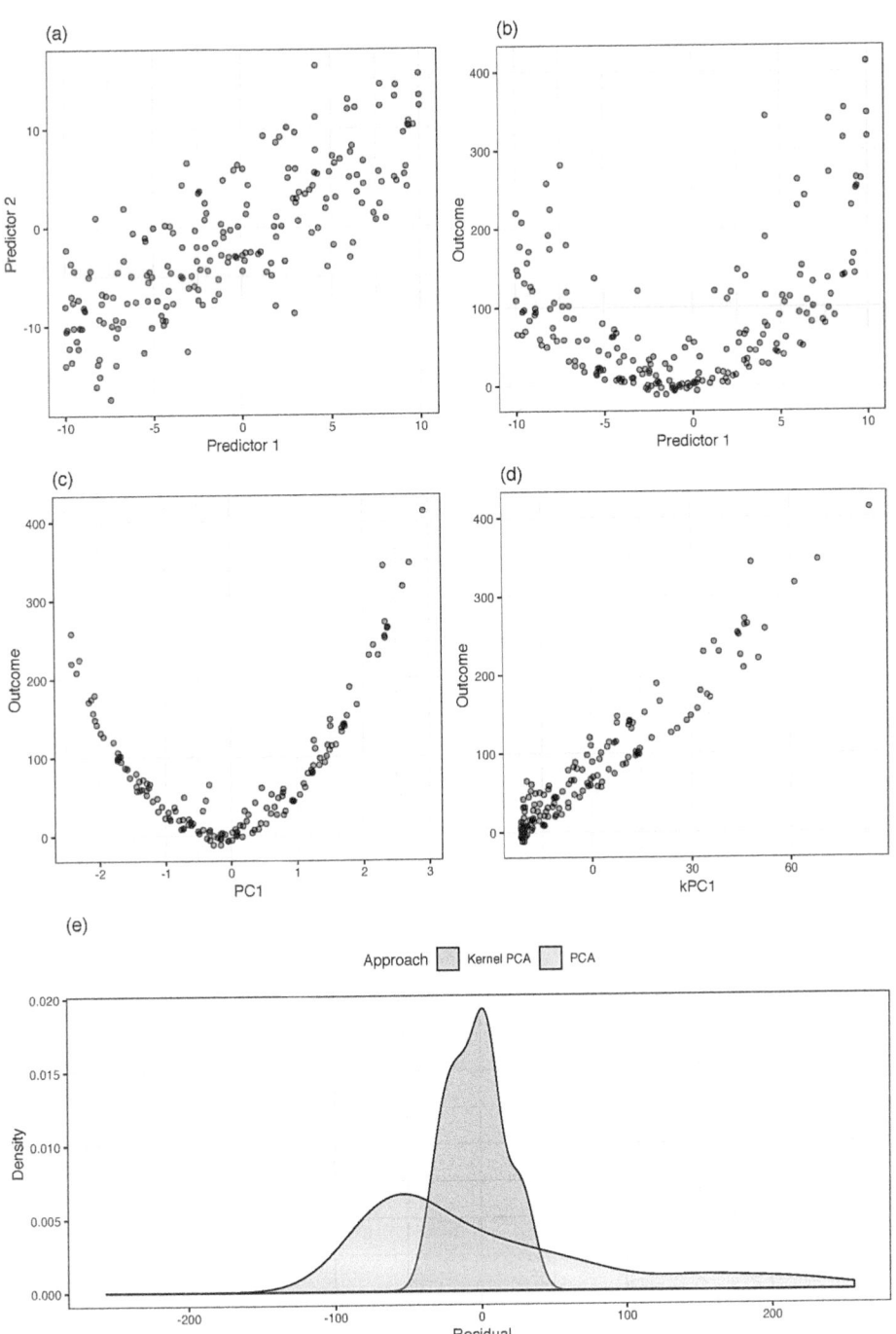

Figure 6.12: A comparison of PCA and kernel PCA. (a) The training set data. (b) The simulated relation between x_1 and the outcome. (c) The PCA results (d) kPCA using a polynomial kernel shows better results. (e) Residual distributions for the two dimension reduction methods.

space using this kernel allows PCA to better find the underlying structure among the predictors. Clearly, this structure has a more predictive relationship with the response than the linear method. The model performance on the test is presented in panel (e) where the distribution of the residuals from the kernel PCA model are demonstrably smaller than those from the PCA model.

For the Chicago data, a radial basis function kernel was used to get new PCA components. The kernel's only parameter (σ) was estimated analytically using the method of Caputo et al. (2002); values between 0.002 and 0.022 seemed reasonable and a value near the middle of the range (0.006) was used to compute the components. Sensitivity analysis was done to assess if other choices within this range led to appreciably different results, which they did not. The second column from the left in Figure 6.9 shows the results. Similar to ordinary PCA, the first kPCA component has a strong association with ridership. However, the kernel-based component has a larger dynamic range on the x-axis. The other components shown in the figure are somewhat dissimilar from regular PCA. While these scores have less association with the outcome (in isolation), it is very possible that, when the scores are entered into a regression model simultaneously, they have predictive utility.

6.3.1.3   Independent Component Analysis

PCA seeks to account for the variation in the data and produces components that are mathematically *orthogonal* to each other. As a result, the components are *uncorrelated*. However, since correlation is a measure of *linear* independence between two variables, they may not be *statistically independent* of one another (i.e., have a covariance of zero). There are a variety of toy examples of scatter plots of uncorrelated variables containing a clear relationship between the variables. One exception occurs when the underlying values follow a Gaussian distribution. In this case, uncorrelated data would also be statistically independent.[62]

Independent component analysis (ICA) is similar to PCA in a number of ways (Lee, 1998). It creates new components that are linear combinations of the original variables but does so in a way that the components are as statistically independent from one another as possible. This enables ICA to be able to model a broader set of trends than PCA, which focuses on orthogonality and linear relationships. There are a number of ways for ICA to meet the constraint of statistical independence and often the goal is to maximize the "non-Gaussianity" of the resulting components. There are different ways to measure non-Gaussianity and the *fastICA* approach uses an information theory called *negentropy* to do so (Hyvarinen and Oja, 2000). In practice, ICA should create components that are dissimilar from PCA unless the predictors demonstrate significant multivariate normality or strictly linear trends. Also, unlike PCA, there is no unique ordering of the components.

---

[62]This does not imply that PCA produces components that follow Gaussian distributions.

It is common to normalize and *whiten* the data prior to running the ICA calculations. Whitening, in this case, means to convert the original values to the full set of PCA components. This may seem counter-intuitive, but this preprocessing doesn't negatively affect the goals of ICA and reduces the computational complexity substantially (Roberts and Everson, 2001). Also, the ICA computations are typically initialized with random parameter values and the end result can be sensitive to these values. To achieve reproducible results, it is helpful to initialize the values manually so that the same results can be re-created at some later time.

ICA was applied to the Chicago data. Prior to estimating the components, the predictors were scaled so that the pre-ICA step of whitening used the centered data to compute the initial PCA data. When scaling was not applied, the ICA components were strongly driven by different individual stations. The ICA components are shown in the third column of Figure 6.9 and appear very different from the other methods. The first component has very little relationship with the outcome and contains a set of outlying values. In this component, most stations had little to no contribution to the score. Many stations that had an effect were near Clark and Lake. While there were 42 stations that negatively affected the component (the strongest being Chicago, State/Lake, and Grand), most have positive coefficients (such as Clark/Division, Irving Park, and Western). For this component, the other important stations were along the Brownline (going northward) as well as the station for Rosemont (just before O'Hare airport). The other ICA components shown in Figure 6.9 had very weak relationships to the outcome. For these data, the PCA approach to finding appropriate rotations of the data was better at locating predictive relationships.

6.3.1.4 Non-negative Matrix Factorization

Non-negative matrix factorization (Gillis, 2017; Fogel et al., 2013) is another linear projection method that is specific to features that are greater than or equal to zero. In this case, the algorithm finds the coefficients of A such that their values are also non-negative (thus ensuring that the new features have the same property). This approach is popular for text data where predictors are word counts, imaging, and biological measures (e.g., the amount of RNA expressed for a specific gene). This tecnique can also be used for the Chicago data.[63]

The method for determining the coefficients is conceptually simple: find the best set of coefficients that make the scores as "close" as possible to the original data with the constraint of non-negativity. Closeness can be measured using mean squared error (aggregated across the predictors) or other measures such as the Kullback–Leibler divergence (Cover and Thomas, 2012). The latter is an information theoretic method to measure the distance between two probability distributions. Like ICA, the numerical solver is initialized using random numbers and the final solution can be sensitive to these values and the order of the components is arbitrary.

---

[63]NNMF can be applied to the log-transformed ridership values because these values are all greater than or equal to zero.

Figure 6.13: Coefficients for the NNMF component 11. The size and color reflect the relative magnitude of the coefficients.

The first five non-negative matrix factorization components shown in Figure 6.9 had modest relationships with ridership at the Clark and Lake station. Of the 20 NNMF components that were estimated, component number 11 had the largest association with ridership. The coefficients for this component are shown in Figure 6.13. The component was strongly influenced by stations along the Pink line (e.g., 54th/Cermak, Damen), the Blue line, and Green lines. The coefficient for O'Hare airport also made a large contribution.

6.3.1.5 Partial Least Squares

The techniques discussed thus far are unsupervised, meaning that the objective of the dimension reduction is based solely on the predictors. In contrast, supervised dimension reduction techniques use the response to guide the dimension reduction of the predictors such that the new predictors are optimally related to the response. Partial least squares (PLS) is a supervised version of PCA that reduces dimension in a way that is optimally related to the response (Stone and Brooks, 1990). Specifically, the objective of PLS is to find linear functions (called latent variables) of the predictors that have optimal covariance with the response. This means that the response guides the dimension reduction such that the scores have the highest possible correlation with the response in the training data. The typical implementation of the optimization problem requires that the data in each of the new dimensions is uncorrelated, which is a similar constraint which is imposed in the PCA optimization problem. Because PLS is a supervised technique, we have found that it takes more PCA components to meet the same level of performance generated by a PLS model (Kuhn and Johnson, 2013). Therefore, PLS can more efficiently reduce dimension, leading to both memory and computational time savings during the model-building process. However, because PLS uses the response to determine the optimal number of latent variables, a rigorous validation approach must be used to ensure that the method does not overfit to the data.

The PLS components for the Chicago weekend data are shown in Figure 6.9. The first component is effectively the same as the first PCA component. Recall that PLS tries to balance dimension reduction in the predictors with maximizing the correlation between the components and the outcome. In this particular case, as shown in the figure, the PCA component has a strong linear relationship with ridership. In effect, the two goals of PLS are accomplished by the unsupervised component. For the other components shown in Figure 6.9, the second component is also very similar to the corresponding PCA version (up to the sign), but the other two components are dissimilar and show marginally better association with ridership.

To quantify how well each method was at estimating the relationship between the outcome and the lagged predictors, linear models were fit using each method. In each case, 20 components were computed and entered into a simple linear regression model. To resample, a similar time-based scheme was used where 800 weekends were the initial analysis set and two sets of weekends were used as an assessment set (which are added to the analysis set for the next iteration). This led to 25 resamples. Within each resample, the projection methods were *recomputed* each time and this replication should capture the uncertainty potentially created by those computations. If we had pre-computed the components for the entire training set of weekends, *and then resampled*, it is very likely that the results would be falsely optimistic. Figure 6.14 shows confidence intervals for the RMSE values over resamples for each method. Although the confidence intervals overlap, there is some indication that PLS and kernel PCA have the potential to outperform the original values. For other models that can be very sensitive to between-predictor correlations (e.g., neural networks), these transformation methods could provide materially better performance when they produce features that are uncorrelated.

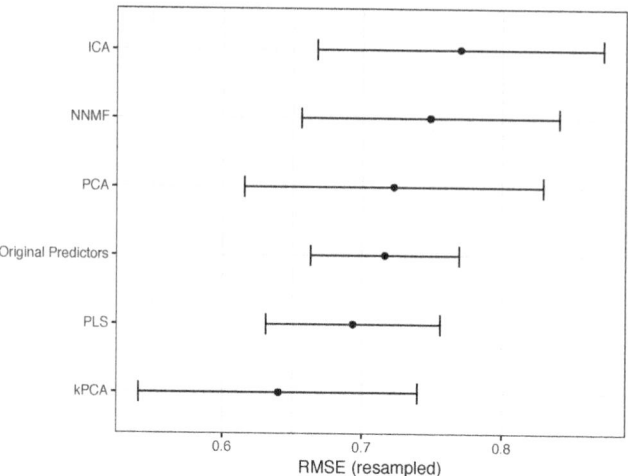

Figure 6.14: Model results for the Chicago weekend data using different dimension reduction methods (each with 20 components).

6.3.2 | Autoencoders

Autoencoders are computationally complex multivariate methods for finding representations of the predictor data and are commonly used in deep learning models (Goodfellow et al., 2016). The idea is to create a nonlinear mapping between the original predictor data and a set of artificial features (that is usually the same size). These new features, which may not have any sensible interpretation, are then used as the model predictors. While this does sound very similar to the previous projection methods, autoencoders are very different in terms of how the new features are derived and also in their potential benefit.

One situation in which to use an autoencoder is when there is an abundance of *unlabeled data* (i.e., the outcome is not yet known). For example, a pharmaceutical company will have access to millions of chemical compounds but a very small percentage of these data have relevant outcome data needed for new drug discovery projects. One common task in chemistry is to create models that use the chemical structure of a compound to predict how good of a drug it might be. The predictors can be calculated for the vast majority of the chemical database and these data, despite the lack of laboratory results, can be used to improve a model.

As an example, suppose that a modeler was going to fit a simple linear regression model. The well-known least squares model would have predictor data in a matrix denoted as X and the outcome results in a vector y. To estimate the regression coefficients, the formula is

$$\hat{\beta} = (X'X)^{-1} \quad X'y.$$

Note that the outcome data are only used on the right-hand part of that equation (e.g., $X'y$). The unlabeled data could be used to produce a very good estimate of the inverse term on the left and saved. When the outcome data are available, $X'y$ can be computed with the relatively small amount of labeled data. The above equation can still be used to produce better parameter estimates when $(X'X)^{-1}$ is created on a large set of data.

The same philosophy can be used with autoencoders. All of the available predictor data can be used to create an equation that estimates (and potentially smooths) the relationships between the predictors. Based on these patterns, a smaller data set of predictors can be more effectively represented by using their "predicted" feature values.

There are a variety of ways that autoencoders can be created. The most (relatively) straightforward method is to use a neural network structure:

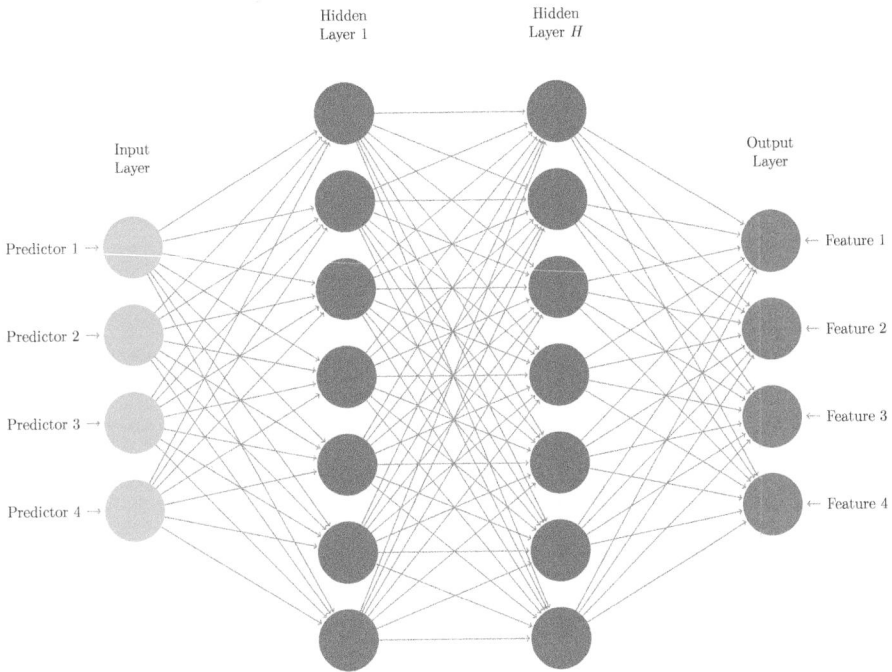

In this architecture, the functions that go from the inputs to the hidden units and between hidden units are nonlinear functions. Commonly used functions are simple sigmoids as well as ReLu functions.[64] The connections to the output features are typically linear. One strategy is to use a relatively small number of nodes in the hidden units to coerce the autoencoder to learn the crucial patterns in the predictors. It may also be that multiple layers with many nodes optimally represent the predictor values.

Each line in the network diagram represents a parameter to be estimated and this number can become excessively large. However, there are successful approaches, such

---

[64]This helps prevent the model from learning a simple identity function from the data (that would provide no real value).

as regularization and dropout methods, to control overfitting as well as a number of fairly efficient deep learning libraries that can ease the computational burden. For this reason, the benefit of this approach occurs when there is a *large* amount of data.

The details of fitting neural network and deep learning models is beyond the scope of this book. Good references for the technical details are Bishop (2011) and Goodfellow et al. (2016), while Chollet and Allaire (2018) is an excellent guide to fitting these types of models.[65] Creating effective autoencoders entails model tuning and many other tasks used in predictive models in general. To measure performance, it is common to use root mean squared error between the observed and predicted (predictor) values, aggregated across variables.

How can autoencoders be incorporated into the modeling process? It is common to treat the relationships estimated by the autoencoder model as "known" values and not re-estimate them during resampling. In fact, it is fairly common in deep learning on images and text data to use pre-trained networks in new projects. These pre-trained sections or blocks of networks can be used as the front of the network. If there is a very large amount of data that goes into this model, and the model fitting process converges, there may be some statistical validity to this approach.

As an example, we used the data from Karthikeyan et al. (2005) where they used chemical descriptors to model the melting point of chemical compounds (i.e., transition from solid to liquid state). They assembled 4401 compounds and 202 numeric predictors. To simulate a drug discovery project just starting up, a random set of 50 data points were used as a training set and another random set of 25 were allocated to a test set. The remaining 4327 data points were treated as unlabeled data that do not have melting point data.

The unlabeled predictors were processed by first eliminating 126 predictors so that there are no pairwise correlations above 0.75. Further, a Yeo-Johnson transformation was used to change the scale of the remaining 76 predictors (to make them less skewed). Finally, all the predictors were centered and scaled. The same preprocessing was applied to the training and test sets.

An autoencoder with two hidden layers, each with 512 units, was used to model the predictors. A hyperbolic tangent activation function was used to go between the original units and the first set of hidden units, as well as between the two hidden layers, and a linear function connected the second set of hidden units and the 76 output features. The total number of model parameters was 341,068. The RMSE on a random 20% validation set of the unlabeled data was used to measure performance as the model was being trained. The model was trained across 500 iterations and Figure 6.15(a) shows that the RMSE drastically decreases in the first 50 iterations, is minimized around 100 iterations, then slowly increases over the remaining iterations. The autoencoder model, based on the coefficients from the 100th iteration, was applied to the training and test set.

---

[65] Code for the example below can be found in the books GitHub repository.

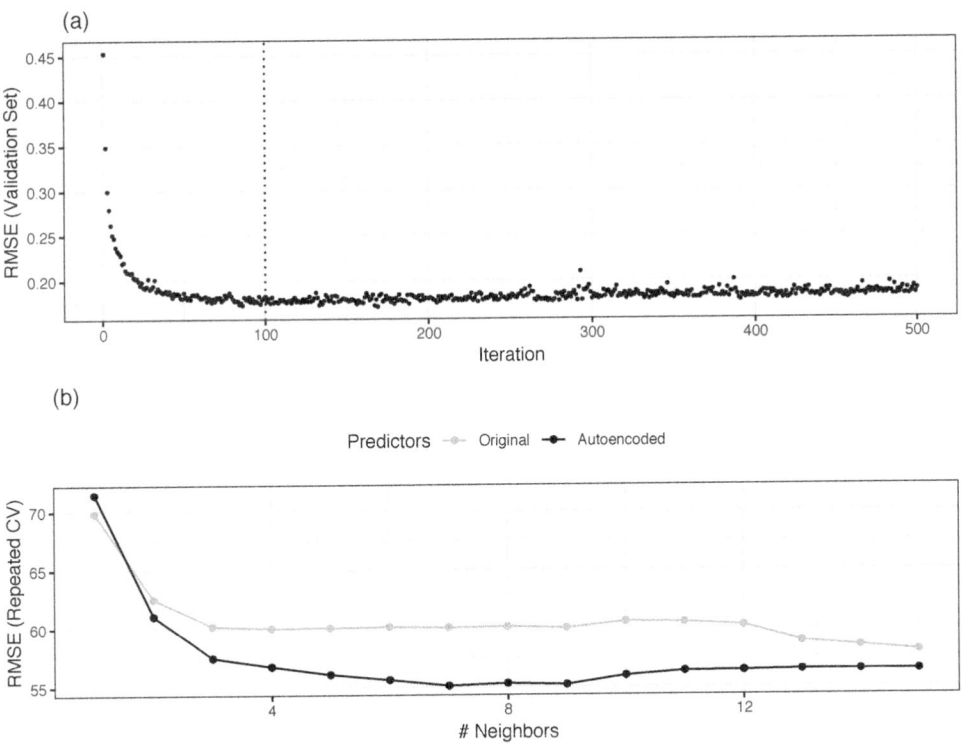

Figure 6.15: The results of fitting an autoencoder to a drug discovery data set. Panel (a) is the holdout data used to measure the MSE of the fitting process and (b) the resampling profiles of K-nearest neighbors with (black) and without (gray) the application of the autoencoder.

A *K*-nearest neighbor model was fit to the original and encoded versions of the training set. Neighbors between one and twenty were tested and the best model was chosen using the RMSE results measured using five repeats of 10-fold cross-validation. The resampling profiles are shown in Figure 6.15(b). The best settings for the encoded data (7 neighbors, with an estimated RMSE of 55.1) and the original data (15 neighbors, RMSE = 58) had a difference of 3 degrees with a 95% confidence interval of (0.7, 5.3) degrees. While this is a moderate improvement, for a drug discovery project, such an improvement could make a substantial impact if the model might be able to better predict the effectiveness of new compounds.

The test set was predicted with both KNN models. The model associated with the raw training set predictors did very poorly, producing an RMSE of 60.4 degrees. This could be because of the selection of training and test set compounds or for some other reason. The KNN model with the encoded values produced results that were consistent with the resampling results, with an RMSE value of 51. The entire process was repeated using different random numbers seeds in order to verify that the results were not produced strictly by chance.

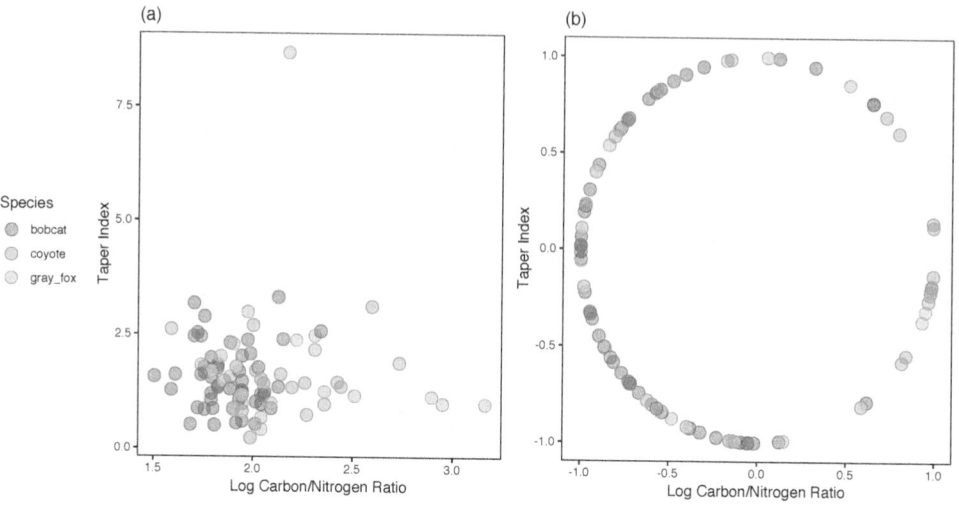

Figure 6.16: Scatterplots of a classification data set before (a) and after (b) the spatial sign transformation.

6.3.3 | Spatial Sign

The spatial sign transformation takes a set of predictor variables and transforms them in a way that the new values have the same distance to the center of the distribution (Serneels et al., 2006). In essence, the data are projected onto a multidimensional sphere using the following formula:

$$x_{ij}^* = \frac{x_{ij}}{\sum_{j=1}^{p} x_{ij}^2}.$$

This approach is also referred to as *global contrast normalization* and is often used with image analysis.

The transformation required computing the norm of the data and, for that reason, the predictors should be centered and scaled prior to the spatial sign. Also, if any predictors have highly skewed distributions, they may benefit from a transformation to induce symmetry, such as the Yeo-Johnson technique.

For example, Figure 6.16(a) shows data from Reid (2015) where characteristics of animal scat (i.e., feces) were used to predict the true species of the animal that donated the sample. This figure displays two predictors on the axes, and samples are colored by the true species. The taper index shows a significant outlier on the *y*-axis. When the transformation is applied to these two predictors, the new values in 6.16(b) show that the data are all equidistant from the center (zero). While the transformed data may appear to be awkward for use in a new model, this transformation can be very effective at mitigating the damage of extreme outliers.

6.3.4 | Distance and Depth Features

In classification, it may be beneficial to make semi-supervised features from the data. In this context, this means that the outcome classes are used in the creation of the new predictors but not in the sense that the features are optimized for predictive accuracy. For example, from the training set, the means of the predictors can be computed for each class (commonly referred to as the *class centroids*). For new data, the distance to the class centroids can be computed for each class and these can be used as features. This basically embeds the quasi-nearest-neighbor technique into the model that uses these features. For example, Figure 6.17(a) shows the class centroids for the scat data using the log carbon-nitrogen ratio and $\delta 13C$ features. Based on these multidimensional averages, it is straightforward to compute a new sample's distance to each class's centers. If the new sample is closer to a particular class, its distance should be small. Figure 6.17(a) shows an example for this approach where

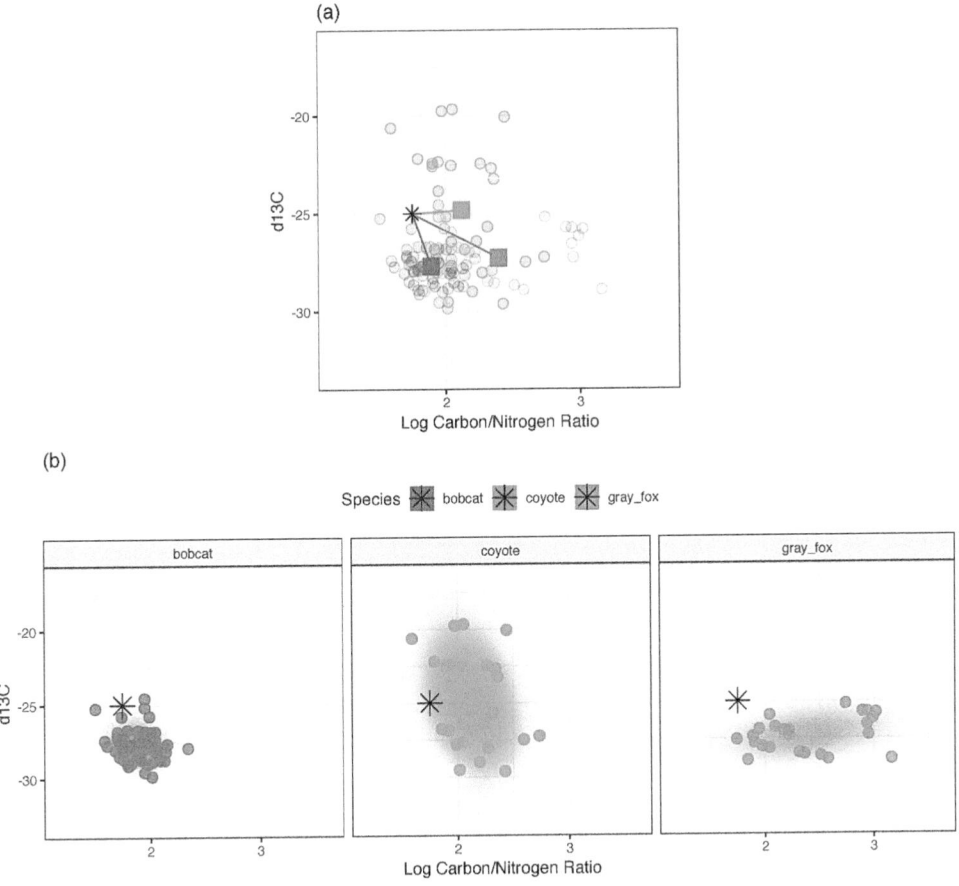

Figure 6.17: Examples of class distance to centroids (a) and depth calculations (b). The asterisk denotes a new sample being predicted and the squares correspond to class centroids.

the new sample (denoted by a ∗) is relatively close to the bobcat and coyote centers. Different distance measures can be used where appropriate. To use this approach in a model, new features would be created for every class that measures the distance of every sample to the corresponding centroids.

Alternatively, the idea of *data depth* can be used (Ghosh and Chaudhuri, 2005; Mozharovskyi et al., 2015). Data depth measures how close a data point is to the center of its distribution. There are many ways to compute depth and there is usually an inverse relationship between class distances and data depths (e.g., *large* depth values are more indicative of being close to the class centroid). If we were to assume bivariate normality for the data previously shown, the probability distributions can be computed within each class. The estimated Gaussian probabilities are shown as shadings in Figure 6.17(b).

In this case, measuring the depth using the probability distributions provides slightly increased sensitivity than the simple Euclidean distance shown in panel (a). In this case, the new sample being predicted resides inside of the probability zone for coyote more than the other species.

There are a variety of depth measures that can be used and many do not require specific assumptions regarding the distribution of the predictors and may also be robust to outliers. Similar to the distance features, class-specific depths can be computed and added to the model.

These types of features can accentuate the ability of some models to predict an outcome. For example, if the true class boundary is a diagonal line, most classification trees would have difficultly emulating this pattern. Supplementing the predictor set with distance or depth features can enable such models in these situations.

6.4   Summary

Numeric features, in their raw form, may or may not be in an effective form that allows a model to find a good relationship with the response. For example, a predictor may be measured, but its squared version is what is truly related to the response. Straightforward transformations such as centering, scaling, or transforming a distribution to symmetry are necessary steps for some models to be able to identify predictive signal. Other transformations such as basis expansions and splines can translate a predictor in its original scale to nonlinear scales that may be informative.

But expanding the predictor space comes at a computational cost. Another way to explore nonlinear relationships between predictors and the response is through the combination of a kernel function and PCA. This approach is very computationally efficient and enables the exploration of much larger dimensional space.

Instead of expanding the predictor space, it may be necessary to reduce the dimension of the predictors. This can be accomplished using unsupervised techniques such as PCA, ICA, NNMF, or a supervised approach like PLS.

Finally, autoencoders, the spatial sign transformation, or distance and depth measures offer novel engineering approaches that can harness information in unlabeled data or dampen the effect of extreme samples.

6.5 | Computing

The website `http://bit.ly/fes-num` contains R programs for reproducing these analyses.

7 | Detecting Interaction Effects

In problems where prediction is the primary purpose, the majority of variation in the response can be explained by the cumulative effect of the important individual predictors. To this point in the book, we have focused on developing methodology for engineering categorical or numeric predictors such that the engineered versions of the predictors have better representations for uncovering and/or improving the predictive relationships with the response. For many problems, additional variation in the response can be explained by the effect of two or more predictors working in conjunction with each other. As a simple conceptual example of predictors working together, consider the effects of water and fertilizer on the yield of a field corn crop. With no water but some fertilizer, the crop of field corn will produce no yield since water is a necessary requirement for plant growth. Conversely, with a sufficient amount of water but no fertilizer, a crop of field corn will produce some yield. However, yield is best optimized with a sufficient amount of water *and* a sufficient amount of fertilizer. Hence water and fertilizer, when combined in the right amounts, produce a yield that is greater than what either would produce alone. More formally, two or more predictors are said to interact if their combined effect is different (less or greater) than what we would expect if we were to add the impact of each of their effects when considered alone. Note that interactions, by definition, are always in the context of *how predictors relate to the outcome*. Correlations between predictors, for example, are not directly related to whether there is an interaction effect or not. Also, from a notational standpoint, the individual variables (e.g., fertilizer and water) are referred to as the *main effect* terms when outside of an interaction.

The predictive importance of interactions were first discussed in Chapter 2 where the focus of the problem was trying to predict an individual's risk of ischemic stroke through the use of predictors based on the images of their carotid arteries. In this example, an interaction between the degree of remodeling of the arterial wall and the arterial wall's maximum thickness was identified as a promising candidate. A visualization of the relationship between these predictors was presented in Figure 2.7 (a), where the contours represent equivalent multiplicative effects between the predictors. Part (b) of the same figure illustrated how the combination of these

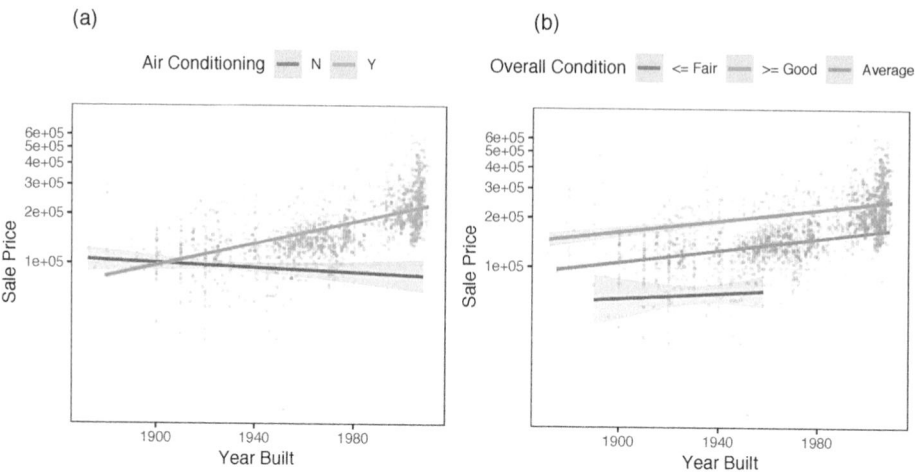

Figure 7.1: Plots of Ames housing variables where the predictors interact (a) and do not interact (b).

predictors generated enhanced separation between patients who did and did not have a stroke.

In the previous example, the important interaction was between two continuous predictors. But important interactions can also occur between two categorical predictors or between a continuous and a categorical predictor, which can be found in the Ames housing data. Figure 7.1(a) illustrates the relationship between the age of a house (x-axis) and sale price of a house (y-axis), categorized by whether or not a house has air conditioning. In this example, the relationship between the age of a house and sales price is increasing for houses with air conditioning; on the other hand, there is no relationship between these predictors for homes without air conditioning. In contrast, part (b) of this figure displays two predictors that do not interact: home age and overall condition. Here the condition of the home does not affect the relationship between home and sales price. The lack of interaction is revealed by the near parallel lines for each condition type.

To better understand interactions, consider a simple example with just two predictors. An interaction between two predictors can be mathematically represented as:

$$y = \beta_0 + \beta_1 x_1 + \beta_2 x_2 + \beta_3 x_1 x_2 + error.$$

In this equation, β_0 represents the overall average response, β_1 and β_2 represent the average rate of change due to x_1 and x_2, respectively, and β_3 represents the incremental rate of change due to the combined effect of x_1 and x_2 that goes beyond what x_1 and x_2 can explain alone. The error term represents the random variation in real data that cannot be explained in the deterministic part of the equation. From data, the β parameters can be estimated using methods like linear regression for a continuous response or logistic regression for a categorical response. Once the parameters have been estimated, the usefulness of the interaction for explaining variation in the response can be determined. There are four possible cases:

1. If β_3 is not significantly different from zero, then the interaction between x_1 and x_2 is not useful for explaining variation of the response. In this case, the relationship between x_1 and x_2 is called *additive*.

2. If the coefficient is meaningfully negative while x_1 and x_2 alone also affect the response, then interaction is called *antagonistic*.

3. At the other extreme, if the coefficient is positive while x_1 and x_2 alone also affect the response, then interaction is called *synergistic*.

4. The final scenario occurs when the coefficient for the interaction is significantly different from zero, but either one or both of x_1 or x_2 *do not* affect the response. In this case, the average response of x_1 across the values of x_2 (or vice versa) has a rate of change that is essentially zero. However, the average value of x_1 at each value of x_2 (or conditioned on x_2) is different from zero. This situation occurs rarely in real data, and we need to understand the visual cues for identifying this particular case. This type of interaction will be called *atypical*.

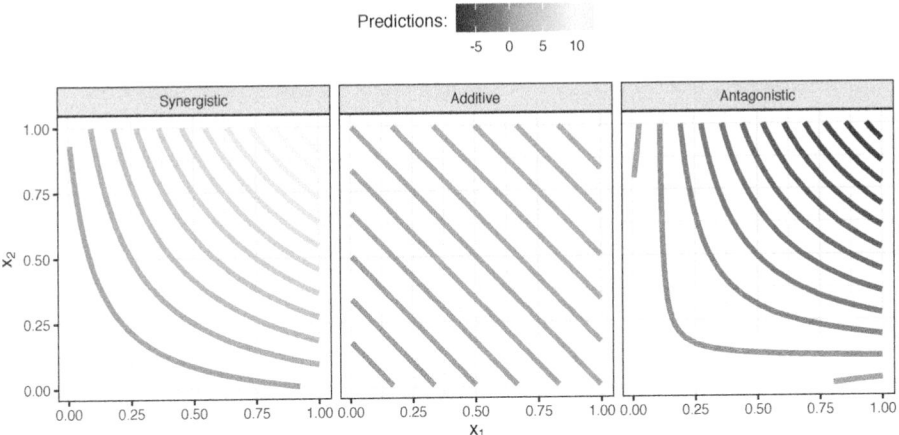

Figure 7.2: Contour plots of the predicted response from a model between two predictors with a synergistic interaction, no interaction (additive), and an antagonistic interaction.

To illustrate the first three types of interactions, simulated data will be generated using the formula from above with coefficients as follows: $\beta_0 = 0$, $\beta_1 = \beta_2 = 1$, and $\beta_3 = -10$, 0, or 10 to illustrate antagonism, no interaction, or synergism, respectively. A random, uniform set of 200 samples were generated for x_1 and x_2 between the values of 0 and 1, and the error for each sample was pulled from a normal distribution. A linear regression model was used to estimate the coefficients and a contour plot of the predicted values from each model was generated (Figure 7.2). In this figure,

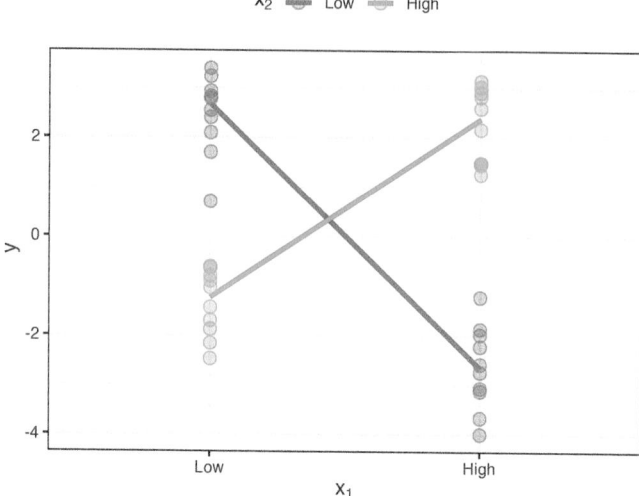

Figure 7.3: An example of an atypical interaction between two categorical factors. In this case each factor alone is not significant, but the interaction between the factors is significant and is characterized by the tell-tale crossing pattern of the average values within groups.

the additive case has parallel contour lines with an increasing response from lower left to upper right. The synergistic case also has an increasing response from lower left to upper right, but the contours are curved, indicating that lower values of each predictor are required to elicit the same response value. The antagonistic case also displays curved contours; here the response decreases from lower left to upper right. This figure reveals that the response profile changes when an interaction between predictors is real and present.

The atypical case is more clearly illustrated when x_1 and x_2 are categorical predictors. To simulate the atypical case, the two predictors will take values of 'low' and 'high', and the response will be generated by $y = 2x_1x_2 + error$, where the low and high values of x_1 and x_2 will be numerically represented by ± 1, respectively. Figure 7.3 displays the relationship between the low and high levels of the factors. Notice that the average response for x_1 at the low level and separately at the high level is approximately zero. However, the average is conditioned on x_2, represented by the two different colors, the responses are opposite at the two levels of x_1. Here the individual effects of each predictor are not significant, but the conditional effect is strongly significant.

When we know which predictors interact, they can be included in a model. However, knowledge of which predictors interact may not be available. A number of recent empirical studies have shown that interactions can be uncovered by more complex modeling techniques. As Elith et al. (2008) noted, tree-based models inherently model interactions between predictors through subsequent recursive splits in the tree. García-Magariños et al. (2009) showed that random forests were effective at identifying unknown interactions between single nucleotide polymorphisms; Lampa et al. (2014) found that a boosted tree model could uncover important interactions in epidemiology problems; and Chen et al. (2008) found that search techniques combined with a support vector machine was effective at identifying gene-gene interactions related to human disease. These findings prompt the question that if sophisticated predictive modeling techniques are indeed able to uncover the predictive information from interactions, then why should we spend any time or effort in pinpointing which interactions are important? The importance comes back to the underlying goal of feature engineering, which is to create features that improve the effectiveness of a model by containing predictively relevant information. By identifying and creating relevant interaction terms, the predictive ability of models that have better interpretability can be improved. Therefore, once we have worked to improve the form of individual predictors, the focus should then turn to searching for interactions among the predictors that could help explain additional variation in the response and improve the predictive ability of the model.

The focus of this chapter will be to explore how to search for and identify interactions between predictors that improve models' predictive performance. In this chapter, the Ames housing data will be the focus. The *base model* that is used consists of a number of variables, including continuous predictors for: general living area, lot area and frontage, years built and sold, pool area, longitude, latitude, and the number of full baths. The base model also contains qualitative predictors for neighborhood, building type (e.g., townhouse, one-family home, etc.), central air, the MS sub-class (e.g., 1 story, 1945 and older, etc.), foundation type, roof style, alley type, garage type, and land contour. The base model consists of main effects and, in this chapter, the focus will be on discovering helpful interactions of these predictors.

7.1 Guiding Principles in the Search for Interactions

First, expert knowledge of the system under study is critical for guiding the process of selecting interaction terms. Neter et al. (1996) recognize the importance of expert knowledge and recommend:

> "It is therefore desirable to identify in advance, whenever possible, those interactions that are most likely to influence the response variable in important ways."

Therefore, expert-selected interactions should be the first to be explored. In many situations, though, expert guidance may not be accessible or the area of research may be so new that there is no a priori knowledge to shepherd interaction term selection. When placed in a situation like this, sound guiding principles are needed that will lead to a reasonable selection of interactions.

One field that has tackled this problem is statistical experimental design. Experimental design employs principles of control, randomization, and replication to construct a framework for gathering evidence to establish causal relationships between inde-

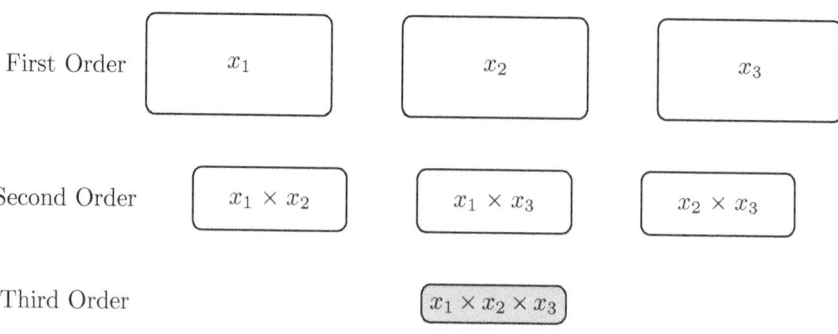

Figure 7.4: An illustration of a set of first-, second-, and third-order terms from a full factorial experiment. The terms are sized based on the interaction hierarchy principle. The three-way interaction term is grayed out due to the effect sparsity principle.

pendent factors (i.e., predictors) and a dependent response. The process begins by determining the population (or units) of interest, the factors that will be systematically changed, and the specific ranges of the factors to be studied. Specific types of experimental designs, called screening designs, generate information in a structured way so that enough data are collected to identify the important factors and interactions among two or more factors. Figure 7.4 conceptually illustrates an experiment with three factors.

While data from an experiment has the ability to estimate all possible interactions among factors, it is unlikely that each of the interactions explain a significant amount of the variation in the measured response. That is, many of these terms are not predictive of the response. It is more likely, however, that only a few of the interactions, if any, are plausible or capture significant amounts of response variation. The challenge in this setting is figuring out which of the many possible interactions are truly important. Wu and Hamada (2011) provide a framework for addressing the challenge of identifying significant interactions. This framework is built on concepts of interaction hierarchy, effect sparsity, and effect heredity. These concepts have been shown to work effectively at identifying the most relevant effects for data generated by an experimental design. Furthermore, this framework can be extended to the predictive modeling framework.

The interaction **hierarchy principle** states that the higher degree of the interaction, the less likely the interaction will explain variation in the response. Therefore, pairwise interactions should be the first set of interactions to search for a relationship with the response, then three-way interactions, then four-way interactions, and so on. A pictorial diagram of the effect hierarchy can be seen in Figure 7.4 for data that has three factors where the size of the nodes represents the likelihood that a term is predictive of the response. Neter et al. (1996) advises caution for including three-way interactions and beyond since these higher-order terms rarely, if ever,

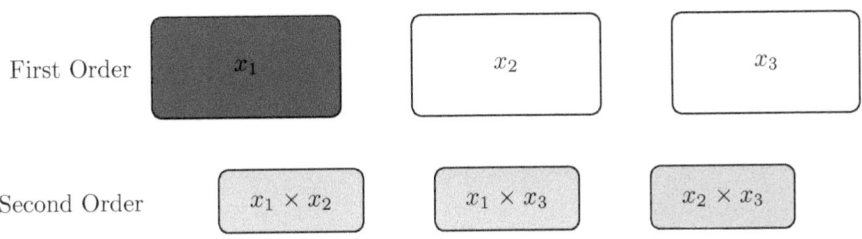

Figure 7.5: An illustration of the effect heredity principle for an example data set with three factors.

capture a significant amount of response variation and are almost impossible to interpret.

The second principle, **effect sparsity**, contends that only a fraction of the possible effects truly explain a significant amount of response variation (Figure 7.4). Therefore, the approach could greatly pare down the possible main effects and interaction terms when searching for terms that optimize the predictive ability of a model.

The effect **heredity principle** is based on principles of genetic heredity and asserts that interaction terms may only be considered if the ordered terms preceding the interaction are effective at explaining response variation. This principle has at least two possible implementations: strong heredity and weak heredity. The *strong heredity* implementation requires that for an interaction to be considered in a model, all lower-level preceding terms must explain a significant amount of response variation. That is, only the $x_1 \times x_2$ interaction would be considered if the main effect of x_1 and the main effect of x_2 each explain a significant amount of response variation.[66] But if only one of the factors is significant, then the interaction will not be considered. *Weak heredity* relaxes this requirement and would consider any possible interaction with the significant factor. Figure 7.5 illustrates the difference between strong heredity and weak heredity for a three-factor experiment. For this example, the x_1 main effect was found to be predictive of the response. The weak heredity principle would explore the x_1x_2 and x_1x_3 interaction terms. However, the x_2x_3 interaction term would be excluded from investigation. The strong heredity principle would not consider any second-order terms in this example since only one main effect was found to be significant.

While these principles have been shown to be effective in the search for relevant and predictive terms to be included in a model, they are not always applicable. For example, complex higher-order interactions have been shown to occur in nature. Bairey et al. (2016) showed that high-order interactions among species can impact ecosystem diversity. The human body is also a haven of complex chemical and biological interactions where multiple systems work together simultaneously to combat pathogens and to sustain life. These examples strengthen our point that

---

[66]For this reason, the interaction shown in Figure 7.3 was labeled as "atypical."

the optimal approach to identifying important interaction terms will come from a combination of expert knowledge about the system being modeled as well as carefully thought-out search routines based on sound statistical principles.

7.2 | Practical Considerations

In the quest to locate predictive interactions, several practical considerations would need to be addressed, especially if there is little or no expert knowledge to suggest which terms should be screened first. A few of the questions to think through before discussing search methodology are: Is it possible with the available data to enumerate and evaluate all possible predictive interactions? And if so, then should all interactions be evaluated? And should interaction terms be created before or after preprocessing the original predictors?

Let's begin with the question of whether or not all of the interaction terms should be completely enumerated. To answer this question, the principles of interaction hierarchy and effect sparsity will be used to influence the approach. While high-order interactions (three-way and above) are possible for certain problems as previously mentioned, they likely occur infrequently (effect sparsity) and may only provide a small improvement on the predictive ability of the model (interaction hierarchy). Therefore, higher-order interactions should be screened only if expert knowledge recommends investigating specific interaction terms. What is more likely to occur is that a fraction of the possible pairwise interaction terms contain predictively relevant information. Even though the recommendation here is to focus on searching through the pairwise interactions, it still may be practically impossible to evaluate a complete enumeration of these terms. More concretely, if there are p predictors, then there are $(p)(p-1)/2$ pairwise interaction terms. As the number of predictors increases, the number of interaction terms increases exponentially. With as few as 100 original predictors, complete enumeration requires a search of 4,950 terms. A moderate increase to 500 predictors requires us to evaluate nearly 125,000 pairwise terms! More strategic search approaches will certainly be required for data that contain even a moderate number of predictors, and will be discussed in subsequent sections.

Another practical consideration is when interaction terms should be created relative to the preprocessing steps. Recall from Chapter 6 that preprocessing steps for individual numeric predictors can include centering, scaling, dimension expansion or dimension reduction. As shown in previous chapters, these steps help models to better uncover the predictive relationships between predictors and the response. What now must be understood is if the order of operations of preprocessing and creation of interaction terms affects the ability to find predictively important interactions. To illustrate how the order of these steps can affect the relationship of the interaction with the response, consider the interaction between the maximum remodeling ratio and the maximum stenosis by area for the stroke data (Chapter 2). For these data, the preprocessing steps were centering, scaling, and individually transformed. Figure 7.6 compares the distributions of the stroke groups when the interaction term is created before the preprocessing steps (a) and after the preprocessing steps

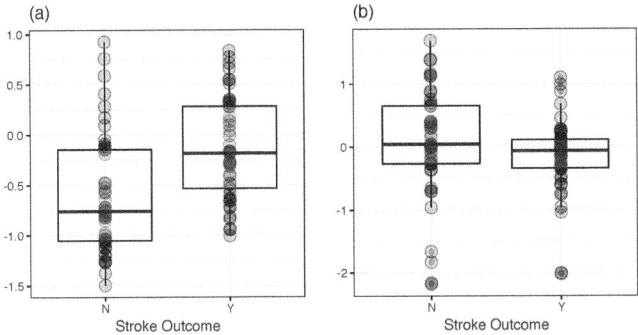

Figure 7.6: An illustration of the impact of the order of operations for preprocessing predictors and creating interaction terms based on the Stroke data. (a) Interactions terms are created, then all terms are preprocessed. (b) The predictors are first preprocessed, then the interaction terms are created based on the preprocessed predictors.

(b). The box plots between the stroke groups make it clear that the interaction's signal, captured by the shift between group distributions, is preserved when the interaction term is first created, followed by the preprocessing steps. However, the interactive predictive signal is almost completely lost when the original predictors are preprocessed prior to creating the interaction term. This case demonstrates that we should be very thoughtful as to at what step interaction terms should be created. In general, interactions are most plausible and practically interpretable on the original scales of measurement. Therefore, the interaction terms *should* probably be created prior to any preprocessing steps. It may also be wise to check the effect of the ordering of these steps.

7.3 The Brute-Force Approach to Identifying Predictive Interactions

For data sets that have a small to moderate number of predictors, all possible pairwise interactions for an association with the response can be evaluated. However, the more interaction terms that can be evaluated, the higher the probability that an interaction that is associated with the response due to random chance is found. More concretely, as more terms that are truly unrelated to the response are evaluated, the chance that at least one of them is associated with the response increases. Terms that have a statistically significant signal simply due to random chance and not due to a true relationship are called false positive findings. An entire sub-field of statistics is devoted to developing methodology for controlling the chance of false positive findings (Dickhaus, 2014). False positive findings can substantially decrease a model's predictive performance, and we need to protect against selecting these types of findings.

7.3.1 | Simple Screening

The traditional approach to screening for important interaction terms is to use nested statistical models. For a linear regression model with two predictors, x_1 and x_2, the main effects model is:

$$y = \beta_0 + \beta_1 x_1 + \beta_2 x_2 + error.$$

The second model with main effects plus an interaction is:

$$y = \beta_0 + \beta_1 x_1 + \beta_2 x_2 + \beta_3 x_1 x_2 + error.$$

These two models are called "nested" since the first model is a subset of the second. When models are nested, a statistical comparison can be made regarding the amount of additional information that is captured by the interaction term. For linear regression, the residual error is compared between these two models and the hypothesis test evaluates whether the improvement in error, adjusted for degrees of freedom, is sufficient to be considered real. The statistical test results in a p-value which reflects the probability that the additional information captured by the term is due to random chance. Small p-values, say less than 0.05, would indicate that there is less than a 5% chance that the additional information captured is due to randomness. It should be noted that the 5% is the rate of false positive findings, and is a historical rule-of-thumb. However, if one is willing to take on more risk of false positive findings for a specific problem, then the cut-off can be set to a higher value.

For linear regression, the *objective function* used to compare models is the statistical likelihood (the residual error, in this case). For other models, such as logistic regression, the objective function to compare nested models would be the binomial likelihood.

An alternative methodology for protecting against false positive findings was presented in Section 2.3. This methodology took the same nested model approach as above, but with two important differences:

1. Resampling was used to create a number of nested model pairs for different versions of the training set. For each resample, the assessment set is used to evaluate the objective function. In the traditional approach, the same data used to create the model are re-predicted and used to evaluate the model (which is understandably problematic for some models). If the model overfits, then computing the objective function from a separate data set allows a potential dissenting opinion that will reflect the overfitting.

2. The other distinction between the resampling approach and the traditional approach is that *any* measure of performance can be used to compare the nested model; it need not be a statistically tractable metric. In Section 2.3, the area under the ROC curve was the objective function to make the comparison. We could have chosen to identify interactions that optimize sensitivity, specificity,

accuracy, or any other appropriate metric. This provides great versatility and enables us to match the screening with an appropriate modeling technique and performance metric.

For both these reasons, but especially the second, the two approaches for evaluating nested models are not guaranteed to make consistent results. This is not problematic *per se*; different objective functions answer different questions and one cannot be said to be objectively better than the other without some context. Friedman (2001) illustrates the potential disconnect between metrics. In his case, different metrics, binomial likelihood and the error rate, were used to choose the optimal number of iterations for a boosted tree. Using the likelihood as the objective function, significantly fewer iterations were chosen when compared to using the misclassification rate. In his view:

> "... the misclassification error rate continues to decrease well after the logistic likelihood has reached its optimum. This degrading the likelihood by overfitting actually *improves* misclassification error rate. Although perhaps counterintuitive, this is not a contradiction; likelihood and error rate measure different aspects of fit quality."

Both the traditional approach and the resampling method from Section 2.3 generate p-values.[67] The more comparisons, the higher the chance of finding a false-positive interaction. There are a range of methods for controlling for false positive findings. One extreme is to not control for false positive findings. We may choose to do this early in our explorations of potential interactions, keeping in mind that there is a good chance of finding one or more false positives. At the other extreme is the Bonferroni correction (Shaffer, 1995) which uses a very strict exponential penalty in order to minimize *any* false positive findings. Because of the severe nature of this penalty, it is not good when we need to perform many statistical tests. The false discovery rate (FDR) procedure, previously mentioned in Section 5.6, falls along this correction spectrum and was developed to minimize false positive findings when there are many tests (Benjamini and Hochberg, 1995). Because complete enumeration can lead to exploring many possible pairwise interactions, the FDR can be a pragmatic compromise between no adjustment and the Bonferroni correction.

One shortfall of this interaction detection approach (previously used in Section 2.3) was that it was overly *reductive*. These candidate models only included the terms involved in the interaction. When the interactions that were discovered were included in a broader model that contains other (perhaps correlated) predictors, their importance to the model may be diminished. As this approach is used in this chapter, the interaction is added to a model that includes *all* of the potentially relevant predictors. This might reduce the number of predictors considered important (since the residual degrees of freedom are smaller) but the discovered interactions are likely to be more reliably important to a larger model.

---

[67] Although, as noted in Section 3.7, there are Bayesian methods that can be used to compare models without generating p-values.

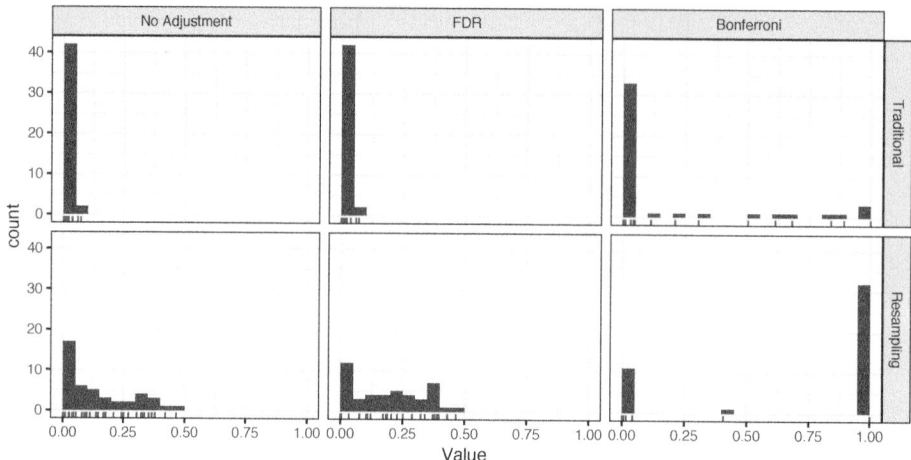

Figure 7.7: A comparison of the distribution of p-values and adjustment method for p-values between traditional and cross-validated estimates for interactions among predictors that produce a reduction in the RMSE.

Let's return to the Ames data example to illustrate evaluating interaction terms. Here nested models will be compared using the traditional approach with the p-values derived through resampling. Recall that a number of predictors in the base model are qualitative. For this analysis, the main effects and interactions for these predictions are done at the group level. This means that a predictor like the type of garage (with four possible values) has three main effects in the model. When this factor is crossed with a quantitative predictor, such as living area, an additional three terms are added. Alternatively, one could look at the effects of more specific interactions (such as a "finished" garage type interacting with another variable). This is very similar to the discussion of factors versus dummy variables for tree-based models in Section 5.7.

Within each of these types of p-value computations, the original p-values will be compared to the FDR and Bonferroni adjusted p-values. Figure 7.7 illustrates the distribution of p-values under each calculation method and adjustment type. For these data, the traditional approach identifies almost every interaction effect to be significant (even after an FDR correction or Bonferroni adjustment). However, the cross-validated approach has a lower rate of significance and is more affected by both types of corrections. Again, this is likely due to different estimation methods being used. Since the traditional approach simply re-predicts the original training set, it is plausible that many of these significant findings are the results of *overfitting* the interactions to the data.

Using the resampling method, if we accepted a 10% false-positive rate, 15 interaction pairs would be declared important. The top five pairs identified by this method, and their associated FDR values, were: latitude × longitude (2e-10), latitude × living area (4e-08), longitude × neighborhood (3e-07), building type × living area (1e-06), and latitude × neighborhood (1e-05). Note that, since the ANOVA method can simultaneously test for the entire set of interaction terms between two variables, these interactions are phrased at the grouped level (i.e, neighborhood instead of a specific neighborhood). For illustrative purposes, almost all of the analyses in subsequent sections decompose any qualitative predictors into dummy variables so that interactions are detected at a more granular level. There is no rule for how to evaluate the interactions; this is generally up to the analyst and the context of the problem.

As will be seen later, the list of significant interactions can be used as a starting point of an investigation using exploratory data analysis. For example, using resampling, the potential interaction with the smallest p-value was between longitude and latitude. From this, visualizations should be used to understand if the effect is practically significant and potentially meaningful. After these activities, the modeler can make a determination about the best approach for encoding the interactions and so on.

7.3.2 | Penalized Regression

Section 7.3.1 presented a way to search through potential interaction terms when complete enumeration is practically possible. This involved evaluating the change in model performance between the model with only the main effects and the model with the main effects and interactions. This approach is simple to implement and can be effective at identifying interaction terms for further evaluation. However, evaluating interaction terms in a one-at-a-time fashion prevents the simplistic models from evaluating the importance of the interaction terms in the presence of the full set of main effects and pairwise interaction terms.

If complete enumeration is practically possible, then another approach to identifying important interactions would be to create and add all interaction terms to the data. By doing this, the data may contain more predictors (original predictors plus interaction terms) than samples. Some models such as trees (or ensembles of trees), neural networks, support vector machines, and K-nearest neighbors, are feasible when the data contains more predictors than samples. However, other techniques like linear and logistic regression cannot be directly used under these conditions. But these models are much more interpretable and could provide good predictive performance if they could be utilized. A goal of feature engineering is to identify features that improve model performance while improving interpretability. Therefore, having a methodology for being able to utilize linear and logistic regression in the scenario where we have more predictors than samples would be useful.

A family of modeling techniques, called penalized models, has been developed to handle the case where there are more predictors than samples yet an interpretable model is desired. To understand the basic premise of these tools, we need to briefly review the objective of linear regression. Suppose that the data consist of p predictors, labeled as x_1, x_2, through x_p. Linear regression seeks to find the coefficients β_1, β_2, through β_p that minimize the sum of the squared distances (or error) from the estimated prediction of the samples to the observed values of the samples. The sum of squared errors can be written as:

$$SSE = \sum_{i=1}^{n}(y_i - \widehat{y}_i)^2$$

where

$$\widehat{y}_i = \widehat{\beta}_1 x_1 + \widehat{\beta}_2 x_2 + \ldots + \widehat{\beta}_p x_p$$

and $\widehat{\beta}_j$ is the regression coefficient estimated from the available data that minimizes SSE. Calculus can be used to solve the minimization problem. But it turns out that the solution requires inverting the covariance matrix of the predictors. If there are more predictors than samples or if one predictor can be written as a combination of one or more of the other predictors, then it is not possible to make the inversion. As predictors become more correlated with each other, the estimated regression coefficients get larger (inflate) and become unstable. Under these conditions, another solution is needed. Hoerl (1970) recognized this challenge, and proposed altering the optimization problem to:

$$SSE_{L_2} = \sum_{i=1}^{n}(y_i - \widehat{y}_i)^2 + \lambda_r \sum_{j=1}^{P} \beta_j^2$$

Because the objective of the equation is to minimize the entire quantity, the λ_r term is called a penalty, and the technique is called ridge regression. As the regression coefficients grow large, the penalty must also increase to enforce the minimization. In essence, the penalty causes the resulting regression coefficients to become smaller and shrink towards zero. Moreover, by adding this simple penalty we can obtain reasonable parameter estimates. However, many of the parameter estimates may not be truly zero, thus making the resulting model interpretation much more difficult. Tibshirani (1996) made a simple and notable modification to the ridge optimization criteria and proposed minimizing the following equation:

$$SSE_{L_1} = \sum_{i=1}^{n}(y_i - \widehat{y}_i)^2 + \lambda_\ell \sum_{j=1}^{P} |\beta_j|$$

This method was termed the least absolute shrinkage and selection operator (lasso). By modifying the penalization term λ_ℓ, the lasso method forces regression coefficients to zero. In doing so, the lasso practically selects model terms down to an optimal subset of predictors and, as such, this method could be used on an entire data set of predictors and interactions to select those that yield the best model performance. Friedman et al. (2010) extended this technique so that it could be applied to the classification setting.

Traditionally, these two different types of penalties have been associated with different tasks; the ridge penalty is mostly associated with combating collinearity between predictors while the lasso penalty is used for the elimination of predictors. In some

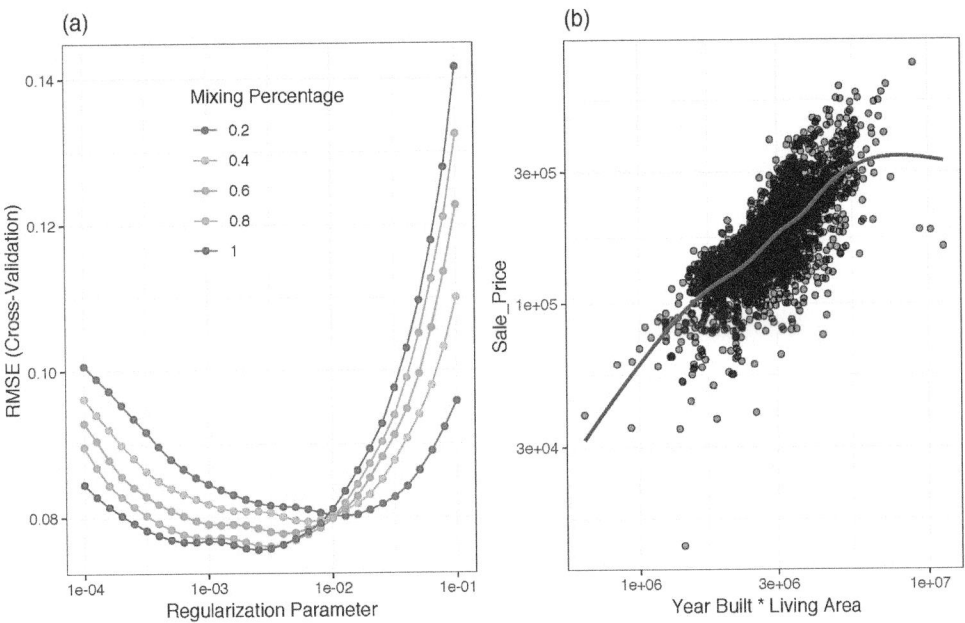

Figure 7.8: (a) The tuning parameter profile for the glmnet method applied to the Ames data. (b) The relationship of the top selected interaction term, Year Built x Living Area, to the outcome data.

data sets, both effects may be required to create an effective model. Zou and Hastie (2005) and Friedman et al. (2010) devised approaches for blending both types of penalties together. The glmnet model has tuning parameters for the total penalty $\lambda = \lambda_r + \lambda_\ell$ and the proportion of λ that is associated with the lasso penalty (usually denoted as α). Selecting $\alpha = 1$ would be a fully lasso penalty model whereas $\alpha = 0.5$ would be an even mix. For regression models, this approach optimizes

$$SSE = \sum_{i=1}^{n}(y_i - \widehat{y}_i)^2 + \lambda \left[(1 - \alpha) \sum_{j=1}^{P} \beta_j^2 + \alpha \sum_{j=1}^{P} |\beta_j| \right]$$

Figure 7.8(a) shows the results of tuning the model using resampling on the Ames data using the base model plus all possible interactions. Each line is the relationship between the estimated RMSE and λ for a given value of α. As the model becomes more lasso-like, the RMSE is reduced. In the end, a pure-lasso model using $\lambda = 0.003$ was best. The estimated RMSE from this model was 0.076 log units. Of the 1,033 predictors, the model selected a subset of 115 terms (only 3 of which were main effects). For these data, the lasso appears to prefer primarily interaction terms in the model; this technique does not enforce the hierarchy principle although Bien et al. (2013) developed a methodology which does for this model.

The interaction effect with the largest coefficient crossed the year built and the living area. This predictor is shown in Figure 7.8(b) where there is a large log-linear effect.

Table 7.1: The top terms (predictors and pairwise interactions) selected by the glmnet procedure.

Predictor 1	Predictor 2	Coefficient
Living Area	Year Built	0.10143
Latitude	Year Built	0.02363
Central Air: Y	Lot Area	0.02281
Lot Frontage	Neighborhood: Northridge Heights	0.01144
Central Air: Y	Neighborhood: Gilbert	-0.01117
Lot Area	Year Built	0.00984
Foundation: PConc	Roof Style: Hip	0.00984
Lot Frontage	Neighborhood: Edwards	-0.00944
Foundation: PConc	Lot Area	0.00893
Alley: No Alley Access	MS SubClass: Two Story PUD 1946 and Newer	-0.00874
Alley: No Alley Access	MS SubClass: Two Story 1946 and Newer	-0.00624
Building Type Duplex	Central Air: Y	-0.00477
Foundation: Slab	Living Area	-0.00454
Land Contour: HLS	Roof Style: Hip	0.00414
Lot Frontage	Pool Area	-0.00409

Table 7.1 shows the top 15 interaction terms. In this set, several model terms show up multiple times (i.e., neighborhood, living area, etc.).

7.4   Approaches when Complete Enumeration Is Practically Impossible

As the number of predictors increases, the number of interaction terms to explore grows exponentially. Just a moderate number of predictors of several hundred, which could be considered small for today's data, generates many more interactions than could be considered from a practical or statistical perspective. On the practical side, the computational burden likewise grows exponentially. With enough computing resources, it still could be possible to evaluate a vast number of interactions. Even if these computations are possible, this is not a statistically wise approach due to the chance of finding false positive results. Because the number of important interactions are usually small, the more insignificant interactions are screened, the greater penalty is incurred for evaluating these interactions. Therefore, truly significant interactions may be missed if all possible interactions are computed and appropriate adjustments are made.

When there are a large number of interaction terms to evaluate, other approaches are necessary to make the search more efficient, yet still effective, at finding important interactions. The remainder of this section will survey several approaches that have been developed to uncover interactions among predictors.

7.4.1 | Guiding Principles and Two-Stage Modeling

In a two-stage modeling approach, the ability of predictors to explain the variation in the response is evaluated using a model that does not directly account for interaction effects. Examples of such models are linear or logistic regression. Once the important predictors are identified, the residuals from this model are calculated. These residuals contain information that the individual predictors themselves cannot explain. The unexplained variation is due to random measurement error, and could also be due to predictor(s) that were not available or measured, or interactions among the observed predictors that were not present in the model. To be more concrete, suppose that the observed data was generated from the equation

$$y = x_1 + x_2 + 10x_1x_2 + error.$$

Suppose also that when the data from this system was collected only the response and x_1 were observed. The best model we could estimate would be limited to

$$y = \beta_1 x_1 + error^*.$$

Since the true response depends on x_2 and the interaction between x_1 and x_2, the error from the imperfect model ($error^*$) contains information about these important terms (that were absent from the model). If it was possible to access x_2 and create the $x_1 \times x_2$ interaction term, then it would be possible to separate out their contribution through a second model

$$error^* = \beta_2 x_2 + \beta_3 x_1 x_2 + error.$$

We cannot do anything to explain random measurement error; it is always inherent to some extent in data. And expert knowledge is necessary for determining if other predictors are relevant for explaining the response and should be included in the model. However, it is still possible to search for interaction terms that help to explain variation in the response. But the focus of this section is the setting where it is practically impossible to search all possible interaction terms.

When there are many predictors, the principles outlined in Section 7.1 can be used to narrow the search. The hierarchy principle would suggest to first look at pairwise interactions for assistance in explaining the residual variation. Next, the sparsity principle would indicate that if there are interactions that are active, then only a relatively few of them are important. And last, the heredity principle would search for these interactions among the predictors that were identified in the first stage. The type of heredity chosen at this point, weak or strong, will determine the number of interaction terms that will be evaluated.

To follow the logic of this process, consider modeling the Ames data. In the first stage, the glmnet model using only the base predictor set (i.e., only main effects) selected 12 predictors.

Since 12 individual predictors were selected, there are 66 combinations of pairs between these predictors. However, once the appropriate dummy variables are created, the second stage glmnet model contains 719 predictor columns as inputs, 660 of which are interactions. The same tuning process was followed and the resulting patterns are very similar to those seen in Figure 7.8(a). This two-stage model was also purely lasso and the best penalty was estimated to be 0.002 with an RMSE of 0.081 log units. Of the entire predictor set, this glmnet model selected 7 main effects and 128 interaction terms.

It is important to note that the process for computing error is different when working with a continuous outcome versus a categorical outcome. The error from the two-stage modeling approach is straightforward to compute when the response is numeric as shown above. In the case of a classification outcome, as with the stroke data, the residuals must be computed differently. The Pearson residual should be used and is defined as

$$\frac{y_i - p_i}{\sqrt{p_i\,(1 - p_i)}}.$$

For categorical response data, y_i is a binary indicator of the response category of the ith sample and p_i is the predicted probability of the ith sample.

7.4.2 | Tree-Based Methods

The methods discussed thus far can be effective at revealing interactions for data with a moderate number of predictors. However, these approaches lose their practical effectiveness as the number of predictors grows. Tree-based methods like recursive partitioning or ensembles of recursive partitioning models can efficiently model data with a large number of predictors. Moreover, tree-based methods have been shown to uncover potential interactions between variables (Breiman et al., 1984). In fact, using tree-based methods for identifying important interactions among predictors has been a popular area of theoretical and empirical research as we will see.

Let's review the basic concepts of recursive partitioning to understand why this technique has the ability to uncover interactions among predictors. The objective of recursive partitioning is to find the optimal split point of a predictor that separates the samples into more homogeneous groups with respect to the response. Once a split is made, the same procedure is applied for the subset of samples at each of the subsequent nodes. This process then continues recursively until a minimum number of samples in a node is reached or until the process meets an optimization criterion. When data are split recursively, each subsequent node can be thought of as representing a localized interaction between the previous node and the current node. In addition, the more frequent the occurrence of subsequent nodes of the same pair of predictors, the more likely that the interaction between the predictors is global interaction across the observed range of the predictors.

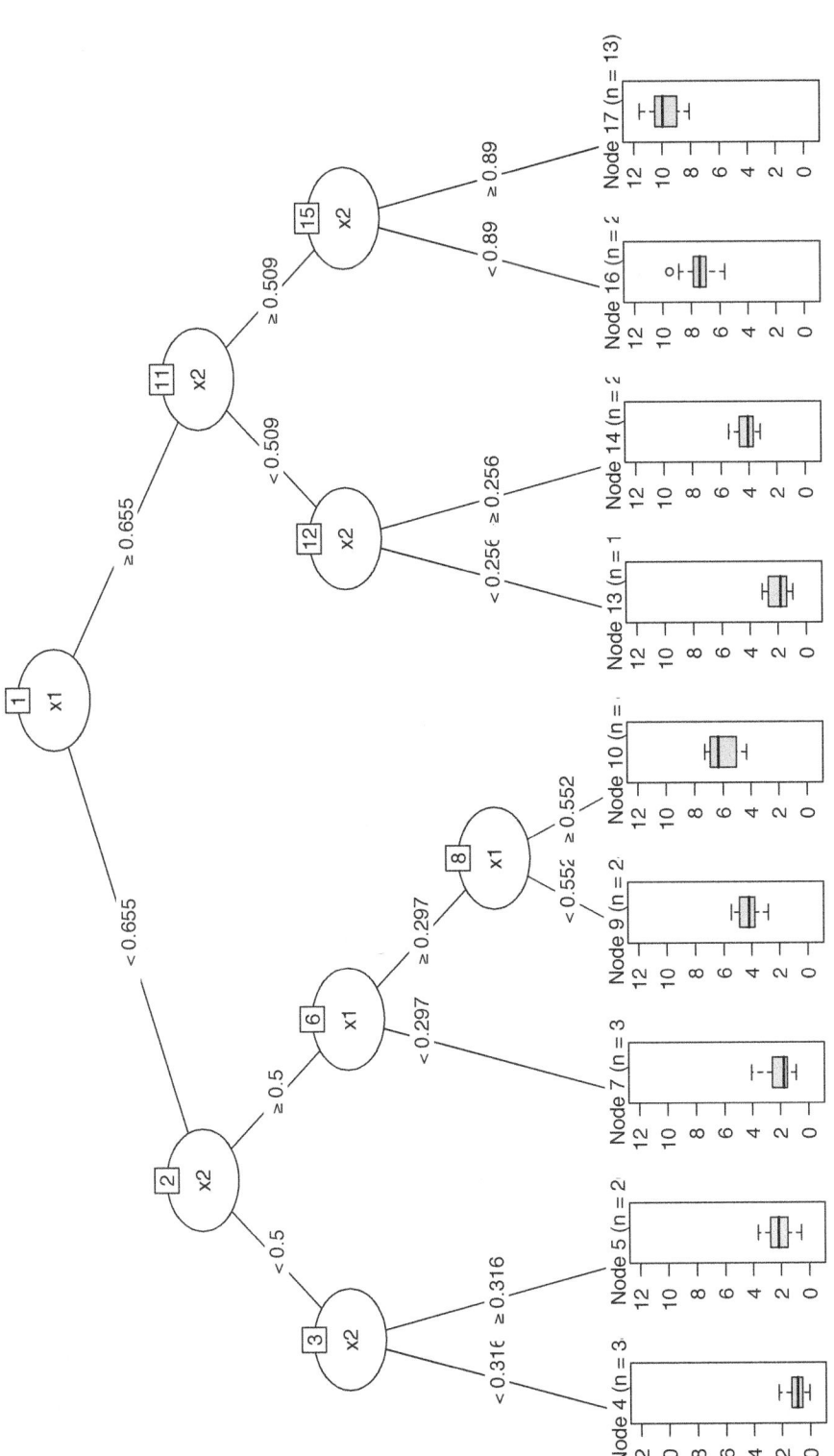

Figure 7.9: A recursive partitioning tree for a synergistic interaction between the two predictors. The interaction is captured by several alternating splits between the two predictors at subsequent levels of the tree.

The simple two-predictor example of a synergistic relationship presented in the introduction will be used to demonstrate how an interaction can be detected through recursive partitioning. The optimal tree for these data is presented in Figure 7.9. Tree-based models are usually thought of as pure interaction due to their prediction equation, which can be written as a set of multiplicative statements. For example, the path to terminal node 4 goes through three splits and can be written as a single rule:

```
node_4 <- I(x1 < 0.655) * I(x2 < 0.5) * I(x2 < 0.316) * 0.939
```

where 0.939 is the mean of the training set data falling into this node (I is a function that is zero or one based on the logic inside the function). The final prediction equation would be computed by adding similar terms from the other eight terminal nodes (which would all be zero except a single rule when predicting). Notice that in Figure 7.9 the terminal nodes have small groups of the data that are homogeneous and range from smaller response values (left) to larger response values (right) on each side of the first split. Achieving homogeneous terminal nodes requires information from both predictors throughout the tree.

But why would a tree need to be this complicated to model the somewhat simple synergistic relationship between x_1 and x_2? Recall from Figure 7.2 that the synergistic relationship induces curved contours for the response. Because the recursive partitioning model breaks the space into rectangular regions, the model needs more regions to adequately explain the response. In essence, the partitioning aspect of trees impedes their ability to represent *smooth, global interactions.*

To see this, let's look at Figure 7.10. In this figure, the predictions from each terminal node are represented by rectangular regions where the color indicates the predicted response value for the region. Overlaid on this graph are the contour lines generated from the linear model where the interaction term has been estimated. The tree-based model is doing well at approximating the synergistic response. But, because the model breaks the space into rectangular regions, it requires many regions to explain the relationship.

Figure 7.10 illustrates that a tree pixelates the relationship between the predictors and the response. It is well known that the coarse nature of the predictions from a tree generally makes this model less accurate than other models. In addition, trees have some potential well-known liabilities. As one example, an individual tree can be highly variable (i.e., the model can substantially change with small changes to the data). Ensembles of trees such as bagging (Breiman, 2001) and boosting (Friedman, 2001) have been shown to mitigate these problems and improve the predictivity of the response.

A brief explanation of these techniques is provided here to aid in the understanding of these techniques. An in-depth explanation of these techniques can be found in Kuhn and Johnson (2013). To build a bagged tree-model, many samples of the original data are independently generated with replacement. For each sample, a tree is built. To predict a new sample, the model averages the predicted response

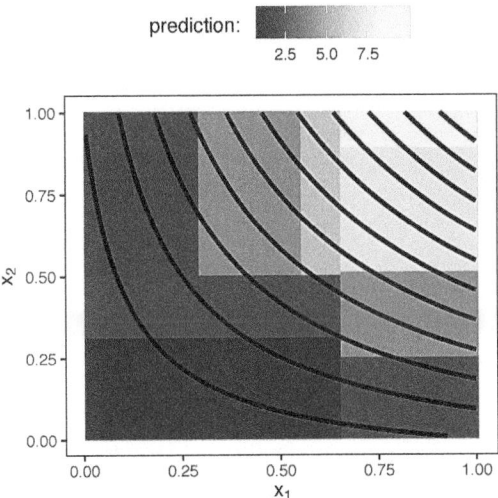

Figure 7.10: An illustration of the predicted response from the recursive partitioning model between two predictors that have a synergistic relationship. The regions are able to approximate the relationship with the response in a pixelated way.

across all of the trees. Boosting also utilizes a sequence of trees, but creates these in a fundamentally different way. Instead of building trees to maximum depth, a boosted model restricts the tree depth. In addition, a boosted tree model utilizes the predictive performance of the previous tree on the samples to reweigh poorly predicted samples. Third, the contribution of each tree in the boosted model is weighted using statistics related to the individual model fit. The prediction for a new sample is a weighted combination across the entire set of trees.

Recent research has shown that these ensemble methods are able to identify important interactions among the predictors. García-Magariños et al. (2009) and Qi (2012) illustrate how random forest, which is a further modification of bagging, can be used to find interactions in biological data, while Basu et al. (2018) uses a variation of the random forest algorithm to identify high-order interactions. For boosting, Lampa et al. (2014) and Elith et al. (2008) found that this methodology could also be used to uncover important interactions. Using a theoretical perspective, Friedman and Popescu (2008) proposed a new statistic for assessing the importance of an interaction in a tree-based model.

The reason why tree ensembles are effective at identifying predictor-response relationships is because they are aggregating many trees from slightly different versions of the original data. The aggregation allows for many more possible response values and better approximating the predictor-response relationships. To see this, a random forest model and a gradient boosting machine model will be built on the two-predictor synergistic example.

Figure 7.11 shows that ensembles of trees are better able to approximate the synergistic relationship than an individual tree. In fact, the cross-validated root mean

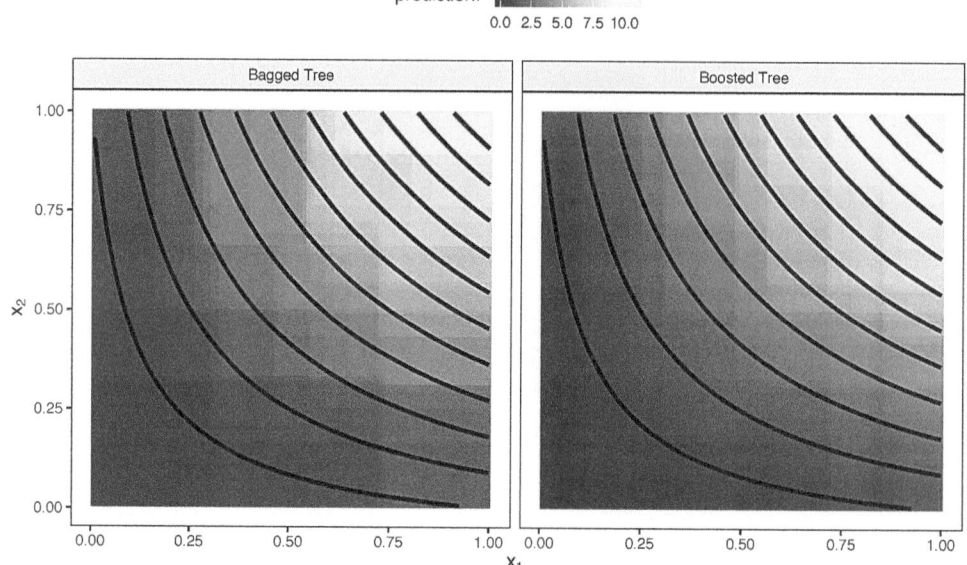

Figure 7.11: An illustration of the predicted response from boosted and bagged tree models when the predictors have a synergistic relationship. These ensemble models can better approximate the relationship between the predictors and the response compared to a single tree.

squared error (RMSE) for boosting is 0.27 and for bagged trees is 0.6. The optimal single tree model is worse with a cross-validated RMSE of 0.9. In this example, smoother models win; the linear regression interaction model's RMSE is 0.05.

Notice that an individual tree or ensemble of trees are less effective and are much more complicated models for uncovering predictive interactions that are globally true across all of the samples. This simple example highlights the main premise of feature engineering. If meaningful predictor relationships can be identified and placed in a more effective form, then a simpler and perhaps more appropriate model to estimate the relationship between the predictors and the response can be used. However, there is a scenario where tree-based methods may have better predictive ability. When important interactions occur within one or more subsets of the samples (i.e., locally), then tree-based methods have the right architecture to model the interactions. Creating features for localized interactions to be included in simpler models is much more tedious. This process would require the creation of a term that is a product of the identifier of the local space (i.e., a dummy predictor) and the predictors that interact within that space, which is essentially an interaction among three terms.

Since trees and ensembles of trees can approximate interactions among predictors, can we harness information from the resulting models to point us to terms that should be included in a simpler linear model? The approach presented by Basu et al.

(2018) uses a form of a random forest model (called feature weighted random forests) that randomly selects features based on weights that are determined by the features' importance. After the ensemble is created, a metric for each co-occurring set of features is calculated which can then be used to identify the top interacting features.

Friedman and Popescu (2008) present an approach based on the theoretical concept of partial dependence (Friedman, 2001) for identifying interactions using ensembles. In short, this approach compares the joint effect of two (or more) predictors with the individual effect of each predictor in a model. If the individual predictor does not interact with any of the other predictors, the difference between the joint effect and the individual effect will be close to zero. However, if there is an interaction between an individual predictor and at least one of the predictors, the difference will be greater than zero. This comparison is numerically summarized in what they term the H statistic. For the 18 predictors in the Ames data (including the random predictor), the H statistics were computed on the entire training set. One issue is that the H statistic, across predictors, can have a fairly bimodal distribution since many of the values are zero (when a predictor isn't involved in any interactions). This data set is fairly small and, for these reasons, 20 bootstrap estimates were computed and summarized using the median H statistic. Both estimates are shown in Figure 7.12. The original statistics and their resampled counterparts agree fairly well. Some H values are fairly close to zero so a cutoff of $H >= 0.001$ was used on the bootstrap estimates to select variables. This would lead us to believe that 6 predictors are involved in at least one interaction. Following the previous two-stage approach, all corresponding two-way interactions of these predictors where added to the main effects in a glmnet model (mimicking the previous analysis). Using resampling, another pure lasso model was selected, this time with an optimal tuning parameter value of $\lambda = 10^{-4}$. From this model, 48 main effects and 68 interactions were selected.

A third and practical way to identify interactions use predictor importance information along with principles of hierarchy, sparsity, and heredity. Simply put, if an interaction among predictors is important, then a tree-based method will use these predictors multiple times throughout multiple trees to uncover the relationship with the response. These predictors will then have non-zero, and perhaps high, importance values. Pairwise interactions among the most important predictors can then be created and evaluated for their relevance in predicting the response. This is very similar to the two-stage modeling approach, except that the base learner is a tree-based model.

Each tree-based method has its own specific algorithm for computing predictor importance. One of the most popular methods is generated by a random forest. Random forest uses bootstrap samples to create many models to be used in the ensemble. Since the bootstrap is being used, each tree has an associated assessment set (historically called the out-of-bag (OOB) samples) that we not used to fit the model. These can be predicted for each tree to understand the quality of the model. Suppose a random forest model contains 1,000 trees. If classification accuracy is

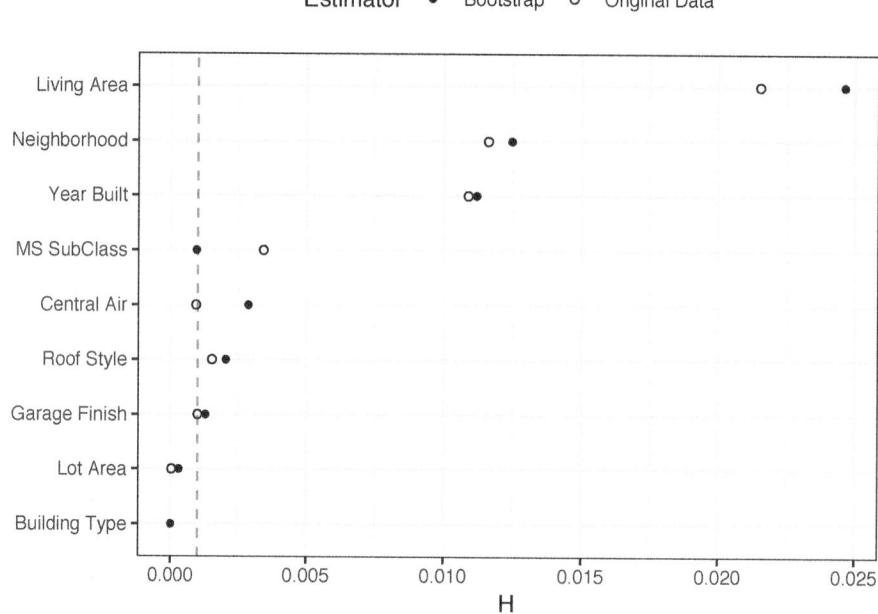

Figure 7.12: A visualization of the H statistic estimates for the Ames data.

computed for the out-of-bag samples for each tree, the average of these accuracy statistics is the OOB estimate. To measure importance for a specific predictor, the values of that predictor are shuffled, and the OOB estimate is recalculated. If the predictor were unimportant, there would not be much difference in the original OOB statistic and the one where the predictor has been made irrelevant. However, if a predictor was crucial, the permuted OOB statistic should be significantly worse. Random forest conducts this analysis for each predictor in the model and a single score is created per predictor where larger values are associated with higher importance.[68]

For the Ames data, the random forest model (using the default value of m_{try}) was fit and the importance scores were estimated. Figure 7.13 displays the top 10 predictors by importance, most of which have all been identified as part of multiple pairwise interactions from the other methods. Of particular importance are the living area and the year built which are the most important predictors and are combined in a scientifically meaningful interaction. This approach is similar in spirit to the two-stage approach discussed above. The primary difference is that the random forest model is simply used for identifying predictors with the potential for being part of important interactions. Interactions among the top predictors identified here could then be included in a simpler model, like logistic regression, to evaluate their importance.

---

[68]These importance scores are discussed further in Section 11.3.

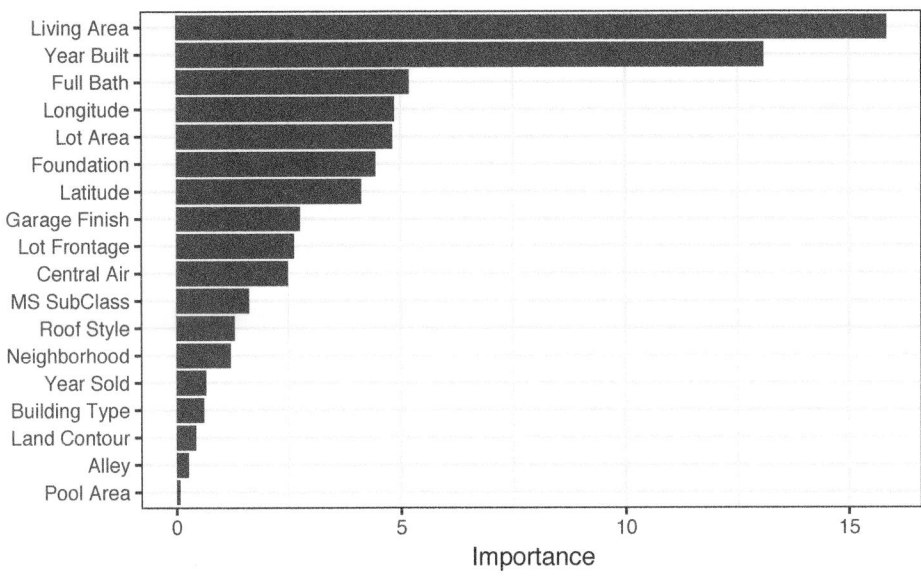

Figure 7.13: Top 10 variable importance scores for a random forest model applied to the Ames data.

7.4.3 | The Feasible Solution Algorithm

As mentioned above, the problem of identifying important interactions can be thought of as identifying the optimal model that includes main effects and interactions. A complete enumeration and evaluation of all of the possible models is practically impossible for even a moderate number of predictors since the number of models is $p!$. When using a linear model such as linear or logistic regression, methods such as forward, backward, and stepwise selection are commonly used to locate an optimal set of predictors (Neter et al., 1996). These techniques are as straightforward to understand as their names imply. The forward selection technique starts with no predictors and identifies the predictor that is most optimally related to the response. In the next step, the next predictor is selected such that it in combination with the first predictor are most optimally related to the response. This process continues until none of the remaining predictors improve the model based on the optimization criteria. The backward selection technique starts with all of the predictors and sequentially removes the predictor that has the least contribution to the optimization criteria. This process continues until the removal of a predictor significantly reduces the model performance. The stepwise procedure adds or removes one predictor at a time at each step based on the optimization criteria until the addition or removal of any of the predictors does not significantly improve the model.[69]

Miller (1984) provided an alternative search technique to the stepwise method for the linear model problem. Instead of removing a predictor, he recommended a

---

[69]This is a feature selection algorithm similar to those discussed in Chapter 10.

substitution approach in which each selected predictor is systematically chosen for substitution. For example, suppose that there are 10 predictors, x_1–x_{10}, and we want to find the best model that contains three predictors. Miller's process begins by randomly selecting three of the ten predictors, say x_2, x_5 and x_9. Model performance is calculated for this trio of predictors. Then the first two predictors are fixed and model performance of these two predictors is computed with the remaining seven other predictors. If none of the other seven predictors give better performance than x_9, then this predictor is kept. If x_9 is not optimal, then it is exchanged with the best predictor. The process then fixes the first and third predictors and uses the same steps to select the optimal second predictor. Then, the second and third predictors are fixed and a search is initiated for the optimal first predictor. This process continues to iterate until it converges on an optimal solution. Because the algorithm is not guaranteed to find the global optimal solution, different random starting sets of predictors should be selected. For each different starting set, the converged solution and the frequency of each solution are determined. Hawkins (1994) extended this algorithm to a robust modeling approach and called the procedure the Feasible Solution Algorithm (FSA).

A primary advantage of this technique is that the search space of the algorithm, which is on the order of $q \times m \times p$ where q is the number of random starts, m is the number of terms in the subset, and p is the number of predictors, is much smaller than the entire search space (p^m). FSA has been shown to converge to an optimal solution, although it is not guaranteed to be the global optimum.

FSA focuses on finding an optimal subset of main effects. However, the original implementation of FSA does not consider searching for interactions among predictors. This can be a considerable drawback to using the algorithm since important interactions do occur among predictors for many problems. Lambert et al. (2018) recognized this deficiency and has provided a generalization to the algorithm to search for interactions. Lambert's approach begins with a base model, which typically includes predictors thought to be important for predicting the response. The order of the desired interaction is chosen (i.e., two-factor, three-factor, etc.). The process of identifying potentially relevant interactions follows the same type of logic as the original FSA. To illustrate the process, let's return to the example from above. Suppose that the base model included three predictors (x_1, x_5, and x_9) and the goal is to identify important pairwise interactions. Lambert's algorithm randomly selects two predictors, say x_1 and x_9, and computes the performance of the model that includes the base terms and this interaction term. The algorithm then fixes the first predictor and systematically replaces the second predictor with the remaining eight (or nine if we want to study quadratic effects) predictors and computes the corresponding model performance. If none of the other predictors give better predictive performance, then x_9 is kept. But if x_9 is not optimal, then it is exchanged with the best predictor. The process then fixes the second predictor and exchanges the first predictor with eight (or nine) of the other predictors. This process continues to iterate until convergence. Then a different random pair of predictors is selected and the process is applied. The converged interactions are tallied across

Table 7.2: Top interaction pairs selected by the Feasible Solution Algorithm for the Ames Data.

Predictor 1	Predictor 2	p-value	RMSE
Foundation: CBlock	Living Area	0.00539	0.0513
Foundation: PConc	Living Area	0.00828	0.0515
Building Type Duplex	Living Area	0.00178	0.0516
Living Area	MS SubClass: Duplex All Styles and Ages	0.00102	0.0516
Roof Style: Hip	Year Built	0.04945	0.0518
Roof Style: Gable	Year Built	0.00032	0.0518
Longitude (spline)	Neighborhood: Northridge Heights	0.00789	0.0518
Living Area	Neighborhood: Northridge Heights	0.09640	0.0519
Full Bath	Land Contour: HLS	0.07669	0.0519
Foundation: CBlock	Lot Area	0.01698	0.0519
MS SubClass: One Story PUD 1946 and Newer	Year Built	0.02508	0.0519
Latitude (spline)	Neighborhood: other	0.03926	0.0520
Latitude (spline)	Neighborhood: College Creek	0.00986	0.0520
Foundation: CBlock	MS SubClass: One Story PUD 1946 and Newer	0.04577	0.0520
Longitude (spline)	Neighborhood: other	0.03762	0.0520

the random starts to identify the potential feasible solutions. This algorithm can be easily generalized to any order of interaction.

FSA was used on the Ames data. Unlike the previous analyses, we chose to look at the specific interaction terms using individual dummy variables to contrast the type of results that can be obtained with the previous results. For example, a randomly generated interaction might involve two dummy variables from different predictors. We constrained the selection so that dummy variables from the same predictor would never be paired. Fifty random starts were used and, during each random start, there were 60 potential swaps of predictors into the randomly generated interactions. Resampling[70] was used to determine if each candidate set of interaction terms was associated with a smaller RMSE value. This determination was made using the base model and the candidate interaction. Across the iterations of the FSA, 31 distinct interaction pairs were identified. Table 7.2 shows the results for the top 15 interactions (indexed by their resulting RMSE). Many of the same variables occur in these lists (e.g., living area, neighborhood, etc.) along with some of the nonlinear terms (e.g., splines for the geocodes). The next step would be to add a selection of these terms to the base model after more exploratory work.

For example, there appears to be a potential interaction between the building type and the living area. What is the nature of their relationship to the outcome? Figure 7.14 shows a scatter plot that is separated for each building type with a linear regression line superimposed in each subgroup. The figure shows very different

---

[70]Using the same repeats of 10-fold cross-validation previously employed.

slopes for several building types. The largest contrast being between Townhouse End Units (`TwnhsE`) and two-family conversions (`TwoFmCon`, originally built as one-family dwelling). It might make sense to encode all five interaction terms or to group some terms with similar slopes together for a more simplistic encoding.

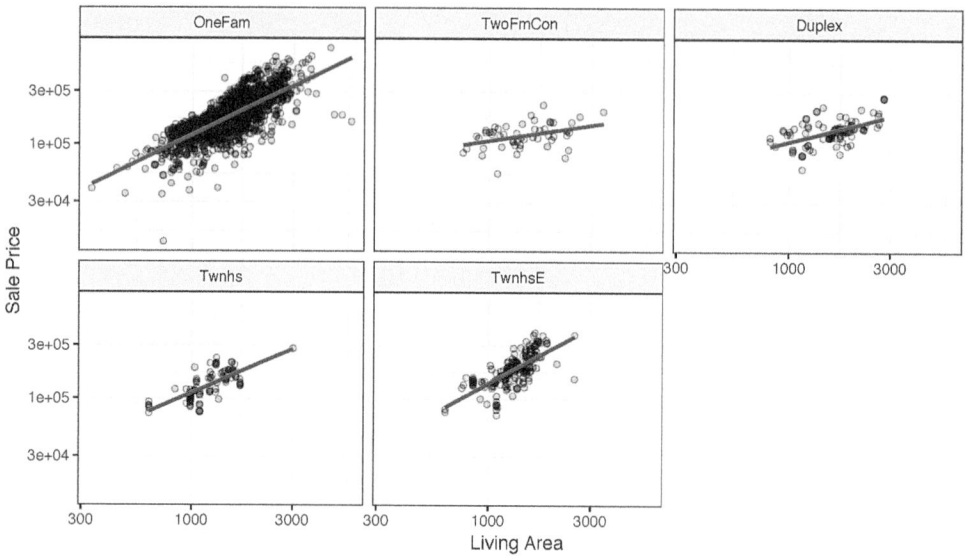

Figure 7.14: An example of heterogeneous slopes for building type discovered using the FSA procedure.

7.5   Other Potentially Useful Tools

There are a few other models that have the ability to uncover predictive interactions that are worthy of mentioning here. Multivariate adaptive regression splines (MARS) is a nonlinear modeling technique for a continuous response that searches through individual predictors and individual values of each predictor to find the predictor and sample value for creating a hinge function that describes the relationship between the predictor and response (Friedman, 1991). The hinge function was described previously in Section 6.2.1 and enables the model to search for nonlinear relationships. In addition to being able to search through individual predictors, MARS can be used to search for products of predictors in order to make nonlinear interactions that isolate portions of the predictor space. The MARS technique also has built-in feature selection. The primary disadvantage of this approach is that it is computationally taxing. MARS has also been extended to classification outcomes, and this method is called flexible discriminant analysis (FDA).

Cubist (Kuhn and Johnson, 2013) is a rule-based regression model that builds an initial tree and decomposes it into sets of rules that are pruned and perhaps eliminated. For each rule, a separate linear model is defined. One example of a rule for the Ames housing data might be:

```
if
```

```
    Year_Built <= 1952
    Central_Air = No
    Longitude <= -93.6255
```

```
then
```

```
    log10 Sale Price = 261.38176 + 0.355 Gr_Liv_Area +
                                    2.88  Longitude +
                                    0.26  Latitude
```

This structure creates a set of disjoint interactions; the set of rules may not cover every combination of values for the predictors that are used in the rules. New data points being predicted may belong to multiple rules and, in this case, the associated linear model predictions are averaged. This model structure is very flexible and has a strong chance of finding local interactions within the entire predictor space. For example, when a property has a small total living area but has a large number of bedrooms, there is usually a *negative* slope in regression models associated with the number of bedrooms. This is certainly not true in general but can be when applied only to properties that are nearly "all bedroom."

7.6   Summary

Interactions between predictors are often overlooked when constructing a predictive model. This may be due to the ability of modern modeling techniques to covertly identify the interactions. Or interactions may be overlooked due to the vast potential additional number of terms these would add to a model. This is especially true if the analyst has no knowledge of which interaction terms may be beneficial to explaining the outcome.

When beginning the search for interactions, expert knowledge about the system will always be most beneficial and can help narrow the search. Algorithmic approaches can also be employed in the search. For data that have a relatively small number of predictors, every pairwise interaction can be completely enumerated. Then resampling approaches or penalized models can be used to locate the interactions that may usefully improve a model.

As the number of predictors grows, complete enumeration becomes practically unfeasible. Instead, methods that can readily identify potentially important interactions without searching the entire space should be used. These include two-stage modeling, tree-based methods, and the feasible solution algorithm.

When the search is complete, the interaction terms that are most likely to improve model performance can be added to a simpler model like linear or logistic regression. The predictive performance can then be estimated through cross-validation and compare to models without these terms to determine the overall predictive improvement.

7.7 | Computing

The website `http://bit.ly/fes-interact` contains R programs for reproducing these analyses.

8 | Handling Missing Data

Missing data are **not** rare in real data sets. In fact, the chance that at least one data point is missing increases as the data set size increases. Missing data can occur any number of ways, some of which include the following.

- Merging of source data sets: A simple example commonly occurs when two data sets are merged by a sample identifier (ID). If an ID is present in only the first data set, then the merged data will contain missing values for that ID for all of the predictors in the second data set.

- Random events: Any measurement process is vulnerable to random events that prevent data collection. Consider the setting where data are collected in a medical diagnostic lab. Accidental misplacement or damage of a biological sample (like blood or serum) would prevent measurements from being made on the sample, thus inducing missing values. Devices that collect actigraphy data can also be affected by random events. For example, if a battery dies or the collection device is damaged then measurements cannot be collected and will be missing in the final data.

- Failures of measurement: Measurements based on images require that an image be in focus. Images that are not in focus or are damaged can induce missing values. Another example of a failure of measurement occurs when a patient in a clinical study misses a scheduled physician visit. Measurements that would have been taken for the patient at that visit would then be missing in the final data.

The goal of feature engineering is to get the predictors into a form which models can better utilize in relating the predictors to the response. For example, projection methods (Section 6.3) or autoencoder transformations (Section 6.3.2) for continuous predictors can lead to a significant improvement in predictive performance. Likewise, a feature engineering maneuver of likelihood encoding for categorical predictors (Section 5.4) could be predictively beneficial. These feature engineering techniques, as well as many others discussed throughout this work, require that the data have no missing values. Moreover, missing values in the original predictors, regardless of any feature engineering, are intolerable in many kinds of predictive models. Therefore, to utilize predictors or feature engineering techniques, we must first address the missingness in the data. Also, the missingness itself may be an important predictor of the response.

In addition to measurements being missing within the predictors, measurements may also be missing within the response. Most modeling techniques cannot utilize samples that have missing response values in the training data. However, a new approach known as semi-supervised learning has the ability to utilize samples with unknown response values in the training set. Semi-supervised learning methods are beyond the scope of this chapter; for further reference see Zhu and Goldberg (2009). Instead, this chapter will focus on methods for resolving missing values within the predictors.

Types of Missing Data

The first and most important question when encountering missing data is *"why* are these values missing?" Sometimes the answer might already be known or could be easily inferred from studying the data. If the data stem from a scientific experiment or clinical study, information from laboratory notebooks or clinical study logs may provide a direct connection to the samples collected or to the patients studied that will reveal why measurements are missing. But for many other data sets, the cause of missing data may not be able to be determined. In cases like this, we need a framework for understanding missing data. This framework will, in turn, lead to appropriate techniques for handling the missing information.

One framework to view missing values is through the lens of the mechanisms of missing data. Three common mechanisms are:

- structural deficiencies in the data,

- random occurrences, or

- specific causes.

A structural deficiency can be defined as a missing component of a predictor that was omitted from the data. This type of missingness is often the easiest to resolve once the necessary component is identified. The Ames housing data provides an example of this type of missingness in the Alley predictor. This predictor takes values of "gravel" or "paved", or is missing. Here, 93.2 percent of homes have a missing value for alley. It may be tempting to simply remove this predictor because most of the values are missing. However, doing this would throw away valuable predictive information of home price since missing, in this case, means that the property has no alley. A better recording for the Alley predictor might be to replace the missing values with "No Alley Access".

A second reason for missing values is due to random occurrences. Little and Rubin (2014) subdivide this type of randomness into two categories:

- Missing completely at random (MCAR): the likelihood of a missing result is equal for all data points (observed or unobserved). In other words, the missing values are independent of the data. This is the best case situation.

- Missing at random (MAR): the likelihood of a missing results is not equal for all data points (observed or unobserved). In this scenario, the probability of a missing result depends on the observed data but not on the unobserved data.

In practice it can be very difficult or impossible to distinguish if the missing values have the same likelihood of occurrence for all data, or have different likelihoods for observed or unobserved data. For the purposes of this text, the methods described herein can apply to either case. We refer the reader to Little and Rubin (2014) for a deeper understanding of these nuanced cases.

A third mechanism of missing data is missingness due to a specific cause (or not missing at random (NMAR) (Little and Rubin, 2014)). This type of missingness often occurs in clinical studies where patients are measured periodically over time. For example, a patient may drop out of a study due to an adverse side effect of a treatment or due to death. For this patient, no measurements will be recorded after the time of drop-out. Data that are not missing at random are the most challenging to handle. The techniques presented here may or may not be appropriate for this type of missingness. Therefore, we must make a good effort to understanding the nature of the missing data prior to implementing any of the techniques described below.

This chapter will illustrate ways for assessing the nature and severity of missing values in the data, highlight models that can be used when missing values are present, and review techniques for removing or imputing missing data.

For illustrations, we will use the Chicago train ridership data (Chapter 4) and the scat data from Reid (2015) (previously seen in Section 6.3.3). The latter data set contains information on animal droppings that were collected in the wild. A variety of measurements were made on each sample including morphological observations (i.e., shape), location/time information, and laboratory tests. The DNA in each sample was genotyped to determine the species of the sample (gray fox, coyote, or bobcat). The goal for these data is to build a predictive relationship between the scat measurements and the species. After gathering a new scat sample, the model would be used to predict the species that produced the sample. Out of the 110 collected scat samples, 19 had one or more missing predictor values.

8.1 Understanding the Nature and Severity of Missing Information

As illustrated throughout this book, visualizing data are an important tool for guiding us towards implementing appropriate feature engineering techniques. The same principle holds for understanding the nature and severity of missing information throughout the data. Visualizations as well as numeric summaries are the first step in grasping the challenge of missing information in a data set. For small to moderate data (100s of samples and 100s of predictors), several techniques are available that allow the visualization of all of the samples and predictors simultaneously for understanding the degree and location of missing predictor values. When the number of samples or predictors become large, the missing information must first be appropriately condensed and then visualized. In addition to visual summaries, numeric summaries are a valuable tool for diagnosing the nature and degree of missingness.

Visualizing Missing Information

When the training set has a moderate number of samples and predictors, a simple method to visualize the missing information is with a *heatmap*. In this visualization the figure has two colors representing the present and missing values. The top of Figure 8.1 illustrates a heatmap of missing information for the animal scat data. The heatmap visualization can be reorganized to highlight commonalities across missing predictors and samples using hierarchical cluster analysis on the predictors and on the samples (Dillon and Goldstein, 1984). This visualization illustrates that most predictors and samples have complete or nearly complete information. The three morphological predictors (diameter, taper, and taper index) are more frequently missing. Also, two samples are missing all three of the laboratory measurements (d13N, d15N, and CN).

A *co-occurrence* plot can further deepen the understanding of missing information. This type of plot displays the frequency of missing predictor combinations (Figure 8.1). As the name suggests, a co-occurrence plot displays the frequencies of the most common missing predictor combinations. The figure makes it easy to see that TI, Tapper, and Diameter have the most missing values, that Tapper and TI are missing together most frequently, and that six samples are missing all three of the morphological predictors. The bottom of Figure 8.1 displays the co-occurrence plot for the scat data.

In addition to exploring the global nature of missing information, it is wise to explore relationships within the data that might be related to missingness. When exploring pairwise relationships between predictors, missing values of one predictor can be called out on the margin of the alternate axis. For example, Figure 8.2 shows the relationship between the diameter and mass of the feces. Hash marks on either axis represent samples that have an observed value for that predictor but have a missing value for the corresponding predictor. The points are colored by whether the sample was a "flat puddle that lacks other morphological traits." Since the missing diameter measurements have this unpalatable quality, it makes sense that some shape attributes cannot be determined. This can be considered to be *structurally missing*, but there still may be a random or informative component to the missingness.

However, in relation to the outcome, the six flat samples were spread across the gray fox ($n = 5$) and coyote ($n = 1$) data. Given the small sample size, it is unclear if this missingness is related to the outcome (which would be problematic). Subject-matter experts would need to be consulted to understand if this is the case.

When the data has a large number of samples or predictors, using heatmaps or co-occurrence plots is much less effective for understanding missingness patterns

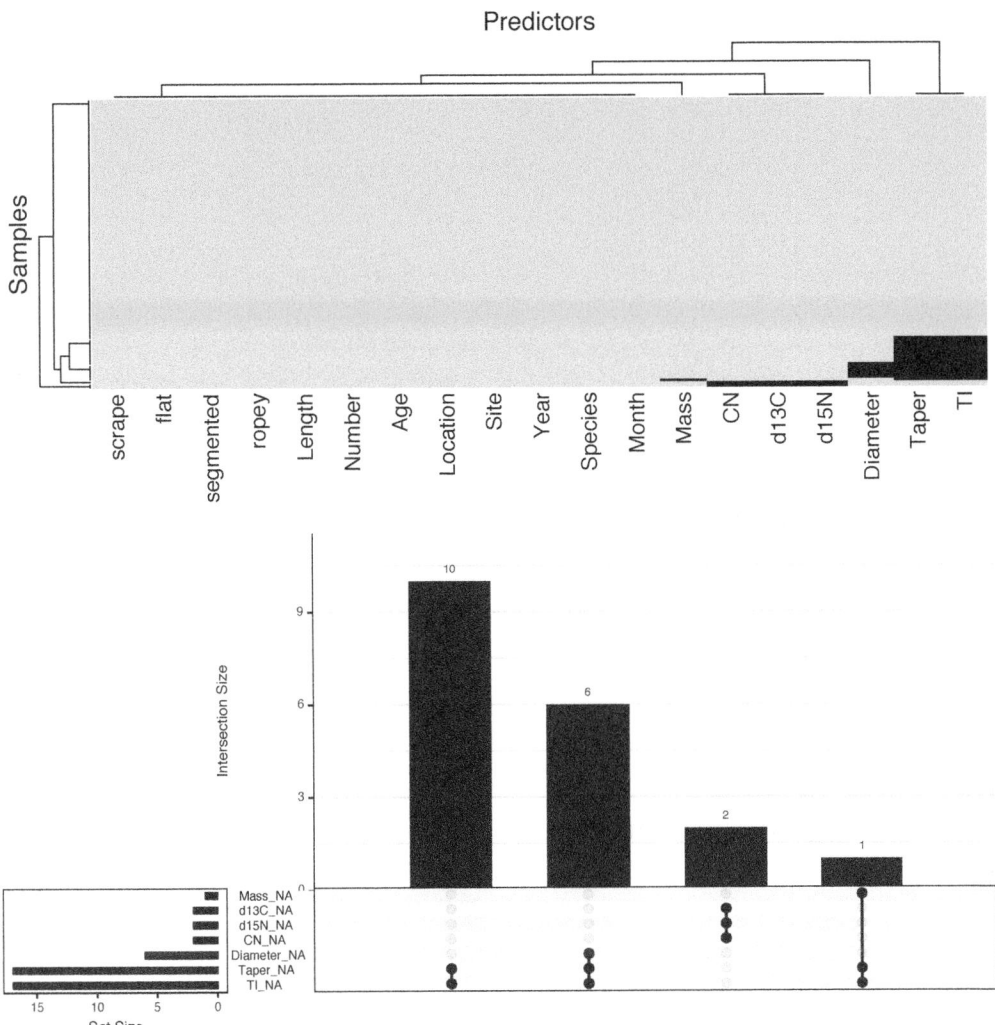

Figure 8.1: Visualizations of missing data patters than examine the entire data set (top) or the co-occurrence of missing data across variables (bottom).

and characteristics. In this setting, the missing information must first be condensed prior to visualization. Principal component analysis (PCA) was first illustrated in Section 6.3 as a dimension reduction technique. It turns out that PCA can also be used to visualize and understand the samples and predictors that have problematic levels of missing information. To use PCA in this setting, the predictor matrix is converted to a binary matrix where a zero represents a non-missing value and a one represents a missing value. Because PCA searches for directions that summarize maximal variability, the initial dimensions capture variation caused by the presence of missing values. Samples that do not have any missing values will be projected onto the same location close to the origin. Alternatively, samples with missing values will be projected away from the origin. Therefore, by simply plotting the scores of the

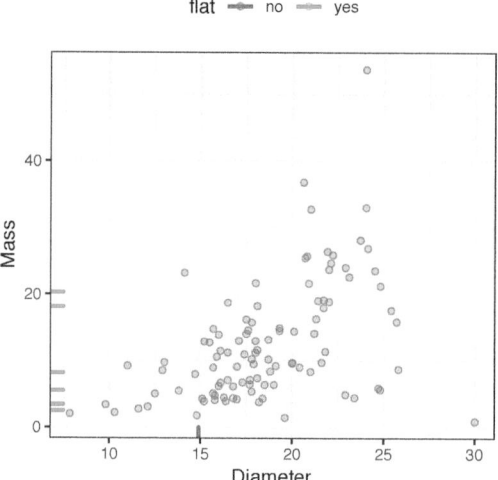

Figure 8.2: A scatter plot of two predictors colored by the value of a third. The rugs on the two axes show where that variable's values occur when the predictor on the other axis is missing.

first two dimensions, we can begin to identify the nature and degree of missingness in large data sets.

The same approach can be applied to understanding the degree of missingness among the predictors. To do this, the binary matrix representing missing values is first transposed so that predictors are now in the rows and the samples are in the columns. PCA is then applied to this matrix and the resulting dimensions now capture variation cause by the presence of missing values across the predictors.

To illustrate this approach, we will return to the Chicago train ridership data which has ridership values for 5,733 days and 146 stations. The missing data matrix can be represented by dates and stations. Applying PCA to this matrix generates the first two components illustrated in Figure 8.3. Points in the plot are sized by the amount of missing stations. In this case each day had at least one station that had a missing value, and there were 8 distinct missing data patterns.

Further exploration found that many of the missing values occurred during September 2013. The cause of these missing values should be investigated further.

To further understand the degree and location of missing values, the data matrix can be transformed and PCA applied again. Figure 8.4 presents the scores for the stations. Here points are labeled by the amount of missing information across stations. In this case, 5 stations have a substantial amount of missing data. The root cause of missing values for these stations should be investigated to better determine an appropriate way for handing these stations.

The stations with excessive missingness and their missing data patterns over time are

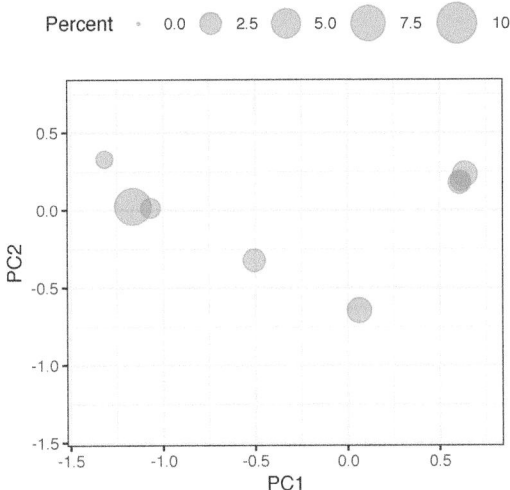

Figure 8.3: A scatter plot of the first two row scores from a binary representation of missing values for the Chicago ridership data to identify patterns of missingness over dates. The size of the symbol depicts the amount of missing data.

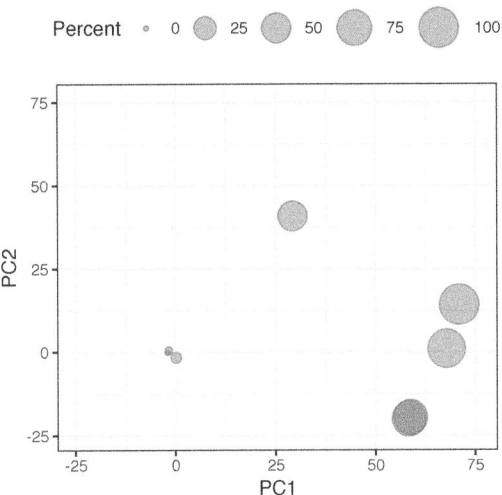

Figure 8.4: A scatter plot of the first two column scores from a binary representation of missing values for the Chicago ridership data to identify patterns is stations. The size of the symbol depicts the amount of missing data.

shown in Figure 8.5. The station order is set using a clustering algorithm while the x-axis is ordered by date. There are nine stations whose data are almost complete except for a single month gap. These stations are all on the Red Line and occur during the time of the Red Line Reconstruction Project that affected stations north of Cermak-Chinatown to the 95th Street station. The Homan station only has one

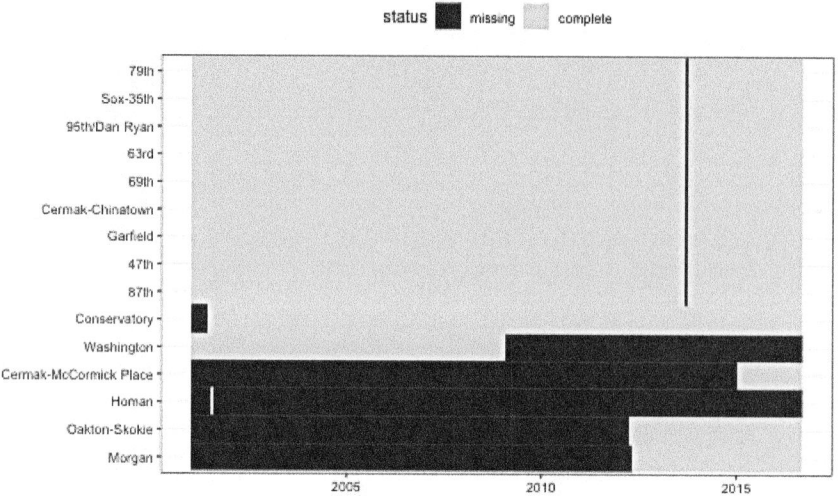

Figure 8.5: Missing data patterns for stations originally in the Chicago ridership data.

Table 8.1: Animal scat predictors by amount of missingness.

Percent Missing (%)	Predictor
0.909	Mass
1.818	d13C, d15N, and CN
5.455	Diameter
15.455	Taper and TI

month of complete data and was replaced by the Conservatory station which opened in 2001, which explains why the Conservatory station has missing data prior to 2001. Another anomaly appears for the Washington-State station. Here the data collection ceased midway through the data set because the station was closed in the mid-2000s. The further investigation of the missing data indicates that these patterns are unrelated to ridership and are mostly due to structural causes.

Summarizing Missing Information

Simple numerical summaries are effective at identifying problematic predictors and samples when the data become too large to visually inspect. On a first pass, the total number or percent of missing values for both predictors and samples can be easily computed. Returning to the animal scat data, Table 8.1 presents the predictors by amount of missing values. Similarly, Table 8.2 contains the samples by amount of missing values. These summaries can then be used for investigating the underlying reasons why values are missing or as a screening mechanism for removing predictors or samples with egregious amounts of missing values.

Table 8.2: Animal scat samples by amount of missingness.

Percent Missing (%)	Sample Numbers
10.5	51, 68, 69, 70, 71, 72, 73, 75, 76, and 86
15.8	11, 13, 14, 15, 29, 60, 67, 80, and 95

8.2 | Models that Are Resistant to Missing Values

Many popular predictive models such as support vector machines, the glmnet, and neural networks, cannot tolerate any amount of missing values. However, there are a few predictive models that can internally handle incomplete data.[71] Certain implementations of tree-based models have clever procedures to accommodate incomplete data. The CART methodology (Breiman et al., 1984) uses the idea of *surrogate splits*. When creating a tree, a separate set of splits are cataloged (using alternate predictors than the current predictor being split) that can approximate the original split logic if that predictor value is missing. Figure 8.6 displays the recursive partitioning model for the animal scat data. All three predictors selected by the tree contain missing values as illustrated in Figure 8.1. The initial split is based on the carbon/nitrogen ratio (CN < 8.7). When a sample has a missing value for CN, then the CART model uses an alternative split based on the indicator for whether the scat was flat or not. These two splits result in the same partitions for 80.6% of the data of the data and could be used if the carbon/nitrogen ratio is missing. Moving further down the tree, the surrogate predictors for d13C and Mass are Mass and d13C, respectively. This is possible since these predictors are not simultaneously missing.

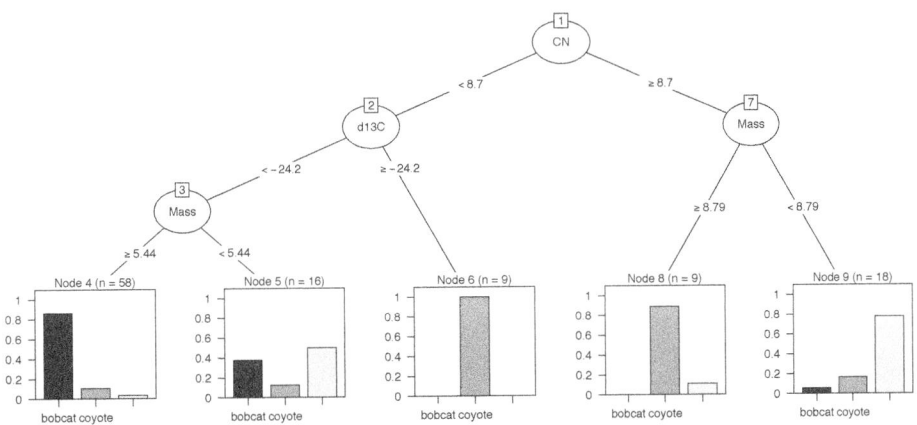

Figure 8.6: A recursive partitioning tree for the animal scat data. All three predictors in this tree contain missing values but can be used due to surrogate predictors.

[71]This is true provided that the mechanism that causes the missing data are not pathological.

C5.0 (Quinlan, 1993; Kuhn and Johnson, 2013) takes a different approach. Based on the distribution of the predictor with missing data, fractional counts are used in subsequent splits. For example, this model's first split for the scat data is `d13C > 24.55`. When this statement is true, all 13 training set samples are coyotes. However, there is a single missing value for this predictor. The counts of each species in subsequent nodes are then *fractional* due to adjusting for the number of missing values for the split variable. This allows the model to keep a running account of where the missing values might have landed in the partitioning.

Another method that can tolerate missing data is Naive Bayes. This method models the class-specific distributions of each predictor *separately*. Again, if the missing data mechanism is not pathological, then these distribution summaries can use the complete observations for each individual predictor and will avoid case-wise deletion.

8.3 | Deletion of Data

When it is desirable to use models that are intolerant to missing data, then the missing values must be extricated from the data. The simplest approach for dealing with missing values is to remove entire predictor(s) and/or sample(s) that contain missing values. However, one must carefully consider a number of aspects of the data prior to taking this approach. For example, missing values could be eliminated by removing all predictors that contain at least one missing value. Similarly, missing values could be eliminated by removing all samples with any missing values. Neither of these approaches will be appropriate for all data as can be inferred from the "No Free Lunch" theorem. For some data sets, it may be true that particular predictors are much more problematic than others; by removing these predictors, the missing data issue is resolved. For other data sets, it may be true that specific samples have consistently missing values across the predictors; by removing these samples, the missing data issue is likewise resolved. In practice, however, specific predictors *and* specific samples contain a majority of the missing information.

Another important consideration is the intrinsic value of samples as compared to predictors. When it is difficult to obtain samples or when the data contain a small number of samples (i.e., rows), then it is not desirable to remove samples from the data. In general, samples are more critical than predictors and a higher priority should be placed on keeping as many as possible. Given that a higher priority is usually placed on samples, an initial strategy would be to first identify and remove predictors that have a sufficiently high amount of missing data. Of course, predictor(s) in question that are known to be valuable and/or predictive of the outcome should not be removed. Once the problematic predictors are removed, then attention can focus on samples that surpass a threshold of missingness.

Consider the Chicago train ridership data as an example. The investigation into these data earlier in the chapter revealed that several of the stations had a vast amount of contiguous missing data (Figure 8.5). Every date (sample) had at least one missing station (predictor), but the degree of missingness across predictors was minimal. However, a handful of stations had excessive missingness. For these stations there is too much contiguous missing data to keep the stations in the analysis set. In this case it would make more sense to remove the handful of stations. It is less clear for the stations on the Red Line as to whether or not to keep these in the analysis sets. Given the high degree of correlation between stations, it was decided that excluding all stations with missing data from all of the analyses in this text would not be detrimental to building an effective model for predicting ridership at the Clark-Lake station. This example illustrates that there are several important aspects to consider when determining the approach for handling missing values.

Besides throwing away data, the main concern with removing samples (rows) of the training set is that it might *bias* the model that relates the predictors to the outcome. A classic example stems from medical studies. In these studies, a subset of patients is randomly assigned to the current standard of care treatment while another subset of patients is assigned to a new treatment. The new treatment may induce an adverse effect for some patients, causing them to drop out of the study thus inducing missing data for future clinical visits. This kind of missing data is clearly not missing at random and the elimination of these data from an analysis would falsely measure the outcome to be better than it would have if their unfavorable results had been included. That said, Allison (2001) notes that if the data are missing completely at random, this may be a viable approach.

8.4 | Encoding Missingness

When a predictor is discrete in nature, missingness can be directly encoded into the predictor as if it were a naturally occurring category. This makes sense for structurally missing values such as the example of alleys in the Ames housing data. Here, it is sensible to change the missing values to a category of "no alley." In other cases, the missing values could simply be encoded as "missing" or "unknown." For example, Kuhn and Johnson (2013) use a data set where the goal is to predict the acceptance or rejection of grant proposals. One of the categorical predictors was grant sponsor which took values such as "Australian competitive grants", "cooperative research centre", "industry", etc.. In total, there were more than 245 possible values for this predictor with roughly 10% of the grant applications having an empty sponsor value. To enable the applications that had an empty sponsor to be used in modeling, empty sponsor values were encoded as "unknown". For many of the models that were investigated, the indicator for an unknown sponsor was one of the most important predictors of grant success. In fact, the odds-ratio that contrasted known versus unknown sponsor was greater than 6. This means that it was much more likely for a grant to be successfully funded if the sponsor predictor was unknown. In fact, in the training set the grant success rate associated with an unknown sponsor was 82.2% versus 42.1% for a known sponsor.

Was encoding the missing values a successful strategy? Clearly, the mechanism that led to the missing sponsor label being identified as strongly associated with grant acceptance was genuinely important. Unfortunately, it is impossible to know *why* this association is so important. The fact that *something* was going on here is important and this encoding helped identify its occurrence. However, it would be troublesome to accept this analysis as final and imply some sort of cause-and-effect relationship.[72] A guiding principle that can be used to determine if encoding missingness is a good idea is to think about how the results would be interpreted if that piece of information becomes important to the model.

8.5   Imputation Methods

Another approach to handling missing values is to *impute* or estimate them. Missing value imputation has a long history in statistics and has been thoroughly researched. Good places to start are Little and Rubin (2014), Van Buuren (2012) and Allison (2001). In essence, imputation uses information and relationships among the non-missing predictors to provide an estimate to fill in the missing value.

Historically, statistical methods for missing data have been concerned with the impact on *inferential models*. In this situation, the characteristics and quality of the imputation strategy have focused on the test statistics that are produced by the model. The goal of these techniques is to ensure that the statistical distributions are tractable and of good enough quality to support subsequent hypothesis testing. The primary approach in this scenario is to use multiple imputations; several variations of the data set are created with different estimates of the missing values. The variations of the data sets are then used as inputs to models and the test statistic replicates are computed for each imputed data set. From these replicate statistics, appropriate hypothesis tests can be constructed and used for decision making.

There are several differences between inferential and predictive models that impact this process:

- In many predictive models, there is no notion of distributional assumptions (or they are often intractable). For example, when constructing *most* tree-based models, the algorithm does not require any specification of a probability distribution to the predictors. As such, many predictive models are incapable of producing inferential results even if that were a primary goal.[73] Given this, traditional multiple imputation methods may not have relevance for these models.

- Many predictive models are computationally expensive. Repeated imputation would greatly exacerbate computation time and overhead. However, there is an interest in capturing the benefit (or detriment) caused by an imputation procedure. To ensure that the variation imparted by the imputation is captured in the training process, we recommend that imputation be performed within the resampling process.

- Since predictive models are judged on their ability to accurately predict yet-to-be-seen samples (including the test set and new unknown samples), as opposed

---

[72] A perhaps more difficult situation would be explaining to the consumers of the model that "We know that this is important but we don't know why!"

[73] Obviously, there are exceptions such as linear and logistic regression, Naive Bayes models, etc.

to statistical appropriateness, it is critical that the imputed values be as close as possible to their true (unobserved) values.

- The general focus of inferential models is to thoroughly understand the relationships between the predictor and response for the available data. Conversely, the focus of predictive models is to understand the relationships between the predictors and the response that are generalizable to yet-to-be-seen samples. Multiple imputation methods do not keep the imputation generator after the missing data have been estimated which provides a challenge for applying these techniques to new samples.

The last point underscores the main objective of imputation with machine learning models: produce the most accurate prediction of the missing data point. Some other important characteristics that a predictive imputation method should have are:

- Within a sample data point, other variables may also be missing. For this reason, an imputation method should be tolerant of other missing data.

- Imputation creates a model embedded within another model. There is a prediction equation associated with every predictor in the training set that might have missing data. It is desirable for the imputation method to be fast and have a compact prediction equation.

- Many data sets often contain both numeric and qualitative predictors. Rather than generating dummy variables for qualitative predictors, a useful imputation method would be able to use predictors of various types as inputs.

- The model for predicting missing values should be relatively (numerically) stable and not be overly influenced by outlying data points.

Virtually any predictive model could be used to impute the data. Here, the focus will be on several that are good candidates to consider.

Imputation does beg the question of how much missing data are too much to impute? Although not a general rule in any sense, 20% missing data within a column might be a good "line of dignity" to observe. Of course, this depends on the situation and the patterns of missing values in the training set.

It is also important to consider that imputation is probably the first step in any preprocessing sequence. Imputing qualitative predictors prior to creating indicator variables so that the binary nature of the resulting imputations can be preserved is a good idea. Also, imputation should usually occur prior to other steps that involve parameter estimation. For example, if centering and scaling is performed on data prior to imputation, the resulting means and standard deviations will inherit the potential biases and issues incurred from data deletion.

K-Nearest Neighbors

When the training set is small or moderate in size, K-nearest neighbors can be a quick and effective method for imputing missing values (Eskelson et al., 2009; Tutz and Ramzan, 2015). This procedure identifies a sample with one or more missing values. Then it identifies the K most similar samples in the training data that are complete (i.e., have no missing values in some columns). Similarity of samples for this method is defined by a distance metric. When all of the predictors are numeric, standard Euclidean distance is commonly used as the similarity metric. After computing the distances, the K closest samples to the sample with the missing value are identified and the average value of the predictor of interest is calculated. This value is then used to replace the missing value of the sample.

Often, however, data sets contain both numeric and categorical predictors. If this is the case, then Euclidean distance is not an appropriate metric. Instead, Gower's distance is a good alternative (Gower, 1971). This metric uses a separate specialized distance metric for both the qualitative and quantitative predictors. For a qualitative predictor, the distance between two samples is 1 if the samples have the same value and 0 otherwise. For a quantitative predictor x, the distance between samples i and j is defined as

$$d(x_i, x_j) = 1 - \frac{|x_i - x_j|}{R_x}$$

where R_x is the range of the predictor. The distance measure is computed for each predictor and the average distance is used as the overall distance. Once the K neighbors are found, their values are used to impute the missing data. The mode is used to impute qualitative predictors and the average or median is used to impute quantitative predictors. K can be a tunable parameter, but values around 5–10 are a sensible default.

For the animal scat data, Figure 8.7 shows the same data from Figure 8.2 but with the missing values filled in using 5 neighbors based on Gower's distance. The new values mostly fall around the periphery of these two dimensions but are within the range of the samples with complete data.

Trees

Tree-based models are a reasonable choice for an imputation technique since a tree can be constructed in the presence of other missing data[74]. In addition, trees generally have good accuracy and will not extrapolate values beyond the bounds of the training data. While a single tree could be used as an imputation technique, it is known to produce results that have low bias but high variance. Ensembles of trees, however, provide a low-variance alternative. Random forests is one such technique and has been studied for this purpose (Stekhoven and Buhlmann, 2011). However, there are a couple of notable drawbacks when using this technique in a predictive modeling setting. First and foremost, the random selection of predictors at each split necessitates a large number of trees (500 to 2000) to achieve a stable, reliable model. Each of these trees is unpruned and the resulting model usually has a large

---

[74]However, many implementations do not have this property.

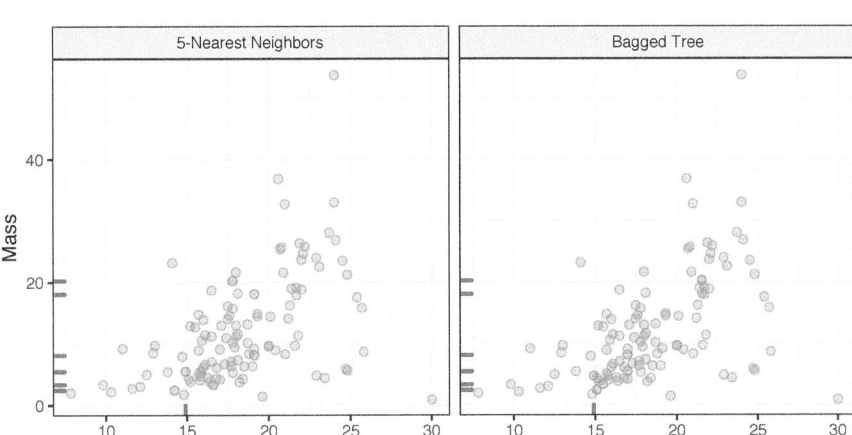

Figure 8.7: A comparison of the K-Nearest Neighbors and bagged tree imputation techniques for the animal scat data. The rugs on the two axes show where that variable's values occurred when the predictor on the other axis was missing.

computational footprint. This can present a challenge as the number of predictors with missing data increases since a separate model must be built and retained for each predictor. A good alternative that has a smaller computational footprint is a bagged tree. A bagged tree is constructed in a similar fashion to a random forest. The primary difference is that in a bagged model, all predictors are evaluated at each split in each tree. The performance of a bagged tree using 25–50 trees is generally in the ballpark of the performance of a random forest model. And the smaller number of trees is a clear advantage when the goal is to find reasonable imputed values for missing data.

Figure 8.7 illustrates the imputed values for the scat data using a bagged ensemble of 50 trees. When imputing the diameter predictor was the goal, the estimated RMSE of the model was 4.16 (or an R^2 of 13.6%). Similarly, when imputing the mass predictor was the goal, the estimated RMSE was 8.56 (or an R^2 of 28.5%). The RMSE and R^2 metrics are not terribly impressive. However, these imputations from the bagged models produce results that are reasonable, are within the range of the training data, and allow the predictors to be retained for modeling (as opposed to case-wise deletion).

Linear Methods

When a complete predictor shows a strong linear relationship with a predictor that requires imputation, a straightforward linear model may be the best approach. Linear models can be computed very quickly and have very little retained overhead. While a linear model does require complete predictors for the imputation model, this is not a fatal flaw since the model coefficients (i.e., slopes) use all of the data for estimation.

Linear regression can be used for a numeric predictor that requires imputation. Similarly, logistic regression is appropriate for a categorical predictor that requires imputation.

The Chicago train ridership data will be used to illustrate the usefulness of a simple technique like linear regression for imputation. It is conceivable that one or more future ridership values might be missing in random ways from the complete data set. To simulate potential missing data, missing values have been induced for ridership at the Halsted stop. As illustrated previously in Figure 4.8, the 14-day lag in ridership within a stop is highly correlated with the current day's ridership. Figure 8.8(a) displays the relationship between these predictors for the Halsted stop in 2010 with the missing values of the current day highlighted on the x-axis. The majority of the data displays a linear relationship between these predictors, with a handful of days having values that are away from the overall trend. The fit of a robust linear model to these data is represented by the dashed line. Figure 8.8(b) compares the imputed values with the original true ridership values. The robust linear model performs adequately at imputing the missing values, with the exception of one day. This day had unusual ridership because it was a holiday. Including holiday as a predictor in the robust model would help improve the imputation.

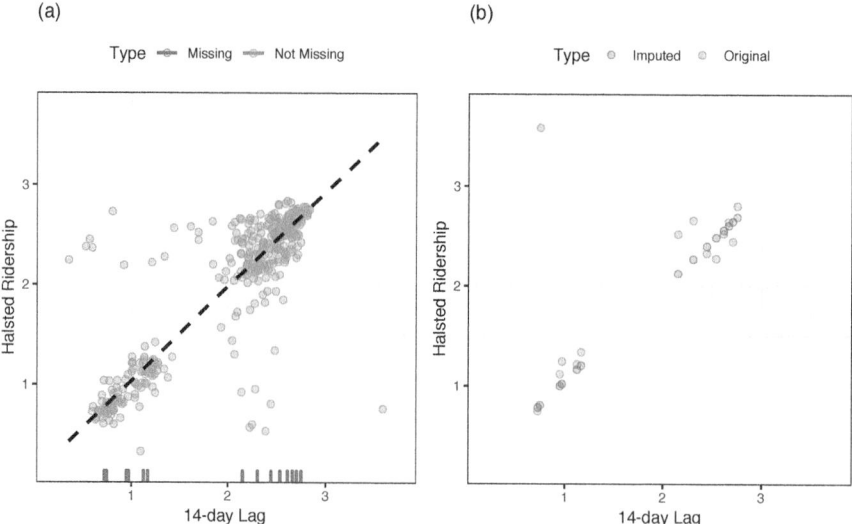

Figure 8.8: An example of linear imputation for ridership at the Halsted stop. (a) 14-day lag ridership versus current-day ridership. (b) Imputed values based on a robust linear model compared to original values.

The concept of linear imputation can be extended to high-dimensional data. Audigier et al. (2016), for example, have developed methodology to impute missing data based on similarity measures and principal components analysis.

8.6 | Special Cases

There are situations where a data point isn't missing but is also *not complete*. For example, when measuring the duration of time until an event, it might be known that the duration is *at least* some time T (since the event has not occurred). These types of values are referred to as *censored*.[75] A variety of statistical methods have been developed to analyze this type of data.

Durations are often *right* censored since the terminating value is not known. In other cases, *left* censoring can occur. For example, laboratory measurements may have a *lower limit of detection*, which means that the measuring instrument cannot reliably quantify values below a threshold X. This threshold is usually determined experimentally in the course of developing the measurement system. When a predictor has values below the lower limit of detection, these values are usually reported as "$< X$". When these data are to be included in a predictive model, there is usually a question of how to handle the censored values. A widely accepted practice is to use the lower limit value of X as the result. While this would not adversely affect some partitioning models, such as trees or rules, it may have a detrimental impact on other models since it assumes that these are the true values. Censored values affect metrics that measure the variability. Specifically, the variability will be underestimated. This effect is similar in a manner to binning in Section 6.2.2, and the model may overfit to a cluster of data points with the same value.

To mitigate the variability issue, left censored values can be imputed using random uniform values between zero and X. In cases where there is good information about the distribution below X, other random value assignment schemes can be used that better represent the distribution. For example, there may be some scientific or physiological reason that the smallest possible value is greater than zero (but less than X). While imputing in this fashion adds random noise to the data, it is likely to be preferable to the potential overfitting issues that can occur by assigning a value of X to the data.

Another atypical situation occurs when the data have a strong time component. In this case, to preserve these characteristics of the data, simple moving average smoothers can be used to impute the data so that any temporal effects are not disrupted. As previously mentioned in Section 6.1, care must be taken on the ends so that the test (or other) data are not used to impute the training set values.

8.7 | Summary

Missing values are common occurrences in data. Unfortunately, most predictive modeling techniques cannot handle any missing values. Therefore, this problem must be addressed prior to modeling. Missing data may occur due to random chance or

---

[75] In cases where the data are not defined outside of lower or upper bounds, the data would be considered *truncated*.

due to a systematic cause. Understanding the nature of the missing values can help to guide the decision process about how best to remove or impute the data.

One of the best ways to understand the amount and nature of missing values is through an appropriate visualization. For smaller data sets, a heatmap or co-occurrence plot is helpful. Larger data sets can be visualized by plotting the first two scores from a PCA model of the missing data indicator matrix.

Once the severity of missing values is known, then a decision needs to be made about how to address these values. When there is a severe amount of missing data within a predictor or a sample, it may be prudent to remove the problematic predictors or samples. Alternatively, if the predictors are qualitative, missing values could be encoded with a category of "missing".

For small to moderate amounts of missingness, the values can be imputed. There are many types of imputation techniques. A few that are particularly useful when the ultimate goal is to build a predictive model are k-nearest neighbors, bagged trees, and linear models. Each of these methods have a relatively small computation footprint and can be computed quickly which enables them to be included in the resampling process of predictive modeling.

8.8 | Computing

The website `http://bit.ly/fes-missing` contains R programs for reproducing these analyses.

9 | Working with Profile Data

Let's review the previous data sets in terms of what has been predicted:

- the price of a house in Iowa
- daily train ridership at Clark and Lake
- the species of a feces sample collected on a trail
- a patient's probability of having a stroke

From these examples, the *unit of prediction* is fairly easy to determine because the data structures are simple. For the housing data, we know the location, structural characteristics, last sale price, and so on. The data structure is straightforward: rows are houses and columns are fields describing them. There is one house per row and, for the most part, we can assume that these houses are statistically independent of one another. In statistical terms, it would be said that the houses are the *independent experimental unit* of data. Since we make predictions on the properties (as opposed to the ZIP Code or other levels of aggregation), the properties are the *unit of prediction*.

This may not always be the case. As a slight departure, consider the Chicago train data. The data set has rows corresponding to specific dates and columns are characteristics of these days: holiday indicators, ridership at other stations a week prior, and so on. Predictions are made daily; this is the *unit of prediction*. However, recall they there are also weather measurements. The weather data was obtained multiple times per day, usually hourly. For example, on January 5, 2001 the first 15 measurements are shown in Table 9.1. These hourly data are not on the unit of prediction (daily). We would expect that, within a day, the hourly measurements are more correlated with each other than they would be with any other random weather measurements taken at random from the entire data set. For this reason, these measurements are statistically dependent.[76]

Since the goal is to make daily predictions, the *profile* of within-day weather measurements should be somehow summarized at the day level in a manner that preserves the potential predictive information. For this example, daily features could include the mean or median of the numeric data and perhaps the range of values within a

---

[76]However, to be fair, the day-to-day measurements have some degree of correlation with one another too.

Table 9.1: An example of hourly weather data in Chicago.

Time	Temp	Humidity	Wind	Conditions
00:53	27.0	92	21.9	Overcast
01:53	28.0	85	20.7	Overcast
02:53	27.0	85	18.4	Overcast
03:53	28.0	85	15.0	Mostly Cloudy
04:53	28.9	82	13.8	Overcast
05:53	32.0	79	15.0	Overcast
06:53	33.1	78	17.3	Overcast
07:53	34.0	79	15.0	Overcast
08:53	34.0	61	13.8	Scattered Clouds
09:53	33.1	82	16.1	Clear
10:53	34.0	75	12.7	Clear
11:53	33.1	78	11.5	Clear
12:53	33.1	78	11.5	Clear
13:53	32.0	79	11.5	Clear
14:53	32.0	79	12.7	Clear

day. Additionally, the rate of change of the features using a slope could be used as a summary. For the qualitative conditions, the percentage of the day that was listed as "Clear", "Overcast", etc. can be calculated so that weather conditions for a specific *day* are incorporated into the data set used for analysis.

As another example, suppose that the stroke data were collected when a patient was hospitalized. If the study contained longitudinal measurements over time we might be interested in predicting the probability of a stroke *over time*. In this case, the unit of prediction is the patient *on a given day* but the *independent experimental unit* is just the patient.

In some cases, there can be multiple hierarchical structures. For online education, there would be interest in predicting what courses a particular student could complete successfully. For example, suppose there were 10,000 students that took the online "Data Science 101" class (labeled DS101). Some were successful and others were not. These data can be used to create an appropriate model *at the student level*. To engineer features for the model, the data structures should be considered. For students, possible predictors might be demographic (e.g., age, etc.) while others would be related to their previous experiences with other classes. If they had successfully completed 10 other courses, it is reasonable to think that they would be more likely to do well than if they did not finish 10 courses. For students who took DS101, a more detailed data set might look like:

Student	Course	Section	Assignment	Start_Time	Stop_Time
Marci	STAT203	1	A	07:36	08:13
			B	13:38	18:01
:	:	:	:	:	:
Marci	CS501	4	Z	10:40	10:59
David	STAT203	1	A	18:26	19:05
:	:	:	:	:	:

The hierarchical structure is *assignment-within-section-within-course-within-student*. If both STAT203 and CS501 are prerequisites for DS101, it would be safe to assume that a relatively complete data set could exist. If that is the case, then very specific features could be created such as, "Is the time to complete the assignment on computing a t-test in STAT203 predictive of their ability to pass or complete the data science course?" Some features could be rolled-up at the course level (e.g., total time to complete course CS501) or at the section level. The richness of this hierarchical structure enables some interesting features in the model but it does pose the question, "Are there good and bad ways of summarizing profile data?"

Unfortunately, the answer to this question depends on the nature of the problem and the profile data. However, there are some aspects that can be considered regarding the nature of variation and correlation in the data.

High content screening (HCS) data in biological sciences is another example (Giuliano et al., 1997; Bickle, 2010; Zanella et al., 2010). To obtain this data, scientists treat cells using a drug and then measure various characteristics of the cells. Examples are the overall shape of the cell, the size of the nucleus, and the amount of a particular protein that is outside the nucleus.

HCS experiments are usually carried out on microtiter plates. For example, the smallest plate has 96 wells that can contain different liquid samples (usually in a 8 by 12 arrangement, such as in Figure 9.1). A sample of cells would be added to each well along with some treatment (such as a drug candidate). After treatment, the cells are imaged using a microscope and a set of images are taken within each well, usually at specific locations (Holmes and Huber, 2019). Let's say that, on average, there are 200 cells in an image and 5 images are taken per well. It is usually safe to pool the images within a cell so that the data structure is then *cell within well within plate*. Each cell is quantified using image analysis so that the underlying features are at the *cellular* level.

However, to make predictions on a drug, the data needs to be summarized to the *wellular* level (i.e., within the well). One method to do this would be to take simple averages and standard deviations of the cell properties, such as the average nucleus size, the average protein abundance, etc. The problem with this approach is that it would break the *correlation structure of the cellular data*. It is common for cell-level features to be highly correlated. As an obvious example, the abundance of a protein in a cell is correlated with the total size of the cell. Suppose that cellular

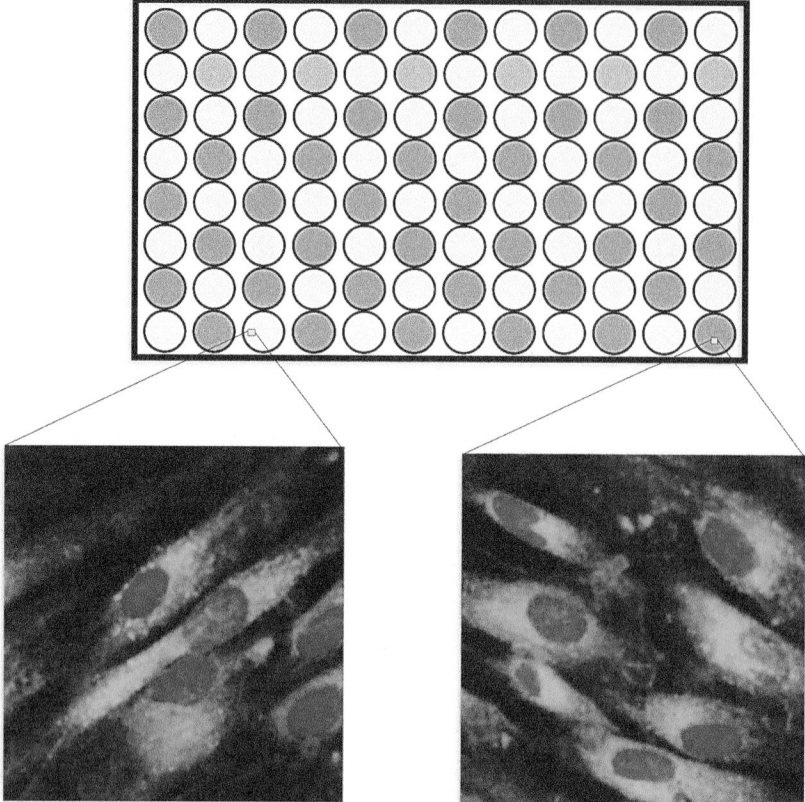

Figure 9.1: A schematic for the micro-titer plates in high content screening. The dark wells might be treated cells while the lightly colored wells would be controls. Multiple images are taken within each well and the cells in each image are isolated and quantified in different ways. The green represents the cell boundary and the blue describes the nucleus.

features X and Y are correlated. If an important feature of these two values is their difference, then there are two ways of performing the calculation. The means, $\bar{X}$ and $\bar{Y}$, could be computed within the well, then the difference between the averages could be computed. Alternatively, the differences could be computed for each cell ($D_i = X_i - Y_i$) and then these could be averaged (i.e., $\bar{D}$). Does it matter? If X and Y are highly correlated, it makes a big difference. From probability theory, we know that the variance of a difference is:

$$Var[X - Y] = Var[X] + Var[Y] - 2Cov[X, Y].$$

If the two cellular features are correlated, the covariance can be large. This can result in a massive reduction of variance if the features are differenced and then averaged at the wellular level. As noted by Wickham and Grolemund (2016):

> "If you think of variation as a phenomenon that creates uncertainty, covariation is a phenomenon that reduces it."

Taking the averages and then differences ignores the covariance term and results in noisier features.

Complex data structures are very prevalent in medical studies. Modern measurement devices such as actigraphy monitors and magnetic resonance imaging (MRI) scanners can generate hundreds of measurements per subject in a very short amount of time. Similarly, medical devices such as an electroencephalogram (EEG) or electrocardiogram (ECG) can provide a dense, continuous stream of measurements related to a patient's brain or heart activity. These measuring devices are being increasingly incorporated into medical studies in an attempt to identify important signals that are related to the outcome. For example, Sathyanarayana et al. (2016) used actigraphy data to predict sleep quality, and Chato and Latifi (2017) examined the relationship between MRI images and survival in brain cancer patients.

While medical studies are more frequently utilizing devices that can acquire dense, continuous measurements, many other fields of study are collecting similar types of data. In the field of equipment maintenance, Caterpillar Inc. collects continuous, real-time data on deployed machines to recommend changes that better optimize performance (Marr, 2017).

To summarize, some complex data sets can have multiple levels of data hierarchies which might need to be collapsed so that the modeling data set is consistent with the unit of prediction. There are good and bad ways of summarizing the data, but the tactics are usually subject-specific.

The remainder of this chapter is an extended case study where the unit of prediction has multiple layers of data. It is a scientific data set where there are a large number of highly correlated predictors, a separate time effect, and other factors. This example, like Figure 1.7, demonstrates that good preprocessing and feature engineering can have a larger impact on the results than the type of model being used.

9.1 | Illustrative Data: Pharmaceutical Manufacturing Monitoring

Pharmaceutical companies use spectroscopy measurements to assess critical process parameters during the manufacturing of a biological drug (Berry et al., 2015). Models built on this process can be used with real-time data to recommend changes that can increase product yield. In the example that follows, Raman spectroscopy was used to generate the data[77] (Hammes, 2005). To manufacture the drug being used for this example, a specific type of protein is required and that protein can be created by a particular type of cell. A batch of cells are seeded into a *bioreactor* which is a device that is designed to help grow and maintain the cells. In production, a large bioreactor would be about 2000 liters and is used to make large quantities of proteins in about two weeks.

---

[77]The data used in this illustration was generated from real data, but has been distinctly modified to preserve confidentiality and achieve illustration purposes.

Many factors can affect product yield. For example, because the cells are living, working organisms, they need the right temperature and sufficient food (glucose) to generate drug product. During the course of their work, the cells also produce waste (ammonia). Too much of the waste product can kill the cells and reduce the overall product yield. Typically, key attributes like glucose and ammonia are monitored daily to ensure that the cells are in optimal production conditions. Samples are collected and off-line measurements are made for these key attributes. If the measurements indicate a potential problem, the manufacturing scientists overseeing the process can tweak the contents of the bioreactor to optimize the conditions for the cells.

One issue is that conventional methods for measuring glucose and ammonia are time consuming and the results may not come in time to address any issues. Spectroscopy is a potentially faster method of obtaining these results if an effective model can be used to take the results of the spectroscopy assay to make predictions on the substances of interest (i.e., glucose and ammonia).

However, it is not feasible to do experiments using many large-scale bioreactors. Two parallel experimental systems were used:

- 15 small-scale (5 liters) bioreactors were seeded with cells and were monitored daily for 14 days.

- Three large-scale bioreactors were also seeded with cells from the same batch and monitored daily for 14 days

Samples were collected each day from all bioreactors and glucose was measured using both spectroscopy and the traditional manner. Figure 9.2 illustrates the design for this experiment. The goal is to create models on the data from the more numerous small-scale bioreactors and then evaluate if these results can accurately predict what is happening in the large-scale bioreactors.

The spectra for several days within one small and one large-scale bioreactor are illustrated in Figure 9.3(a). These spectra exhibit similar patterns in the profiles with notable peaks occurring in similar wavelength regions across days. The heights of the intensities, however, vary within the small bioreactor and between the small and large bioreactors. To get a sense for how the spectra change over the two-week period, examine Figure 9.3(b). Most of the profiles have similar overall patterns across the wavenumbers, with increasing intensity across time. This same pattern is true for the other bioreactors, with additional shifts in intensity that are unique to each bioreactor.

9.2 │ What Are the Experimental Unit and the Unit of Prediction?

Nearly 2600 wavelengths are measured each day for two weeks for each of 15 small-scale bioreactors. This type of data forms a *hierarchical* structure in which wavelengths are measured within each day and within each bioreactor. Another way

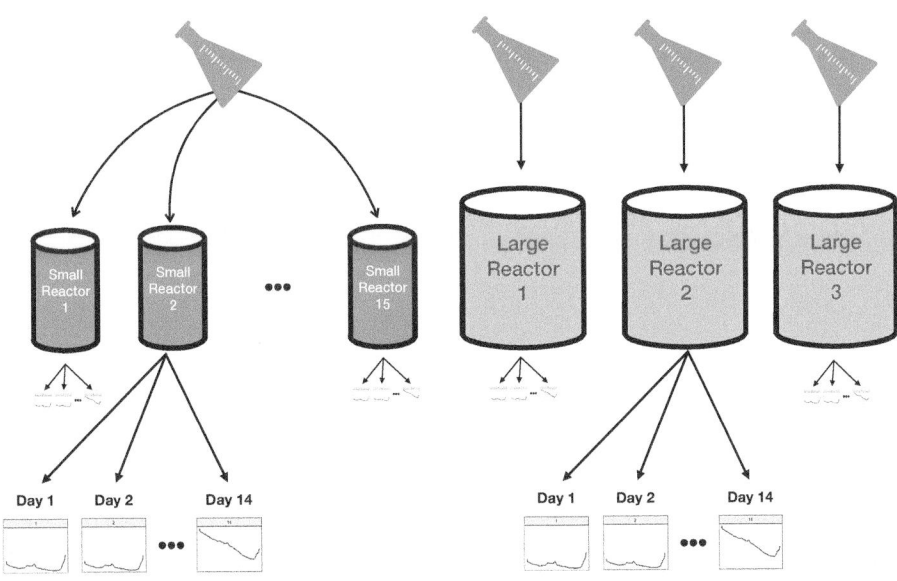

Figure 9.2: A schematic for the experimental design for pharmaceutical manufacturing.

to say this is that the wavelength measurements are nested within-day which is further nested within bioreactor. A characteristic of nested data structures is that the measurements within each nesting are more related to each other than the measurements between nestings. Here, the wavelengths within a day are more related or correlated to each other than the wavelengths between days. And the wavelengths between days are more correlated to each other than the wavelengths between bioreactors.

These intricate relationships among wavelengths within a day and bioreactor can be observed through a plot of the autocorrelations. An autocorrelation is the correlation between the original series and each sequential lagged version of the series. In general, the autocorrelation decreases as the lag value increases, and can increase at later lags with trends in seasonality. The autocorrelation plot for the wavelengths of the first bioreactor and several days are displayed in Figure 9.4 (a) (which is representative of the relationships among wavelengths across other bioreactor and day combinations). This figure indicates that the correlations between wavelengths are different across days. Later days tend to have higher between-wavelength correlations but, in all cases, it takes many hundreds of lags to get the correlations below zero.

The next level up in the hierarchy is within bioreactor across days. Autocorrelations can also be used here to understand the correlation structure at this level of the hierarchy. To create the autocorrelations, the intensities across wavelengths will be

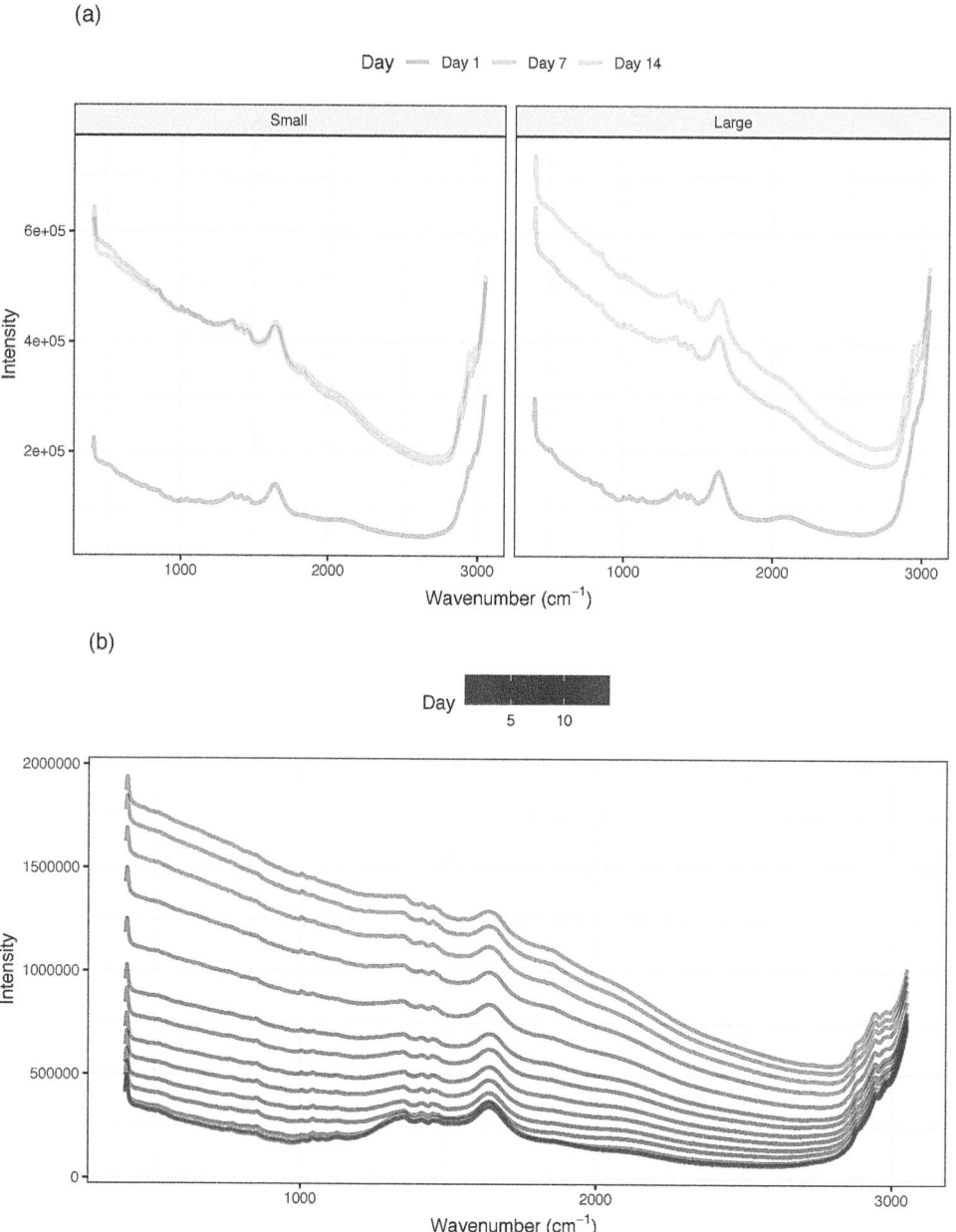

Figure 9.3: (a) Spectra for a small- and large-scale bioreactor on days 1, 7, and 14. (b) The spectra for a single reactor over 14 days.

averaged within a bioreactor and day. The average intensities will then be lagged across day and the correlations between lags will be created. Here again, we will use small-scale bioreactor 1 as a representative example. Figure 9.4 (b) shows the autocorrelations for the first 13 lagged days. The correlation for the first lag is greater than 0.95, with correlations tailing off fairly quickly.

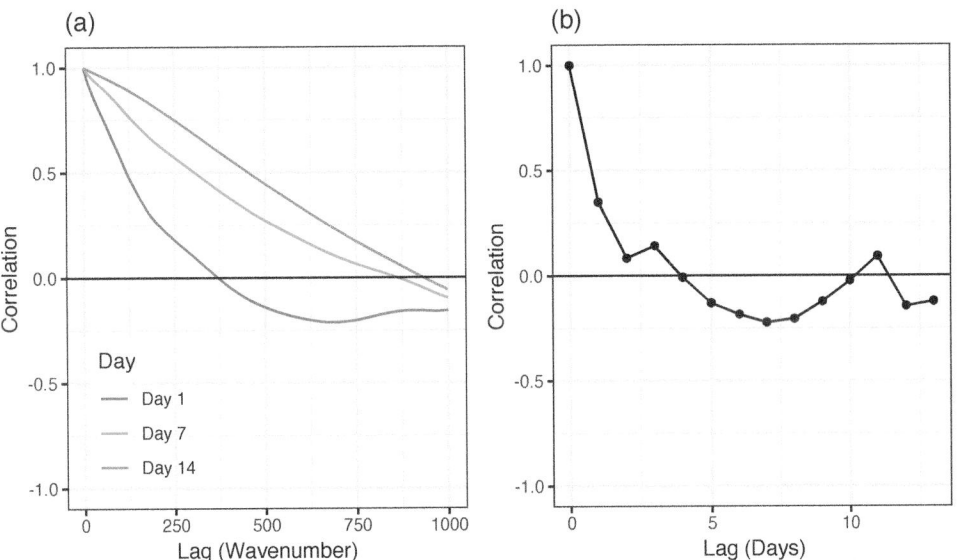

Figure 9.4: (a) Autocorrelations for selected lags of wavelengths for small-scale bioreactor 1 on the first day. (b) Autocorrelations for lags of days for average wavelength intensity for small-scale bioreactor 1.

The top-most hierarchical structure is the bioreactor. Since the reactions occurring within one bioreactor do not affect what is occurring within a different reactor, the data at these levels are independent of one another. As discussed above, data within bioreactor are likely to be correlated with one another. What is the unit of prediction? Since the spectra are all measured simultaneously, we can think of this level of the hierarchy to be *below* the unit of prediction; we would not make a prediction at a specific wavelength. The use case for the model is to make a prediction for a bioreactor for a specific number of days that the cells have been growing. For this reason, the unit of prediction is day within bioreactor.

Understanding the units of experimentation will guide the selection of the cross validation method (Section 3.4) and is crucial for getting an honest assessment of a model's predictive ability on new days. Consider, for example, if each day (within each bioreactor) was taken to be the independent experimental unit and V-fold cross-validation was used as the resampling technique. In this scenario, days within the same bioreactor will likely be in both the analysis and assessment sets (Figure 3.5). Given the amount of correlated data within day, this is a bad idea since it will lead to artificially optimistic characterizations of the model.

A more appropriate resampling technique is to leave out all of the data for one or more bioreactors out *en mass*. A day effect will be used in the model, so the collection of data corresponding to different days should move with each bioreactor as they are allocated with the analysis or assessment sets. In the last section, we will compare these two methods of resampling.

The next few sections describe sequences of preprocessing methods that are applied to this type of data. While these methods are most useful for spectroscopy, they illustrate how the preprocessing and feature engineering can be applied to different layers of data.

9.3   Reducing Background

A characteristic seen in Figures 9.3 is that the daily data across small- and large-scale bioreactors are shifted on the intensity axis where they tend to trend down for higher wavelengths. For spectroscopy data, intensity deviations from zero are called baseline drift and are generally due to noise in the measurement system, interference, or fluorescence (Rinnan et al., 2009) and are not due to the actual chemical substance of the sample. Baseline drift is a major contributor to measurement variation; for the small-scale bioreactors in Figure 9.3, the vertical variation is much greater than variation due to peaks in the spectra which are likely to be associated with the amount of glucose. The excess variation due to spurious sources that contribute to background can have a detrimental impact on models like principle component regression and partial least squares, which are driven by predictor variation.

For this type of data, if background was completely eliminated, then intensities would be zero for wavenumbers that had no response to the molecules present in the sample. While specific steps can be taken during the measurement process to reduce noise, interference and fluorescence, it is almost impossible to experimentally remove all background. Therefore, the background patterns must be approximated and this approximation must be removed from the observed intensities.

A simple and clever way to approximate the background is to use a polynomial fit to the lowest intensity values across the spectrum. An algorithm for this process is:

Algorithm 9.1: Background correction using polynomial functions.

---

1 Select the polynomial degree, d, and maximum number of iterations, m;

2 **while** *The number of selected points is greater than d and the number of iterations is less than m* **do**

3 | Fit a polynomial of degree d to the data;

4 | Calculate the residuals (observed minus fitted);

5 | Subset the data to those points with negative residuals;

6 **end**

7 Subtract the estimated baseline from the original profile;

For most data, a polynomial of degrees 3 to 5 for approximating the baseline is usually sufficient.

Figure 9.5 illustrates the original spectra, a final polynomial fit of degree $d = 5$, and the corrected intensities across wavelengths. While there are still areas of the spectra

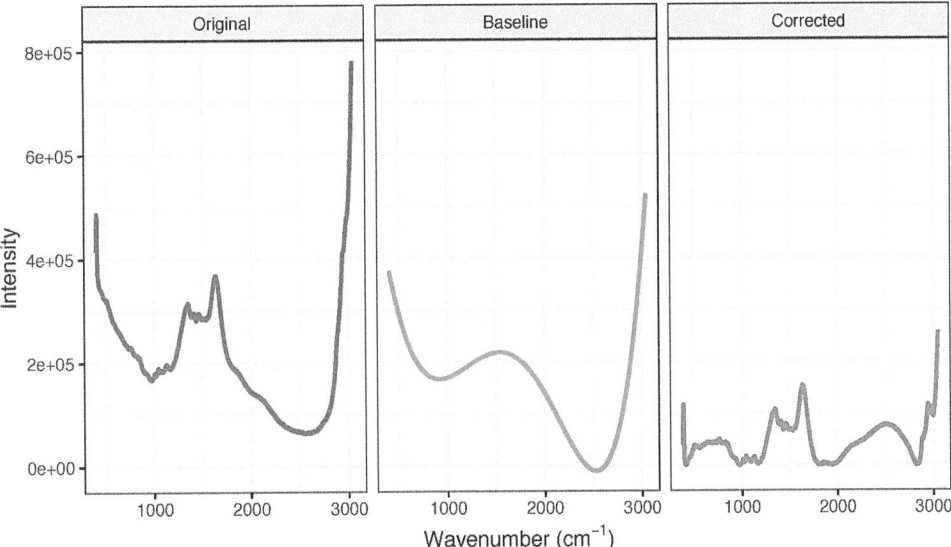

Figure 9.5: A polynomial baseline correction for the small-scale bioreactor 1, day 1 intensities.

that are above zero, the spectra have been warped so that low-intensity regions are near zero.

9.4 | Reducing Other Noise

The next step in processing profile data is to reduce extraneous noise. Noise appears in two forms for these data. First, the amplitudes vary greatly from spectrum to spectrum across bioreactors which is likely due to measurement system variation rather than variation due to the types and amounts of molecules within a sample. What is more indicative of the sample contents are the relative peak amplitudes across spectra. To place the spectra on similar scales, the baseline adjusted intensity values are standardized within each spectrum such that the overall mean of a spectrum is zero and the standard deviation is 1. The spectroscopy literature calls this transformation the standard normal variate (SNV). This approach ensures that the sample contents are directly comparable across samples. One must be careful, however, when using the mean and standard deviation to standardize data since each of these summary statistics can be affected by one (or a few) extreme, influential value. To prevent a small minority of points from affecting the mean and standard deviation, these statistics can be computed by excluding the most extreme values. This approach is called *trimming*, and provides more robust estimates of the center and spread of the data that are typical of the vast majority of the data.

Figure 9.6 compares the profiles of the spectrum with the lowest variation and highest variation for (a) the baseline corrected data across all days and small-scale bioreactors. This figure demonstrates that the amplitudes of the profiles can vary

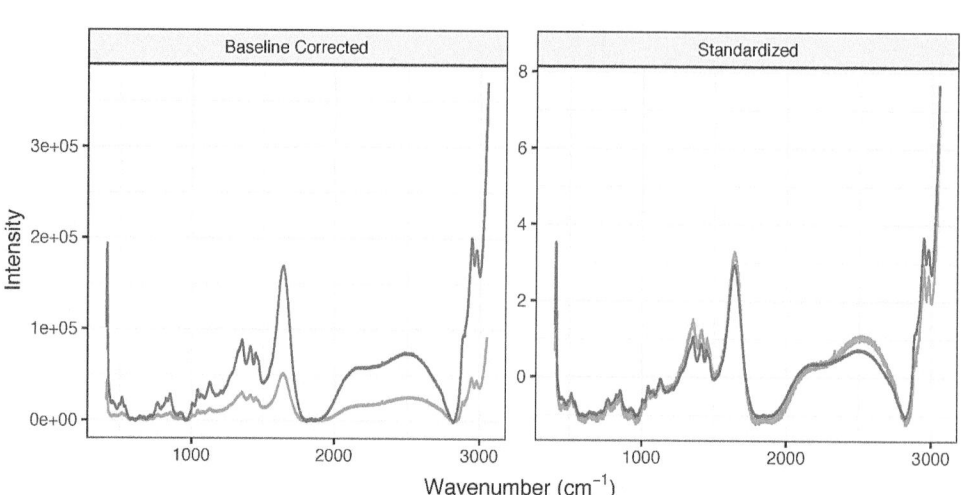

Figure 9.6: (a) The baseline-corrected intensities for the spectra that are the most and least variable. (b) The spectra after standardizing each to have a mean of 0 and standard deviation of 1.

greatly. After standardizing (b), the profiles become more directly comparable which will allow more subtle changes in the spectra due to sample content to be identified as signal which is related to the response.

A second source of noise is apparent in the intensity measurements for each wavelength within a spectrum. This can be seen in the jagged nature of the profile illustrated in either the "Original" or "Corrected" panels of Figure 9.5. Reducing this type of noise can be accomplished through several different approaches such as smoothing splines and moving averages. To compute the moving average of size k at a point p, the k values symmetric about p are averaged together which then replace the current value. The more points are considered for computing the moving average, the smoother the curve becomes.

To more distinctly see the impact of applying a moving average, a smaller region of the wavelengths will be examined. Figure 9.7 focuses on wavelengths 950 through 1200 for the first day of the first small-scale bioreactor and displays the standardized spectra, as well as the moving averages of 5, 10, and 50. The standardized spectra is quite jagged in this region. As the number of wavelengths increases in the moving average, the profile becomes smoother. But what is the best number of wavelengths to choose for the calculation? The trick is to find a value that represents the general structure of the peaks and valleys of the profile without removing them. Here, a moving average of 5 removes most of the jagged nature of the standardized profiles but still traces the original profile closely. On the other hand, a value of 50 no longer represents the major peaks and valleys of the original profile. A value of 15 for these data seems to be a reasonable number to retain the overall structure.

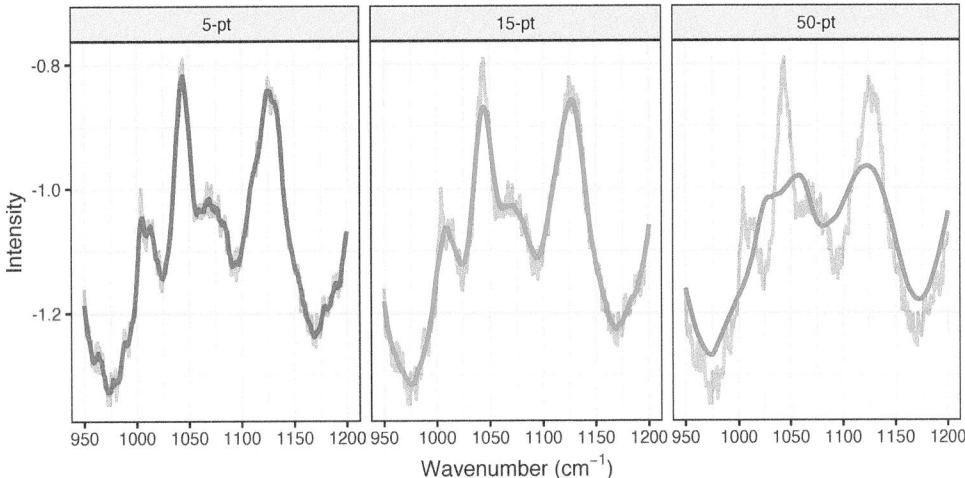

Figure 9.7: The moving average of lengths 5, 15, and 50 applied to the first day of the first small-scale bioreactor for wavelengths 950 through 1200.

It is important to point out that the appropriate number of points to consider for a moving average calculation will be different for each type of data and application. Visual inspection of the impact of different number of points in the calculation like that displayed Figure 9.7 is a good way to identify an appropriate number for the problem of interest.

9.5 | Exploiting Correlation

The previous steps of estimating and the adjusting baseline and reducing noise help to refine the profiles and enhance the true signal that is related to the response within the profiles. These steps, however, do not reduce the between-wavelength correlation within each sample which is still a problematic characteristic for many predictive models.

Reducing between-wavelength correlation can be accomplished using several previously described approaches in Section 6.3 such as principal component analysis (PCA), kernel PCA, or independent component analysis. These techniques perform dimension reduction on the predictors across *all* of the samples. Notice that this is a different approach than the steps previously described; specifically, the baseline correction and noise reduction steps are performed within each sample. In the case of traditional PCA, the predictors are condensed in such a way that the variation across samples is maximized with respect to all of the predictor measurements. As noted earlier, PCA is ignorant to the response and may not produce predictive features. While this technique may solve the issue of between-predictor correlations, it isn't guaranteed to produce effective models.

When PCA is applied to the small-scale data, the amount of variation in intensity

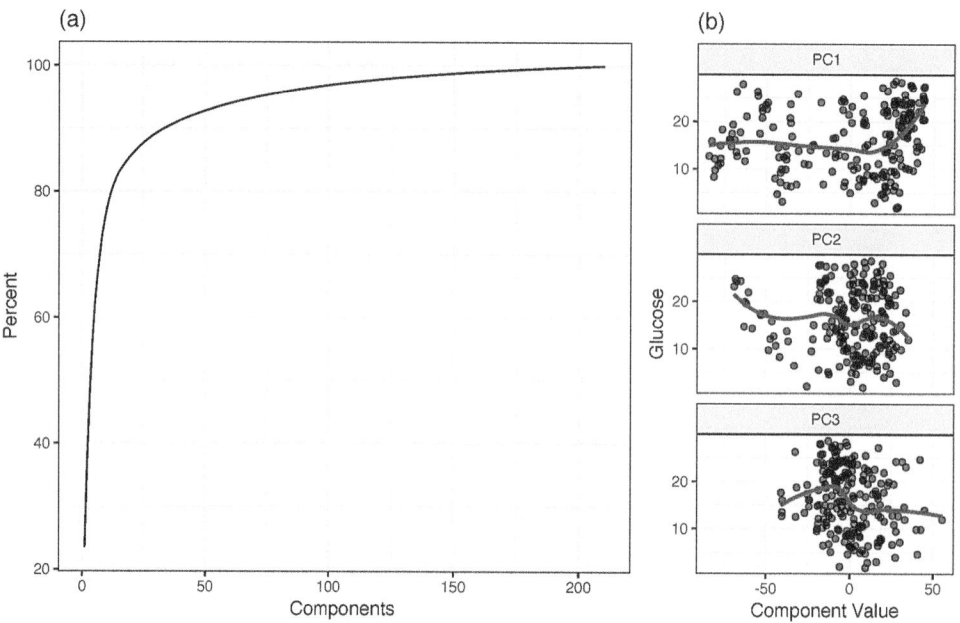

Figure 9.8: PCA dimension reduction applied across all small-scale bioreactor data. (a) A scree plot of the cumulative variability explained across components. (b) Scatterplots of Glucose and the first three principal components.

values across wavelengths can be summarized with a scree plot (Figure 9.8(a)). For these data, 11 components explain approximately 80 percent of the predictor variation, while 33 components explain nearly 90 percent of the variation. The relationships between the response and each of the first three new components are shown in Figure 9.8(b). The new components are uncorrelated and are now an ideal input to any predictive model. However, none of the first three components have a strong relationship with the response. When faced with highly correlated predictors, and when the goal is to find the optimal relationship between the predictors and the response, a more effective alternative technique is partial least squares (described in Section 6.3.1.5).

A second approach to reducing correlation is through first-order differentiation within each profile. To compute first-order differentiation, the response at the $(p-1)^{st}$ value in the profile is subtracted from the response at the pth value in the profile. This difference represents the rate of change of the response between consecutive measurements in the profile. Larger changes correspond to a larger movement and could potentially be related to the signal in the response. Moreover, calculating the first-order difference makes the new values relative to the previous value and reduces the relationship with values that are 2 or more steps away from the current value. This means that the autocorrelation across the profile should greatly be reduced. This is directly related to the equation shown in the introduction to this chapter; large positive covariances, such as those seen between wavelengths, can drastically reduce noise and variation in the differences.

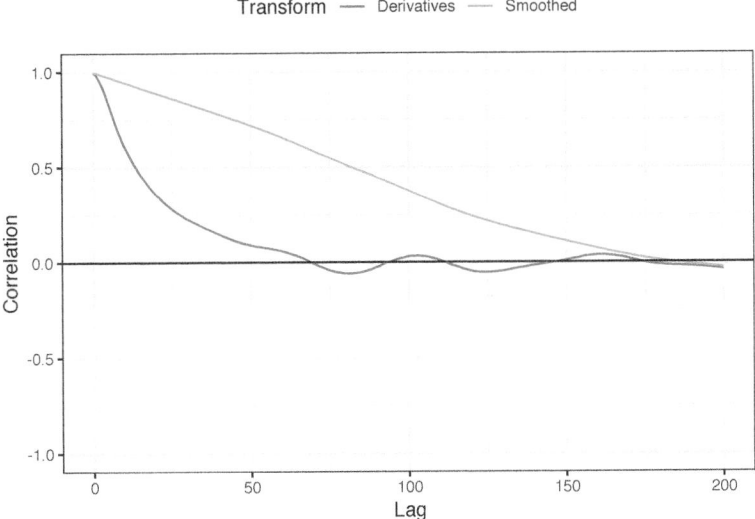

Figure 9.9: Autocorrelations before and after taking derivatives of the spectra.

Figure 9.9 shows the autocorrelation values across the first 200 lags within the profile for the first day and bioreactor before and after the derivatives were calculated. The autocorrelations drop dramatically, and only the first 3 lags have correlations greater than 0.95.

The original profile of the first day of the first small-scale bioreactor and the profile of the baseline corrected, standardized, and first-order differenced version of the same bioreactor are compared in Figure 9.10. These steps show how the within-spectra drift has been removed and most of the trends that are unrelated to the peaks have been minimized.

While the number of lagged differences that are highly correlated is small, these will still pose a problem for predictive models. One solution would be to select every m^{th} profile, where m is chosen such that the autocorrelation at that lag falls below a threshold such as 0.9 or 0.95. Another solution would be to filter out highly correlated differences using the correlation filter (Section 2) across all profiles.

9.6 Impacts of Data Processing on Modeling

The amount of preprocessing could be considered a tuning parameter. If so, then the goal would be to select the best model and the appropriate amount of signal processing. The strategy here is to use the small-scale bioreactor data, with their corresponding resampled performance estimates, to choose the best combination of analytical methods. Once we come up with one or two candidate combinations of preprocessing and modeling, the models built with the small-scale bioreactor spectra are used to predict the analogous data from the large-scale bioreactor. In other words, the small bioreactors are the training data while the large bioreactors are the

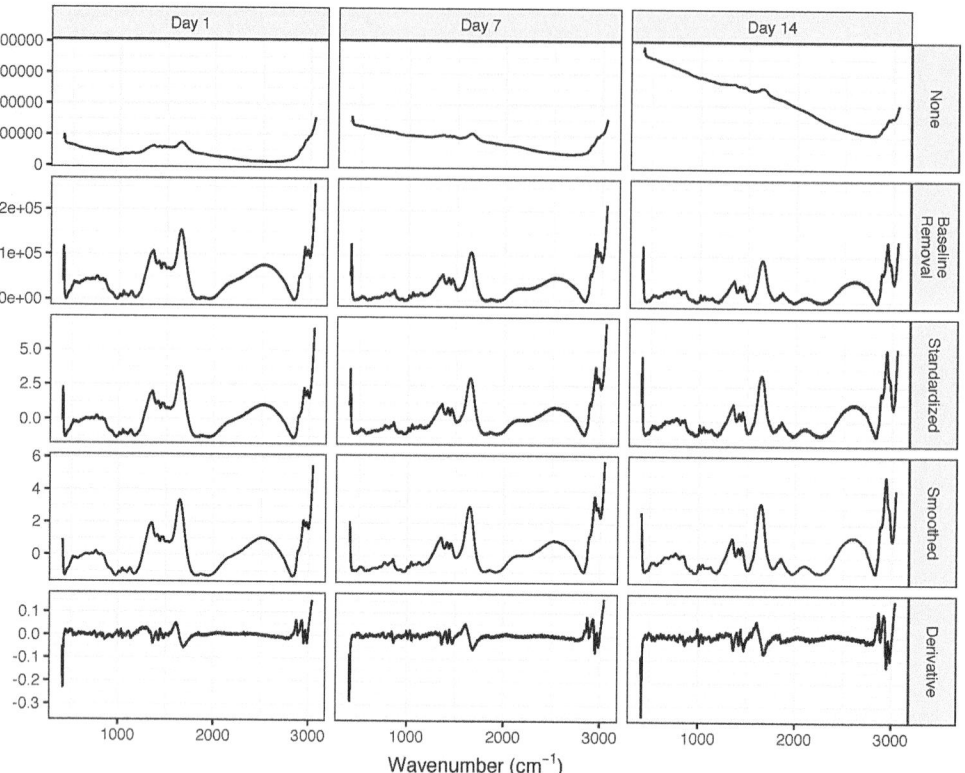

Figure 9.10: Spectra for the first day of the first small-scale bioreactor where the preprocessing steps have been sequentially applied.

test data. Hopefully, a model built on one data set will be applicable to the data from the production-sized reactors.

The training data contain 15 small bioreactors each with 14 daily measurements. While the number of experimental units is small, there still are several reasonable options for cross-validation. The first option one could consider in this setting would be leave-one-bioreactor-out cross-validation. This approach would place the data from 14 small bioreactors in the analysis set and use the data from one small bioreactor for the assessment set. Another approach would be to consider grouped V-fold cross-validation but use the bioreactor as the experimental unit. A natural choice for V in this setting would be 5; each fold would place 12 bioreactors in the analysis set and 3 in the assessment set. As an example, Table 9.2 shows an example of such an allocation for these data.

In the case of leave-one-bioreactor-out and grouped V-fold cross-validation, each sample is predicted exactly once. When there are a small number of experimental units, the impact of one unit on model tuning and cross-validation performance increases. Just one unusual unit has the potential to alter the optimal tuning parameter selection as well as the estimate of model predictive performance. Averaging performance across more cross-validation replicates, or repeated cross-validation, is

Table 9.2: An example of grouped *V*-fold cross-validation for the bioreactor data.

Resample	Heldout Bioreactor
1	5, 9, and 13
2	4, 6, and 11
3	3, 7, and 15
4	1, 8, and 10
5	2, 12, and 14

an effective way to dampen the effect of an unusual unit. But how many repeats are necessary? Generally, 5 repeats of grouped *V*-fold cross-validation are sufficient but the number can depend on the computational burden that the problem presents as well as the sample size. For smaller data sets, increasing the number of repeats will produce more accurate performance metrics. Alternatively, for larger data sets, the number of repeats may need to be less than 5 to be computationally feasible. For these data, 5 repeats of grouped 5-fold cross-validation were performed.

When profile data are in their raw form, the choices of modeling techniques are limited. Modeling techniques that incorporate simultaneous dimension reduction and prediction, like principal component regression and partial least squares, are often chosen. Partial least squares, in particular, is a very popular modeling technique for this type of data. This is due to the fact that it condenses the predictor information into a smaller region of the predictor space that is optimally related to the response. However, PLS and PCR are only effective when the relationship between the predictors and the response follows a straight line or plane. These methods are not optimal when the underlying relationship between predictors and response is non-linear.

Neural networks and support vector machines cannot directly handle profile data. But their ability to uncover non-linear relationships between predictors and a response make them very desirable modeling techniques. The preprocessing techniques presented earlier in this chapter will enable these techniques to be applied to profile data.

Tree-based methods can also tolerate the highly correlated nature of profile data. The primary drawback of using these techniques is that the variable importance calculations may be misled due to the high correlations among predictors. Also, if the trends in the data are truly linear, these models will have to work harder to approximate linear patterns.

In this section, linear, non-linear, and tree-based models will be trained. Specifically, we will explore the performance of PLS, Cubist, radial basis function SVMs, and feed-forward neural networks. Each of these models will be trained on the small-scale bioreactor analysis set. Then each model will be trained on the sequence of preprocessing of baseline-correction, standardization, smoothing, and first-order

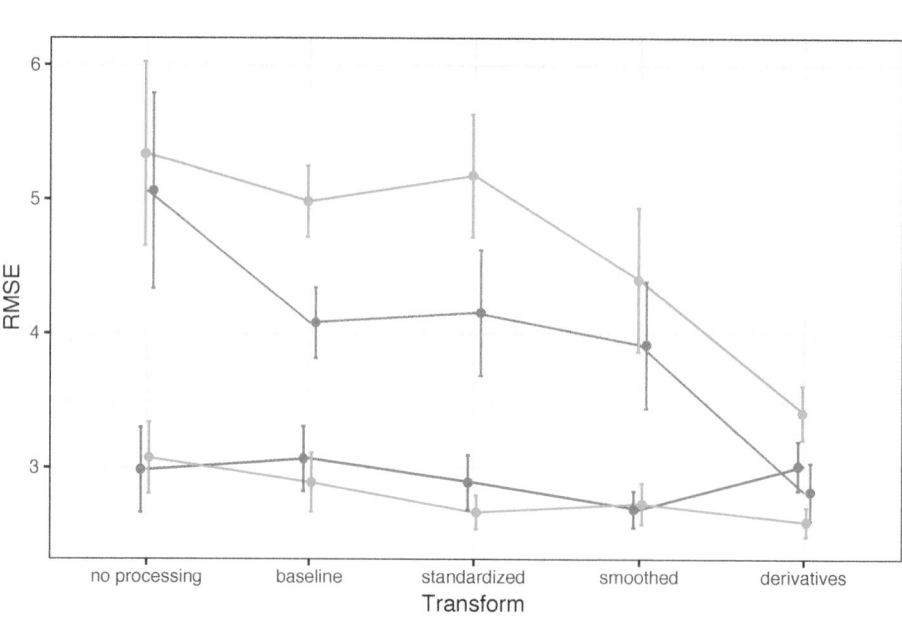

Figure 9.11: Cross-validation performance using different preprocessing steps for profile data across several models.

derivatives. Model-specific preprocessing steps will also be applied within each sequence of the profile preprocessing. For example, centering and scaling each predictor is beneficial for PLS. While the profile preprocessing steps significantly reduce correlation among predictors, some high correlation still remains. Therefore, highly correlated predictor removal will be an additional step for the SVM and neural network models.

The results from repeated cross-validation across models and profile preprocessing steps are presented in Figure 9.11. This figure highlights several important results. First, overall average model performance of the assessment sets dramatically improves for SVM and neural network models with profile preprocessing. In the case of neural networks, the cross-validated RMSE for the raw data was 5.34. After profile preprocessing, cross-validated RMSE dropped to 3.41. Profile preprocessing provides some overall improvement in predictive ability for PLS and Cubist models. But what is more noticeable for these models is the reduction in variation of RMSE. The standard deviation of the performance in the assessment sets for PLS with no profile preprocessing was 3.08 and was reduced to 2.6 after profile preprocessing.

Based on these results, PLS using derivative features appears to be the best combination. Support vector machines with derivatives also appears promising, although the Cubist models showed good results prior to computing the derivatives. To go forward to the large-scale bioreactor test set, the derivative-based PLS model and

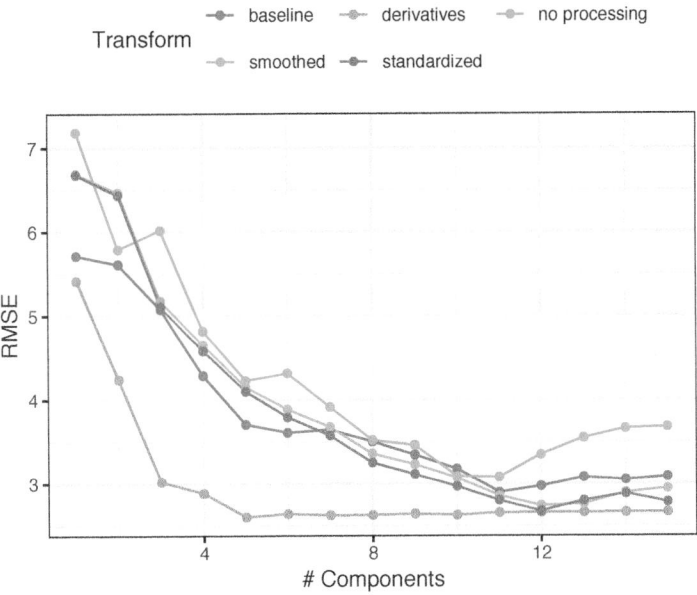

Figure 9.12: The tuning parameter profiles for partial least squares across profile preprocessing steps.

Cubist models with smoothing will be evaluated. Figure 9.12 shows the PLS results in more detail where the resampled RMSE estimates are shown against the number of retained components for each of the preprocessing methods. Performance is equivalent across many of the preprocessing methods but the use of derivatives clearly had a positive effect on the results. Not only were the RMSE values smaller but only a small number of components were required to optimize performance. Again, this is most likely driven by the reduction in correlation between the wavelength features after differencing was used.

Taking the two candidate models to the large-scale bioreactor test set, Figure 9.13 shows a plot of the observed and predicted values, colored by day, for each preprocessing method. Clearly, up until standardization, the models appreciably under-predicted the glucose values (especially on the initial days). Even when the model fits are best, the initial few days of the reaction appear to have the most noise in prediction. Numerically, the Cubist model had slightly smaller RMSE values (2.07 for Cubist and 2.14 for PLS). However, given the simplicity of the PLS model, this approach may be preferred.

The modeling illustration was based on the knowledge that the experimental unit was the bioreactor rather than the individual day within the bioreactor. As discussed earlier, it is important to have a firm understanding of the unit for utilizing an appropriate cross-validation scheme. For these data, we have seen that the daily measurements within a bioreactor are more highly correlated than measurements between bioreactors. Cross-validation based on daily measurements would then likely

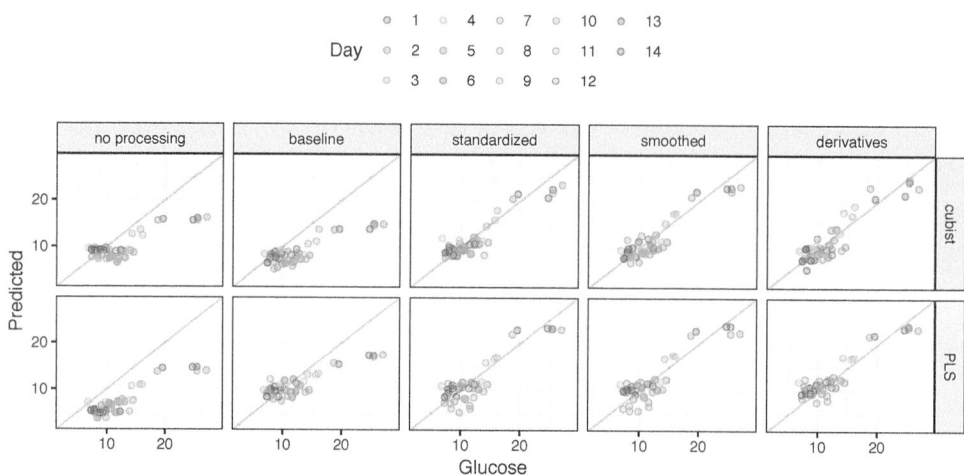

Figure 9.13: A comparison of the observed and predicted glucose values for the large-scale bioreactor data.

lead to better hold-out performance. But this performance would be *misleading* since, in this setting, new data would be based on entirely new bioreactors. Additionally, repeated cross-validation was performed on the small-scale data, where the days were inappropriately considered as the experimental unit. The hold-out performance comparison is provided in Figure 9.14. Across all models and all profile preprocessing steps, the hold-out RMSE values are artificially lower when day is used as the experimental unit as compared to when bioreactor is used as the unit. Ignoring or being unaware that bioreactor was the experimental unit would likely lead one to be overly optimistic about the predictive performance of a model.

9.7 Summary

Profile data is a particular type of data that can stem from a number of different structures. This type of data can occur if a sample is measured repeatedly over time, if a sample has many highly related/correlated predictors, or if sample measurements occur through a hierarchical structure. Whatever the case, the analyst needs to be keenly aware of what the experimental unit is. Understanding the unit informs decisions about how the profiles should be preprocessed, how samples should be allocated to training and test sets, and how samples should be allocated during resampling.

Basic preprocessing steps for profiled data can include estimating and adjusting the baseline effect, reducing noise across the profile, and harnessing the information contained in the correlation among predictors. An underlying goal of these steps is to remove the characteristics that prevent this type of data from being used with most predictive models while simultaneously preserving the predictive signal between the profiles and the outcome. No one particular combination of steps will work for

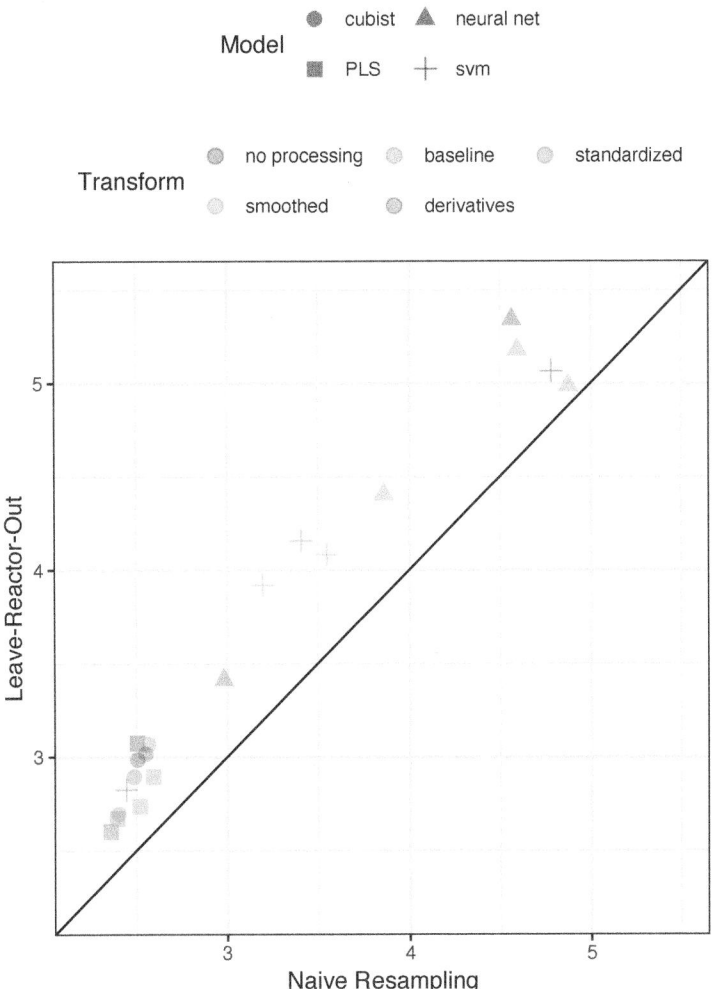

Figure 9.14: Cross-validation performance comparison using bioreactors, the experimental units, or naive resampling of rows.

all data. However, putting the right combination of steps together can produce a very effective model.

9.8 | Computing

The website `http://bit.ly/fes-profile` contains R programs for reproducing these analyses.

10 | Feature Selection Overview

Chapters 5 through 9 have provided tools for engineering features (or predictors) to put them in a form that enables models to better find the predictive signal relative to the outcome. Some of these techniques, like 1-1 transformations or dimensions reduction methods, lead to a new predictor set that has as many or fewer predictors than the original data. Other transformations, like basis expansions, generate more features than the original data. The hope is that some of the newly engineered predictors capture a predictive relationship with the outcome. But some may not be relevant to the outcome. Moreover, many of the original predictors also may not contain predictive information. For a number of models, predictive performance is degraded as the number of uninformative predictors increases. Therefore, there is a genuine need to appropriately select predictors for modeling.

The remaining chapters will focus on *supervised* feature selection where the choice of which predictors to retain is guided by their effect on the outcome. This chapter will provide an introduction to feature selection, the general nomenclature for these methods, and some notable pitfalls.

10.1 | Goals of Feature Selection

In practice, we often have found that collaborators desire to have a model that has the best predictive ability *and* is interpretable. But there is often a trade-off between predictive performance and interpretability, and it is generally not possible to maximize both at the same time (Shmueli, 2010). A misunderstanding of this trade-off leads to the belief that simply filtering out uninformative predictors will help elucidate which factors are influencing the outcome. Then an explanation can be constructed as to why the remaining predictors are related to the outcome. This rationale is problematic for several reasons. First, consider the case when the number of predictors is much greater than the number of samples. In this scenario, there are likely to be many mutually exclusive subsets of predictors that result in models of nearly equivalent predictive performance (e.g., local optima). To find the best global solution, i.e. the subset of predictors that has best performance, would require evaluating all possible predictor subsets and may be computationally infeasible. But even if it is possible to find the global optimum with the available data, the identified subset may not be the true global optimum (if one exists) due to the inherent noise in the available predictors and outcome.

Many models are complex in the ways that they relate the predictors to the outcome. For such models it is nearly impossible to decipher the relationship between any individual predictor and the outcome. One approach that attempts to gain insight for an individual predictor in a complex model is to fix all other selected predictors to a single value, then observe the effect on the outcome by varying the predictor of interest. This approach, called a partial dependence plot, is overly simplistic and only provides a small sliver of insight on the predictor's true impact.

Here we would like to refocus the motivations for removing predictors from a model. The primary motivations should be to either mitigate a specific problem in the interplay between predictors and a model, or to reduce model complexity. For example:

- Some models, notably support vector machines and neural networks, are sensitive to irrelevant predictors. As will be shown below, superfluous predictors can lead to reduced predictive performance in some situations.

- Other models like linear or logistic regression are vulnerable to correlated predictors (see Chapter 6). Removing correlated predictors will reduce multi-collinearity and thus enable these types of models to be fit.

- Even when a predictive model is *insensitive* to extra predictors, it makes good scientific sense to include the minimum possible set that provides acceptable results. In some cases, removing predictors can reduce the cost of acquiring data or improve the throughput of the software used to make predictions.

The working premise here is that it is generally better to have fewer predictors in a model. For the remaining chapters, the goal of feature selection will be re-framed as follows:

> Reduce the number of predictors as far as possible without compromising predictive performance.

There are a variety of methods to reduce the predictor set. The next section provides an overview of the general classes of feature selection techniques.

10.2 | Classes of Feature Selection Methodologies

Feature selection methodologies fall into three general classes: intrinsic (or implicit) methods, filter methods, and wrapper methods. Intrinsic methods have feature selection naturally incorporated with the modeling process, whereas filter and wrapper methods work to marry feature selection approaches with modeling techniques. The most seamless and important of the three classes for reducing features are intrinsic methods. Some examples include:

- Tree- and rule-based models. These models search for the best predictor and split point such that the outcomes are more homogeneous within each new partition (recall Section 5.7). Therefore, if a predictor is not used in any split, it is functionally independent of the prediction equation and has been excluded from the model. Ensembles of trees have the same property, although some algorithms, such as random forest, deliberately force splits on irrelevant predictors when constructing the tree. This results in the over-selection of predictors (as shown below).

- Multivariate adaptive regression spline (MARS) models. These models create new features of the data that involve one or two predictors at a time (Section 6.2.1). The predictors are then added to a linear model in sequence. Like trees, if a predictor is not involved in at least one MARS feature, it is excluded from the model.

- Regularization models. The regularization approach penalizes or shrinks predictor coefficients to improve the model fit. The lasso (Section 7.3) uses a type of penalty that shrinks coefficients to absolute zero. This forces predictors to be excluded from the final model.

The advantage of implicit feature selection methods is that they are relatively fast since the selection process is embedded within the model fitting process; no external feature selection tool is required. Also, implicit methods provide a direct connection between selecting features and the *objective function*. The objective function is the statistic that the model attempts to optimize (e.g., the likelihood in generalized linear models or the impurity in tree-based models). The direct link between selection and modeling objective makes it easier to make informed choices between feature sparsity and predictive performance.

The main downside to intrinsic feature selection is that it is model-dependent. If the data are better fit by a non-intrinsic feature selection type of model, then predictive performance may be sub-optimal when all features are used. In addition, some intrinsic models (like single tree models) use a *greedy* approach to feature selection. Greedy approaches usually identify a narrow set of features that have sub-optimal predictive performance. A more detailed discussion of this problem appears later in this chapter.

If a model does not have intrinsic feature selection, then some sort of search procedure is required to identify feature subsets that improve predictive performance. There are two general classes of techniques for this purpose: *filters* and *wrappers*.

Filter methods conduct an initial supervised analysis of the predictors to determine which are important and then only provide these to the model. Here, the search is performed just once. The filter methods often consider each predictor separately although this is not a requirement. An example of filtering occurred in Section 5.6 where we considered a set of keywords derived from the OkCupid text fields. The relationship between the occurrence of the keyword and the outcome was assessed using an odds-ratio. The rationale to keep or exclude a word (i.e., the filtering rule) was based on statistical significance as well as the magnitude of the odds-ratio. The words that made it past the filter were then added to the logistic regression model. Since each keyword was considered separately, they are unlikely to be capturing independent trends in the data. For instance, the words `programming` and `programmer` were both selected using the filtering criteria and represent nearly the same meaning with respect to the outcome

Filters are simple and tend to be fast. In addition, they can be effective at capturing the large trends (i.e., individual predictor-outcome relationship) in the data. However, as just mentioned, they are prone to over-selecting predictors. In many cases, some measure of statistical significance is used to judge "importance" such as a raw or multiplicity adjusted p-value. In these cases, there may be a disconnect between the objective function for the filter method (e.g., significance) and what the model requires (predictive performance). In other words, a selection of predictors that meets a filtering criterion like statistical significance may not be a set that improves predictive performance.[78] Filters will be discussed in more detail in Chapter 11.

Wrapper methods use iterative search procedures that repeatedly supply predictor subsets to the model and then use the resulting model performance estimate to guide the selection of the *next subset* to evaluate. If successful, a wrapper method will iterate to a smaller set of predictors that has better predictive performance than the original predictor set. Wrapper methods can take either a *greedy* or non-greedy approach to feature selection. A greedy search is one that chooses the search path based on the direction that seems *best at the time* in order to achieve the best immediate benefit. While this can be an effective strategy, it may show immediate benefits in predictive performance that stall out at a locally best setting. A non-greedy search method would re-evaluate previous feature combinations and would have the ability to move in a direction that is initially unfavorable if it appears to have a potential benefit *after* the current step. This allows the non-greedy approach to escape being trapped in a local optimum.

An example of a greedy wrapper method is backwards selection (otherwise known as *recursive feature elimination* or RFE). Here, the predictors are initially ranked by some measure of importance. An initial model is created using the complete predictor set. The next model is based on a smaller set of predictors where the least important have been removed. This process continues down a prescribed path (based on the ranking generated by the importances) until a very small number of predictors are in the model. Performance estimates are used to determine when too many features have been removed; hopefully a smaller subset of predictors can result in an improvement. Notice that the RFE procedure is greedy in that it considers the variable ranking as the search direction. It does not re-evaluate the search path at any point or consider subsets of mixed levels of importance. This approach to feature selection will likely fail if there are important interactions between predictors where only one of the predictors is significant only in the presence of the other(s). RFE will be discussed in more detail in Chapter 11 and in Section 10.4 below.

Examples of non-greedy wrapper methods are genetic algorithms (GA) and simulated

---

[78] An example of this was discussed in the context of simple screening in Section 7.3.

annealing (SA). The SA method is non-greedy since it incorporates randomness into the feature selection process. The random component of the process helps SA to find new search spaces that often lead to more optimal results. Both of these techniques will be discussed in detail in Chapter 12.

Wrappers have the potential advantage of searching a wider variety of predictor subsets than simple filters or models with built-in feature selection. They have the most potential to find the globally best predictor subset (if it exists). The primary drawback is the computational time required for these methods to find the optimal or near optimal subset. The additional time can be excessive to the extent of being counter-productive. The computational time problem can be further exacerbated by the type of model with which it is coupled. For example, the models that are in most need of feature selection (e.g., SVMs and neural networks) can be very computationally taxing themselves. Another disadvantage of wrappers is that they have the most potential to overfit the predictors to the training data and require *external validation* (as discussed below).

Figure 10.1 visually contrasts greedy and non-greedy search methods using the Goldstein-Price equation:

$$f(x, y) = [1 + (x + y + 1)^2(19 - 14x + 3x^2 - 14y + 6xy + 3y^2)] \times$$
$$[30 + (2x - 3y)^2(18 - 32x + 12x^2 + 48y - 36xy + 27y^2)].$$

The minimum of this function, where both x and y are between -2 and 2, is located at $x = 0$ and $y = -1$. The function's outcome is relatively flat in the middle with large values along the top and lower-right edges. Panel (a) shows the results of two greedy methods that use gradient decent to find the minimum. The first starting value (in orange, top right) moves along a path defined by the gradient and stalls at a local minimum. It is unable to re-evaluate the search. The second attempt, starting at $x = -1$ and $y = 1$, also follows a different greedy path and, after some course-corrections, finds a solution close to the best value at iteration 131. Each change in the path was to a new direction of immediate benefit. Panel (b) shows a global search methods called *simulated annealing*. This is a controlled but random approach that will be discussed in greater detail later. It generates a new candidate point using random deviations from the current point. If the new point is better, the process continues. If it is a worse solution, it may accept it with a certain probability. In this way, the global search does not always proceed in the current best direction and tends to explore a broader range of candidate values. In this particular search, a value near the optimum is determined more quickly (at iteration 64).

Given the advantages and disadvantages of the different types of feature selection methods, how should one utilize these techniques? A good strategy which we use in practice is to start the feature selection process with one or more intrinsic methods to see what they yield. Note that it is unrealistic to expect that models using intrinsic feature selection would select the same predictor subset, especially if linear and nonlinear methods are being compared. If a non-linear intrinsic method has good

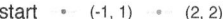

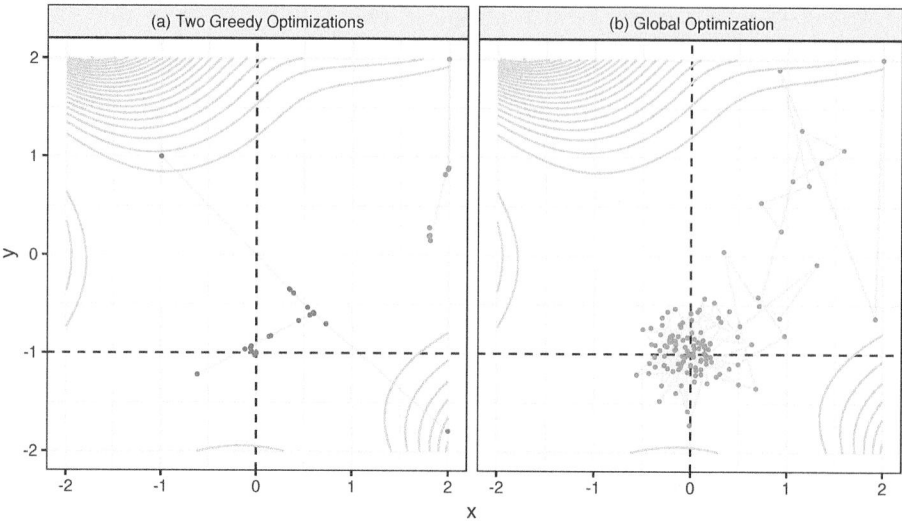

Figure 10.1: Examples of greedy and global search methods when solving functions with continuous inputs.

predictive performance, then we could proceed to a wrapper method that is combined with a non-linear model. Similarly, if a linear intrinsic method has good predictive performance, then we could proceed to a wrapper method combined with a linear model. If multiple approaches select *large* predictor sets then this may imply that reducing the number of features may not be feasible.

Before proceeding, let's examine the effect of irrelevant predictors on different types of models and how the effect changes over different scenarios.

10.3 Effect of Irrelevant Features

How much do extraneous predictors hurt a model? Predictably, that depends on the type of model, the nature of the predictors, as well as the ratio of the size of the training set to the number of predictors (a.k.a the $p : n$ ratio). To investigate this, a simulation was used to emulate data with varying numbers of irrelevant predictors and to monitor performance. The simulation system is taken from Sapp et al. (2014) and consists of a nonlinear function of the 20 relevant predictors:

$$y = x_1 + \sin(x_2) + \log(|x_3|) + x_4^2 + x_5 x_6 + I(x_7 x_8 x_9 < 0) + I(x_{10} > 0) +$$
$$x_{11} I(x_{11} > 0) + \sqrt{|x_{12}|} + \cos(x_{13}) + 2x_{14} + |x_{15}| + I(x_{16} < -1) +$$
$$x_{17} I(x_{17} < -1) - 2x_{18} - x_{19} x_{20} + \epsilon. \tag{10.1}$$

Each of these predictors was generated using independent standard normal random

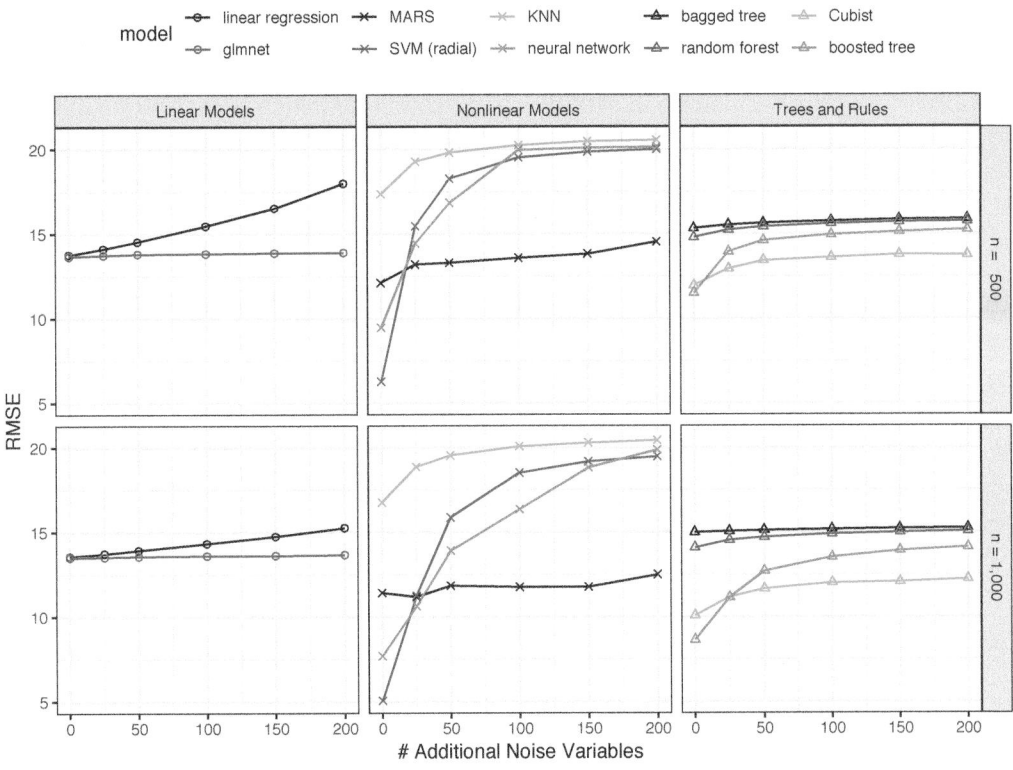

Figure 10.2: RMSE trends for different models and simulation configurations.

variables and the error was simulated as random normal with a zero mean and standard deviation of 3. To evaluate the effect of extra variables, varying numbers of random standard normal predictors (with no connection to the outcome) were added. Between 10 and 200 extra columns were appended to the original feature set. The training set size was either $n = 500$ or $n = 1,000$. The root mean squared error (RMSE) was used to measure the quality of the model using a large simulated test set since RMSE is a good estimate of irreducible error.

A number of models were tuned and trained for each of the simulated data sets including, linear, nonlinear, and tree/rule-based models.[79] The details and code for each model can be found in a GitHub repo[80] and the results are summarized below.

Figure 10.2 shows the results. The linear models (ordinary linear regression and the glmnet) showed mediocre performance overall. Extra predictors eroded the performance of linear regression while the lasso penalty used by the glmnet resulted in stable performance as the number of irrelevant predictors increased. The effect

---

[79]Clearly, this data set requires nonlinear models. However, the simulation is being used to look at the relative impact of irrelevant predictors (as opposed to absolute performance). For linear models, we assumed that the interaction terms x_5x_6 and $x_{19}x_{20}$ would be identified and included in the linear models.

[80]https://github.com/topepo/FES_Selection_Simulation.

on linear regression was smaller as the training set size increases from 500 to 1,000. This reinforces the notation that the $p : n$ ratio may be the underlying issue.

Nonlinear models had different trends. K-nearest neighbors had overall poor performance that was moderately impacted by the increase in p. MARS performance was good and showed good resistance to noise features due to the intrinsic feature selection used by that model. Again, the effect of extra predictors was smaller with the larger training set size. Single layer neural networks and support vector machines showed the overall best performance (getting closest to the best possible RMSE of 3.0) when no extra predictors were added. However, both methods were drastically affected by the inclusion of noise predictors to the point of having some of the worst performance seen in the simulation.

For trees, random forest and bagging showed similar but mediocre performance that was not affected by the increase in p. The rule-based ensemble Cubist and boosted trees fared better. However, both models showed a moderate decrease in performance as irrelevant predictors were added; boosting was more susceptible to this issue than Cubist.

These results clearly show that there are a number of models that may require a reduction of predictors to avoid a decrease in performance. Also, for models such as random forest or the glmnet, it appears that feature selection may be useful to find a smaller subset of predictors without affecting the model's efficacy.

Since this is a simulation, we can also assess how models with built-in feature selection performed at finding the *correct predictor subset*. There are 20 relevant predictors in the data. Based on which predictors were selected, a *sensitivity*-like proportion can be computed that describes the rate at which the true predictors were retained in the model. Similarly, the number of irrelevant predictors that were selected gives a sense of the *false positive* rate (i.e., one minus specificity) of the selection procedure. Figure 10.3 shows these results using a plot similar to how ROC curves are visualized.

The three tree ensembles (bagging, random forest, and boosting) have excellent true positive rates; the truly relevant predictors are almost always retained. However, they also show poor results for irrelevant predictors by selecting many of the noise variables. This is somewhat related to the nature of the simulation where most of the predictors enter the system as smooth, continuous functions. This may be causing the tree-based models to try too hard to achieve performance by training deep trees or larger ensembles.

Cubist had a more balanced result where the true positive rate was greater than 70%. The false positive rate is around 50% for a smaller training set size and *increased* when the training set was larger and less than or equal to 100 extra predictors. The overall difference between the Cubist and tree ensemble results can be explained by the factor that Cubist uses an ensemble of linear regression models that are fit on small subsets of the training set (defined by rules). The nature of the simulation system might be better served by this type of model.

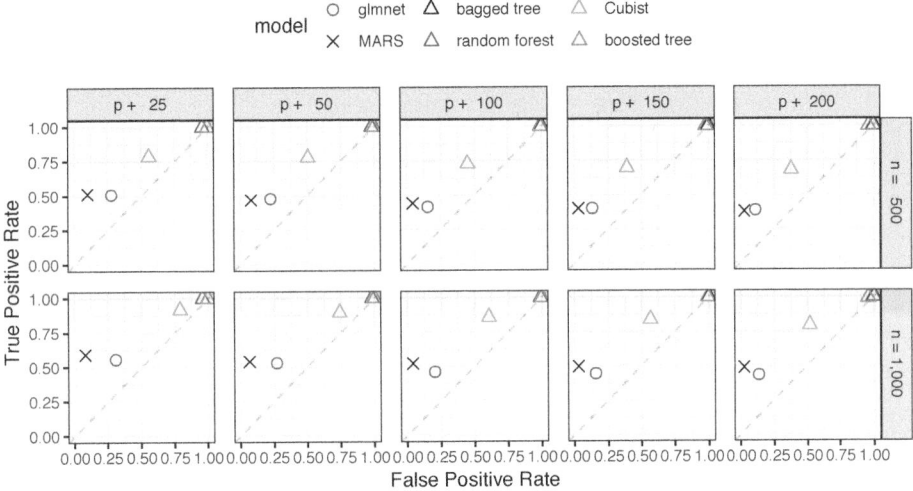

Figure 10.3: ROC-like plots for the feature selection results in the simulated data sets. The value p represents the number of relevant predictors, while the additional number represents the number of irrelevant predictors.

Recall that the main random forest tuning parameter, m_{try}, is the number of randomly selected variables that should be evaluated at each split. While this encourages diverse trees (to improve performance), it can necessitate the splitting of irrelevant predictors. Random forest often splits on the majority of the predictors (regardless of their importance) and will vastly over-select predictors. For this reason, an additional analysis might be needed to determine which predictors are truly essential based on that model's variable importance scores. Additionally, trees in general poorly estimate the importance of highly correlated predictors. For example, adding a duplicate column that is an exact copy of another predictor will numerically dilute the importance scores generated by these models. For this reason, important predictors can have low rankings when using these models.

MARS and the glmnet also made trade-offs between sensitivity and specificity in these simulations that tended to favor specificity. These models had false positive rates under 25% and true positive rates that were usually between 40% and 60%.

10.4 | Overfitting to Predictors and External Validation

Section 3.5 introduced a simple example of the problem of overfitting the available data during selection of model tuning parameters. The example illustrated the risk of finding tuning parameter values that over-learn the relationship between the predictors and the outcome in the training set. When models over-interpret patterns in the training set, the predictive performance suffers with new data. The solution to this problem is to evaluate the tuning parameters on a data set that is not used to estimate the model parameters (via validation or assessment sets).

An analogous problem can occur when performing feature selection. For many data sets it is possible to find a subset of predictors that has good predictive performance on the training set but has poor performance when used on a test set or other new data set. The solution to this problem is similar to the solution to the problem of overfitting: feature selection needs to be part of the resampling process.

Unfortunately, despite the need, practitioners often combine feature selection and resampling inappropriately. The most common mistake is to only conduct resampling *inside* of the feature selection procedure. For example, suppose that there are five predictors in a model (labeled as A though E) and that each has an associated measure of importance derived from the training set (with A being the most important and E being the least). Applying backwards selection, the first model would contain all five predictors, the second would remove the least important (E in this example), and so on. A general outline of one approach is:

Algorithm 10.1: An inappropriate resampling scheme for recursive feature elimination.

---

1 Rank the predictors using the training set;

2 **for** *a feature subset of size 5 to 1* **do**

3   **for** *each resample* **do**

4   Fit model with the feature subset on the analysis set;

5   Predict the assessment set;

6   **end**

7   Determine the best feature subset using resampled performance;

8   Fit the best feature subset using the entire training set;

9 **end**

There are two key problems with this procedure:

1. Since the feature selection is *external* to the resampling, resampling cannot effectively measure the impact (good or bad) of the selection process. Here, resampling is not being exposed to the variation in the selection process and therefore cannot measure its impact.

2. The same data are being used to measure performance and to guide the direction of the selection routine. This is analogous to fitting a model to the training set and then re-predicting the same set to measure performance. There is an obvious bias that can occur if the model is able to closely fit the training data. Some sort of *out-of-sample* data are required to accurately determine how well the model is doing. If the selection process results in overfitting, there are no data remaining that could possibly inform us of the problem.

As a real example of this issue, Ambroise and McLachlan (2002) reanalyzed the results of Guyon et al. (2002) where high-dimensional classification data from a RNA expression microarray were modeled. In these analyses, the number of samples was low (i.e., less than 100) while the number of predictors was high (2K to 7K). Linear support vector machines were used to fit the data and backwards elimination (a.k.a. RFE) was used for feature selection. The original analysis used leave-one-out (LOO) cross validation with the scheme outlined above. In these analyses, LOO resampling reported error rates very close to zero. However, when a set of samples were held out and only used to measure performance, the error rates were measured to be 15%–20% higher. In one case, where the $p : n$ ratio was at its worst, Ambroise and McLachlan (2002) shows that a zero LOO error rate could still be achieved *even when the class labels were scrambled.*

A better way of combining feature selection and resampling is to make feature selection a **component of the modeling process**. Feature selection should be incorporated the same way as preprocessing and other engineering tasks. What we mean is that that an appropriate way to perform feature selection is to do this inside of the resampling process.[81] Using the previous example with five predictors, the algorithm for feature selection within resampling is modified to:

Algorithm 10.2: Appropriate resampling for recursive feature elimination.

---

1   Split data into analysis and assessment sets;

2   **for** *each resample* **do**

3   Rank the predictors using the analysis set;

4   **for** *a feature subset of size 5 to 1* **do**

5   Fit model with the feature subset on the analysis set;

6   Predict the assessment set;

7   **end**

8   Average the resampled performance for each model and feature subset size;

9   Choose the model and feature subset with the best performance;

10   Fit a model with the best feature subset using the entire training set;

11 **end**

In this procedure, the same data set is **not** being used to determine both the optimal subset size and the predictors within the subset. Instead, different versions of the data are used for each purpose. The external resampling loop is used to decide *how far* down the removal path that the selection process should go. This is then used to decide the subset size that should be used in the final model on the entire training set. Basically, the subset size is treated as a tuning parameter. In addition, separate data are then used to determine the *direction* that the selection path should take. The direction is determined by the variable rankings computed on analysis sets (inside of resampling) or the entire training set (for the final model).

Performing feature selection within the resampling process has two notable implications. The first implication is that the process provides a more realistic estimate

---

[81]Even if it is a single (hopefully large) validation set.

of predictive performance. It may be hard to conceptualize the idea that different sets of features may be selected during resampling. Recall, resampling forces the modeling process to use different data sets so that the variation in performance can be accurately measured. The results reflect different realizations of the entire modeling process. If the feature selection process is unstable, it is helpful to understand how noisy the results could be. In the end, resampling is used to measure the overall performance of the modeling process and tries to estimate what performance *will* be when the final model is fit to the training set with the best predictors. The second implication is an increase in computational burden.[82] For models that are sensitive to the choice of tuning parameters, there may be a need to *retune* the model when each subset is evaluated. In this case, a separate, nested resampling process is required to tune the model. In many cases, the computational costs may make the search tools practically infeasible.

In situations where the selection process is unstable, using multiple resamples is an expensive but worthwhile approach. This could happen if the data set were small, and/or if the number of features were large, or if an extreme class imbalance exists. These factors might yield different feature sets when the data are slightly changed. On the other extreme, large data sets tend to greatly reduce the risk of overfitting to the predictors during feature selection. In this case, using separate data splits for feature ranking/filtering, modeling, and evaluation can be both efficient and effective.

10.5   A Case Study

One of our recent collaborations highlights the perils of incorrectly combining feature selection and resampling. In this problem, a researcher had collected 75 samples from each of the two classes. There were approximately 10,000 predictors for each data point. The ultimate goal was to attempt to identify a subset of predictors that had the ability to classify samples into the correct response with an accuracy of at least 80%.

The researcher chose to use 70% of the data for a training set, 10-fold cross-validation for model training, and the implicit feature selection methods of the glmnet and random forest. The best cross-validation accuracy found after a broad search through the tuning parameter space of each model was just under 60%, far from the desired performance. Reducing the dimensionality of the problem might help, so principal component analysis followed by linear discriminant analysis and partial least squares discriminant analysis were tried next. Unfortunately, the cross-validation accuracy of these methods was even worse.

The researcher surmised that part of the challenge in finding a predictive subset of variables was the vast number of irrelevant predictors in the data. The logic was then to first identify and select predictors that had a univariate signal with the response.

---

[82]Global search methods described in Chapter 12 also exacerbate this problem.

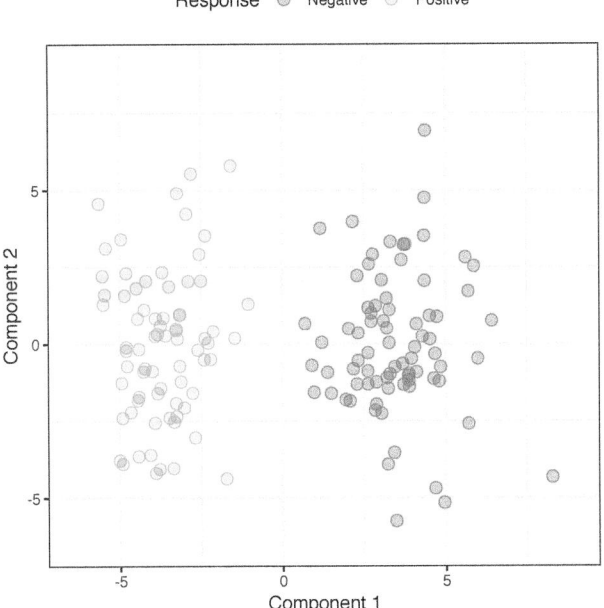

Figure 10.4: The first two principal components applied to the top features identified using a feature selection routine outside of resampling. The predictors and response on which this procedure was performed were randomly generated, indicating that feature selection outside of resampling leads to spurious results.

A t-test was performed for each predictor, and predictors were ranked by significance of separating the classes. The top 300 predictors were selected for modeling.

To demonstrate that this selection approach was able to identify predictors that captured the best predictors for classifying the response, the researcher performed principal component analysis on the 300 predictors and plotted the samples projected onto the first two components colored by response category. This plot revealed near perfect classification between groups, and the researcher concluded that there was good evidence of predictors that could classify the samples.

Regrettably, because feature selection was performed outside of resampling, the apparent signal that was found was simply due to random chance. In fact, the univariate feature selection process used by the researcher can be shown to produce perfect separation between two groups in completely random data. To illustrate this point we have generated a 150-by-10,000 matrix of random standard normal numbers ($\mu = 0$ and $\sigma = 1$). The response was also a random selection such that 75 samples were in each class. A t-test was performed for each predictor and the top 300 predictors were selected. The plot of the samples onto the first two principal components colored by response status is displayed in Figure 10.4.

Clearly, having good separation using this procedure does not imply that the selected features have the ability to predict the response. Rather, it is as likely that the apparent separation is due to random chance and an improper feature selection routine.

10.6 | Next Steps

In the following chapters, the topics that were introduced here will be revisited with examples. In the next chapter, greedy selection methods will be described (such as simple filters and backwards selection). Then the subsequent chapter describes more complex (and computationally costly) approaches that are not greedy and more likely to find a better local optimum.

10.7 | Computing

The website `http://bit.ly/fes-selection` contains R programs for reproducing these analyses.

11 | Greedy Search Methods

In this chapter, greedy search methods such as simple univariate filters and recursive feature elimination are discussed. Before proceeding, another data set is introduced that will be used in this chapter to demonstrate the strengths and weaknesses of these approaches.

11.1 | Illustrative Data: Predicting Parkinson's Disease

Sakar et al. (2019) describes an experiment where a group of 252 patients, 188 of whom had a previous diagnosis of Parkinson's disease, were recorded speaking a particular sound three separate times. Several signal processing techniques were then applied to each replicate to create 750 numerical features. The objective was to use the features to classify patients' Parkinson's disease status. Groups of features within these data could be considered as following a profile (Chapter 9), since many of the features consisted of related sets of fields produced by each type of signal processor (e.g., across different sound wavelengths or sub-bands). Not surprisingly, the resulting data have an extreme amount of multicollinearity; about 10,750 pairs of predictors have absolute rank correlations greater than 0.75.

Since each patient had three replicated data points, the feature results used in our analyses are the averages of the replicates. In this way, the patient is both the independent experimental unit and the unit of prediction.

To illustrate the tools in this chapter, we used a stratified random sample based on patient disease status to allocate 25% of the data to the test set. The resulting training set consisted of 189 patients, 138 of whom had the disease. The performance metric to evaluate models was the area under the ROC curve.

11.2 | Simple Filters

The most basic approach to feature selection is to screen the predictors to see if any have a relationship with the outcome prior to including them in a model. To do this, a numeric scoring technique is required to quantify the strength of the relationship. Using the scores, the predictors are ranked and filtered with either a threshold or by taking the top p predictors. Scoring the predictors can be done separately for each predictor, or can be done simultaneously across all predictors (depending on the technique that is used). If the predictors are screened separately, there is a large variety of scoring methods. A summary of some popular and effective scoring techniques is provided below and is organized based on the type of predictor and outcome. The list is not meant to be exhaustive, but is meant to be a starting place. A similar discussion in contained in Chapter 18 of Kuhn and Johnson (2013).

When screening individual *categorical predictors*, there are several options depending on the type of outcome data:

- When the outcome is categorical, the relationship between the predictor and outcome forms a contingency table. When there are three or more levels for the predictor, the degree of association between predictor and outcome can be measured with statistics such as χ^2 (chi-squared) tests or exact methods (Agresti, 2012). When there are exactly two classes for the predictor, the odds-ratio can be an effective choice (see Section 5.6).

- When the outcome is numeric, and the categorical predictor has two levels, then a basic t-test can be used to generate a statistic. ROC curves and precision-recall curves can also be created for each predictor and the area under the curves can be calculated.[83] When the predictor has more than two levels, the traditional ANOVA F-statistic can be calculated.

When the *predictor is numeric*, the following options exist:

- When the outcome is categorical, the same tests can be used in the case above where the predictor is categorical and the outcome is numeric. The roles are simply reversed in the t-test, curve calculations, F-test, and so on. When there are a large number of tests or if the predictors have substantial multicollinearity, the correlation-adjusted t-scores of Opgen-Rhein and Strimmer (2007) and Zuber and Strimmer (2009) are a good alternative to simple ANOVA statistics.

- When the outcome is numeric, a simple pairwise correlation (or rank correlation) statistic can be calculated. If the relationship is nonlinear, then the maximal information coefficient (MIC) values (Reshef et al., 2011) or A statistics (Murrell et al., 2016), which measure strength of the association between numeric outcomes and predictors, can be used.

- Alternatively, a generalized additive model (GAM) (Wood, 2006) can fit non-linear smooth terms to a set of predictors simultaneously and measure their importance using a p-value that tests against a null hypothesis that there is no trend of each predictor. An example of such a model with a categorical outcome was shown in Figure 4.15(b).

A summary of these simple filters can be found in Figure 11.1.

---

[83]ROC curves are typically calculated when the input is continuous and the outcome is binary. For the purpose of simple filtering, the input-output direction is swapped in order to quantify the relationship between the numeric outcome and categorical predictor.

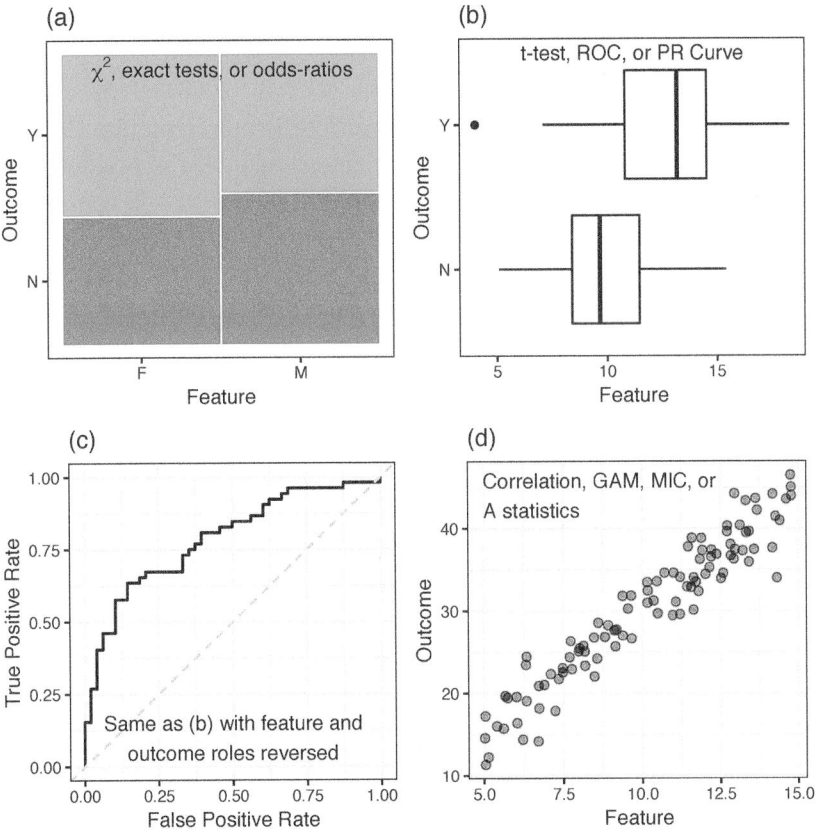

Figure 11.1: A comparison of the simple filters that can be applied to feature and outcome type combinations: (a) categorical feature and categorical outcome, (b) continuous feature and categorical outcome, (c) categorical feature and continuous outcome, and (d) continuous feature and continuous outcome.

Generating summaries of how effective individual predictors are at classifying or correlating with the outcome is straightforward when the predictors are all of the same type. But most data sets contain a mix of predictor types. In this setting it can be challenging to arrive at a ranking of the predictors since their screening statistics are in different units. For example, an odds-ratio and a t-statistic are not compatible since they are on different scales. In many cases, each statistic can be converted to a p-value so that there is a commonality across the screening statistics.

Recall that a p-value stems from the statistical framework of hypothesis testing. In this framework, we assume that there is no association between the predictor and the outcome. The data are then used to refute the assumption of no association. The p-value is then the probability that we observe a more extreme value of the statistic if, in fact, no association exists between the predictor and outcome.

Each statistic can be converted to a p-value, but this conversion is easier for some statistics than others. For instance, converting a t-statistic to a p-value is a well-known process, provided that some basic assumptions are true. On the other hand, it is not easy to convert an AUC to a p-value. A solution to this problem is by using a *permutation* method (Good, 2013; Berry et al., 2016). This approach can be applied to any statistic to generate a p-value. Here is how a randomization approach works: for a selected predictor and corresponding outcome, the predictor is randomly permuted, but the outcome is not. The statistic of interest is then calculated on the permuted data. This process disconnects the observed predictor and outcome relationship, thus creating no association between the two. The same predictor is randomly permuted many times to generate a distribution of statistics. This distribution represents the distribution of no association (i.e., the null distribution). The statistic from the original data can then be compared to the distribution of no association to get a probability, or p-value, of coming from this distribution. An example of this is given in Kuhn and Johnson (2013) for Relief scores (Robnik-Sikonja and Kononenko, 2003).

In March of 2016, the American Statistical Association released a statement[84] regarding the use of p-values (and null-hypothesis-based test) to discuss their utility and misuse. For example, one obvious issue is that the application of an arbitrarily defined threshold to determine "significance" outside of any practical context can be problematic. Also, it is important to realize that when a statistic such as the area under the ROC curve is converted into a p-value, the level of evidence required for "importance" is being set very low. The reason for this is that the null hypothesis is *no association.* One might think that, if feature A has a p-value of 0.01 and feature B has a p-value of 0.0001, that the original ROC value was larger for B. This may not be the case since the p-value is a function of signal, noise, and other factors (e.g., the sample size). This is especially true when the p-values are being generated from different data sets; in these cases, the sample size may be the prime determinant of the value.

For our purposes, the p-values are being used algorithmically to search for appropriate subsets, especially since it is a convenient quantity to use when dealing with different predictor types and/or importance statistics. As mentioned in Section 5.6 and below in Section 11.4, it is very important to not take p-values generated in such processes as meaningful or formal indicators of statistical significance.

Good summaries of the issues are contained in Yaddanapudi (2016), Wasserstein and Lazar (2016), and the entirety of Volume 73 of *The American Statistician.*

Simple filters are effective at identifying individual predictors that are associated with the outcome. However, these filters are very susceptible to finding predictors that have strong associations in the available data but do not show any association with new data. In the statistical literature, these selected predictors are labeled as *false positives.* An entire sub-field of statistics has been devoted to developing approaches for minimizing the chance of false positive findings, especially in the context of hypothesis testing and p-values. One approach to reducing false positives

---

[84]https://www.amstat.org/asa/files/pdfs/P-ValueStatement.pdf

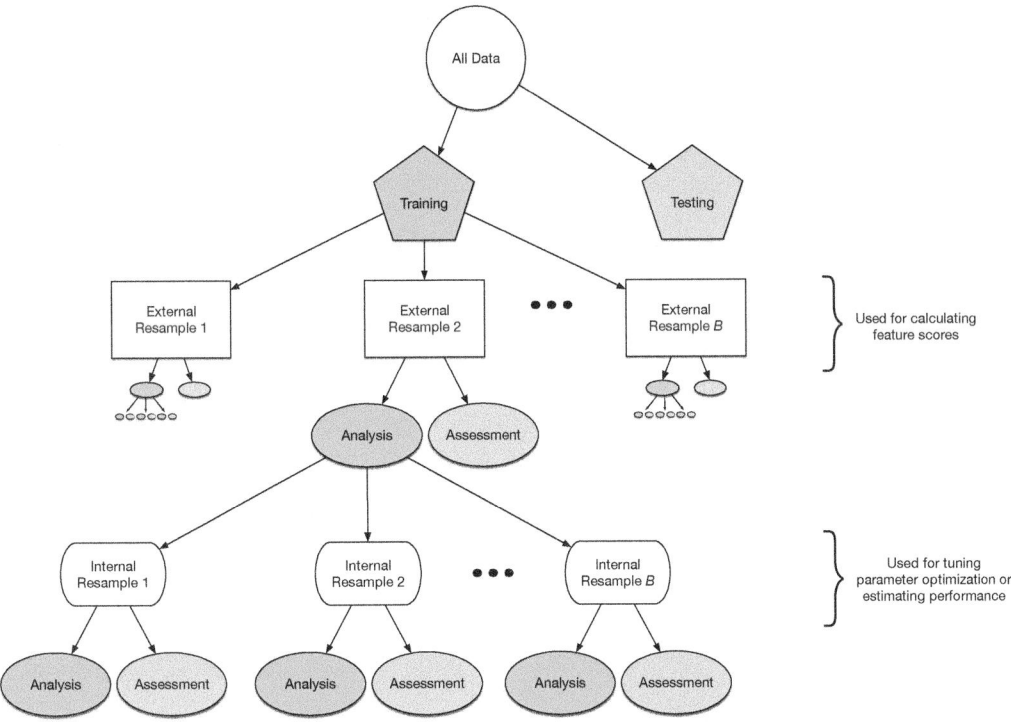

Figure 11.2: A diagram of external and internal cross-validation for simple filters.

is to adjust the p-values to effectively make them larger and thus less significant (as was shown in previous chapters).

In the context of predictive modeling, false positive findings can be minimized by using an independent set of data to evaluate the selected features. This context is exactly parallel to the context of identifying optimal model tuning parameters. Recall from Section 3.4 that cross-validation is used to identify an optimal set of tuning parameters such that the model does not overfit the available data. The model building process now needs to accomplish two objectives: to identify an effective subset of features, and to identify the appropriate tuning parameters such that the selected features and tuning parameters do not overfit the available data. When using simple screening filters, selecting both the subset of features and model tuning parameters cannot be done in the same layer of cross-validation, since the filtering must be done independently of the model tuning. Instead, we must incorporate another layer of cross-validation. The first layer, or external layer, is used to filter features. Then the second layer (the "internal layer") is used to select tuning parameters. A diagram of this process is illustrated in Figure 11.2.

As one can see from this figure, conducting feature selection can be computationally costly. In general, the number of models constructed and evaluated is $I \times E \times T$, where I is the number if internal resamples, E is the number of external resamples, and T is the total number of tuning parameter combinations.

11.2.1 | Simple Filters Applied to the Parkinson's Disease Data

The Parkinson's disease data has several characteristics that could make modeling challenging. Specifically, the predictors have a high degree of multicollinearity, the sample size is small, and the outcome is imbalanced (74.6% of patients have the disease). Given these characteristics, partial least squares discriminant analysis (PLSDA, Barker and Rayens (2003)) would be a good first model to try. This model produces linear class boundaries which may constrain it from overfitting to the majority class. The model tuning parameters for PLSDA is the number of components to retain.

The second choice we must make is the criteria for filtering features. For these data we will use the area under the ROC curve to determine if a feature should be included in the model. The initial analysis of the training set showed that there were 5 predictors with an area under the ROC curve of at least 0.80 and 21 were between 0.75 and 0.80.

For the internal resampling process, an audio feature was selected if it had an area under the ROC curve of at least 0.80. The selected features within the internal resampling node are then passed to the corresponding external sampling node where the model tuning process is conducted.

Once the filtering criteria and model are selected, then the type of internal and external resampling methods are chosen. In this example, we used 20 iterations of bootstrap resampling for internal resampling and 5 repeats of 10-fold cross-validation for external resampling. The final estimates of performance are based on the external resampling hold-out sets. To preprocess the data, a Yeo-Johnson transformation was applied to each predictor, then the values were centered and scaled. These operations were conducted within the resampling process.

During resampling the number of predictors selected ranged from 2 to 12 with an average of 5.7. Some predictors were selected regularly; 2 of them passed the ROC threshold in each resample. For the final model, the previously mentioned 5 predictors were selected. For this set of predictors, the PLSDA model was optimized with 4 components. The corresponding estimated area under the ROC curve was 0.827.

Looking at the selected predictors more closely, there appears to be considerable redundancy. Figure 11.3 shows the rank correlation matrix of these predictors in a heatmap. There are very few pairs with absolute correlations less than 0.5 and many extreme values. As mentioned in Section 6.3, partial least squares models are good at *feature extraction* by creating new variables that are linear combinations of the original data. For this model, these PLS components are created in a way that summarizes the maximum amount of variation in the predictors while simultaneously minimizing the misclassification among groups. The final model used 4 components, each of which is a combination of 5 of the original predictors that survived the filter. Looking at Figure 11.3, there is a relatively clean separation into distinct blocks/clusters. This indicates that there are probably only a handful of underlying pieces of information in the filtered predictors. A separate principal component analysis was used to gauge the magnitude of the between-predictor correlations. In this analysis, only 1 feature was needed to capture 90% of the total variation. This reinforces that there are only a small number of *underlying effects* in these predictors. This is the reason that such a small number of components were found to be optimal.

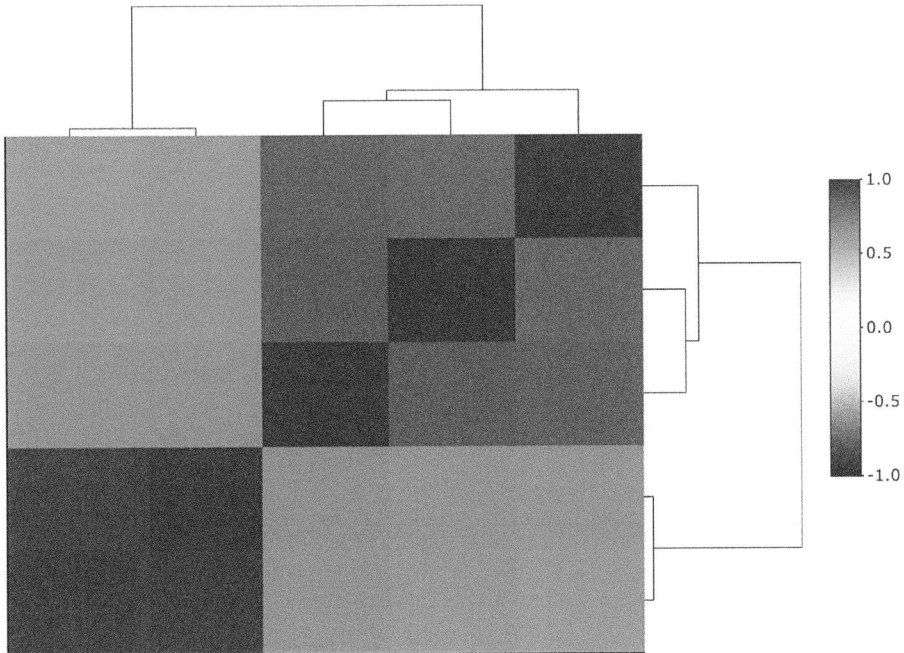

Figure 11.3: The correlation matrix of the predictors chosen by the filtering process based on a threshold on the area under the ROC curve.

Did this filtering process help the model? If the same resamples are used to fit the PLS model using the entire predictor set, the model still favored a small number of projections (4 components) and the area under the ROC curve was estimated to be a little better than the full model (0.812 versus 0.827 when filtered). The p-value from a paired t-test ($p = 0.446$) showed no evidence that the improvement is real. However, given that only 0.7% of the original variables were used in this analysis, the simplicity of the new model might be very attractive.

One *post hoc* analysis that is advisable when conducting feature selection is to determine if the particular subset that is found is any better than a random subset of the same size. To do this, 100 random subsets of 5 predictors were created and PLS models were fitted with the same external cross-validation routine. Where does our specific subset fit within the distribution of performance that we see from a random subset? The area under the ROC curve using the above approach (0.827) was better than 97 percent of the values generated using random subsets of the same size. The largest AUC from a random subset was 0.832. This result grants confidence that the filtering and modeling approach found a subset with credible signal. However, given the amount of correlation between the predictors, it is unlikely that this is the only subset of this size that could achieve equal performance.

While partial least squares is very proficient at accommodating predictor sets with a high degree of collinearity, it does raise the question about what the minimal variable set might be that would achieve nearly optimal predictive performance. For this example, we could reduce the subset size by increasing the stringency of the filter (i.e., $AUC > 0.80$). Changing the threshold will affect both the sparsity of the solution and the performance of the model. This process is the same as backwards selection which will be discussed in the next section.

To summarize, the use of a simple screen prior to modeling can be effective and relatively efficient. The filters should be included in the resampling process to avoid optimistic assessments of performance. The drawback is that there may be some redundancy in the selected features and the subjectivity around the filtering threshold might leave the modeler wanting to understand how many features *could* be removed before performance is impacted.

11.3 | Recursive Feature Elimination

As previously noted, recursive feature elimination (RFE, Guyon et al. (2002)) is basically a backward selection of the predictors. This technique begins by building a model on the entire set of predictors and computing an importance score for each predictor. The least important predictor(s) are then removed, the model is re-built, and importance scores are computed again. In practice, the analyst specifies the number of predictor subsets to evaluate as well as each subset's size. Therefore, the subset size is a *tuning parameter* for RFE. The subset size that optimizes the performance criteria is used to select the predictors based on the importance rankings. The optimal subset is then used to train the final model.

Section 10.4 described in detail the appropriate way to estimate the subset size. The selection process is resampled in the same way as fundamental tuning parameters from a model, such as the number of nearest neighbors or the amount of weight decay in a neural network. The resampling process *includes* the feature selection routine and the *external* resamples are used to estimate the appropriate subset size.

Not all models can be paired with the RFE method, and some models benefit more from RFE than others. Because RFE requires that the initial model uses the full predictor set, some models cannot be used when the number of predictors exceeds the number of samples. As noted in previous chapters, these models include multiple linear regression, logistic regression, and linear discriminant analysis. If we desire to use one of these techniques with RFE, then the predictors must first be winnowed down. In addition, some models benefit more from the use of RFE than others. Random forest is one such case (Svetnik et al., 2003) and RFE will be demonstrated using this model for the Parkinson's disease data.

Backwards selection is frequently used with random forest models for two reasons. First, as noted in Chapter 10, random forest tends not to exclude variables from the prediction equation. The reason is related to the nature of model ensembles. Increased performance in ensembles is related to the diversity in the constituent models; averaging models that are effectively the same does not drive down the variation in the model predictions. For this reason, random forest coerces the trees to contain sub-optimal splits of the predictors using a random sample of predictors.[85] The act of restricting the number of predictors that could possibly be used in a split increases the likelihood that an irrelevant predictor will be chosen. While such a predictor may not have much direct impact on the performance of the model, the prediction equation is functionally dependent on that predictor. As our simulations showed, tree ensembles may use every possible predictor at least once in the ensemble. For this reason, random forest can use some *post hoc* pruning of variables that are not essential for the model to perform well. When many irrelevant predictors are included in the data and the RFE process is paired with random forest, a wide range of subset sizes can exhibit very similar predictive performances.

The second reason that random forest is used with RFE is because this model has a well-known internal method for measuring feature importance. This was described previously in Section 7.4 and can be used with the first model fit within RFE, where the entire predictor set is used to compute the feature rankings.

One notable issue with measuring importance in trees is related to multicollinearity. If there are highly correlated predictors in a training set that are useful for predicting the outcome, then which predictor is chosen is essentially a random selection. It is common to see that a set of highly redundant and useful predictors are *all* used in the splits across the ensemble of trees. In this scenario, the predictive performance of the ensemble of trees is unaffected by highly correlated, useful features. However, the redundancy of the features dilutes the importance scores. Figure 11.4 shows an example of this phenomenon. The data were simulated using the same system described in Section 10.3 and a random forest model was tuned using 2,000 trees. The largest importance was associated with feature x_4. Additional copies of this feature were added to the data set and the model was refit. The figure shows the decrease in importance for x_4 as copies are added.[86] There is a clear decrease in importance as redundant features are added to the model. For this reason, we should be careful about labeling predictors as unimportant when their permutation scores are not large since these may be masked by correlated variables. When using RFE with random forest, or other tree-based models, we advise experimenting with filtering out highly correlated features prior to beginning the routine.

For the Parkinson's data, backwards selection was conducted with random forests,[87]

---

[85] Recall that the size of the random sample, typically denoted as m_{try}, is the main tuning parameter.

[86] Each model was replicated five times using different random number seeds.

[87] While m_{try} is a tuning parameter for random forest models, the default value of $m_{try} \approx sqrt(p)$ tends to provide good overall performance. While tuning this parameter may have led to better performance, our experience is that the improvement tends to be marginal.

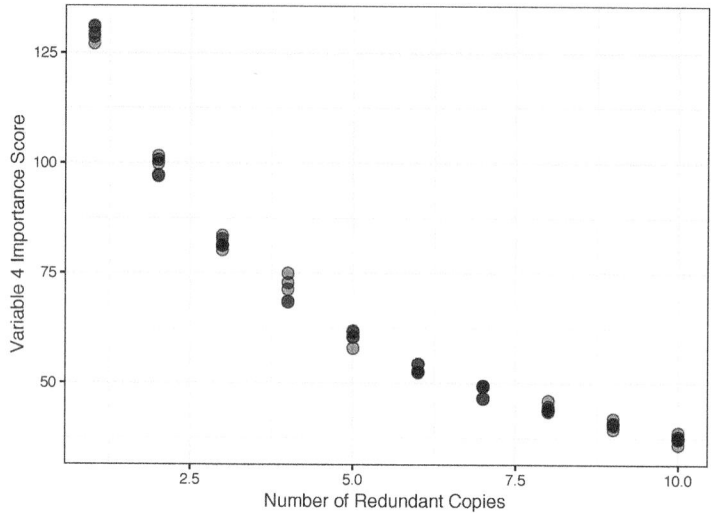

Figure 11.4: The dilution effect of random forest permutation importance scores when redundant variables are added to the model.

and each ensemble contained 10,000 trees.[88] The model's importance scores were used to rank the predictors. Given the number of predictors and the large amount of correlation between them, the subset sizes were specified on the $\log_{10}$ scale so that the models would more thoroughly investigate smaller sizes. The same resampling scheme was used as the simple filtering analysis. Given our previous comments regarding correlated predictors, the RFE analysis was conducted with and without a correlation filter that excluded enough predictors to coerce all of the absolute pairwise correlations to be smaller than 0.50. The performance profiles, shown in the left-hand panel of Figure 11.5, illustrate that the models created using all predictors show an edge in model performance; the ROC curve AUC was 0.064 larger with a confidence interval for the difference being (0.037, 0.091). Both curves show a fairly typical pattern; performance stays fairly constant until relevant features are removed. Based on the unfiltered models, the numerically best subset size is 377 predictors, although the filtered model could probably achieve similar performance using subsets with around 100 predictors.

What would happen if the rankings were based on the ROC curve statistics instead of the random forest importance scores? On one hand, important predictors that share a large correlation might be ranked higher. On the other hand, the random forest rankings are based on the simultaneous presence of all features. That is, all of the predictors are being considered at once. For this reason, the ranks based on the simultaneous calculation might be better. For example, if there are important predictor interactions, random forest would account for these contributions, thus increasing the importance scores.

---

[88]In general, we do not advise using this large ensemble size for every random forest model. However, in data sets with relatively small (but wide) training sets, we have noticed that increasing the number of trees to 10K–15K is required to get accurate and reliable predictor rankings.

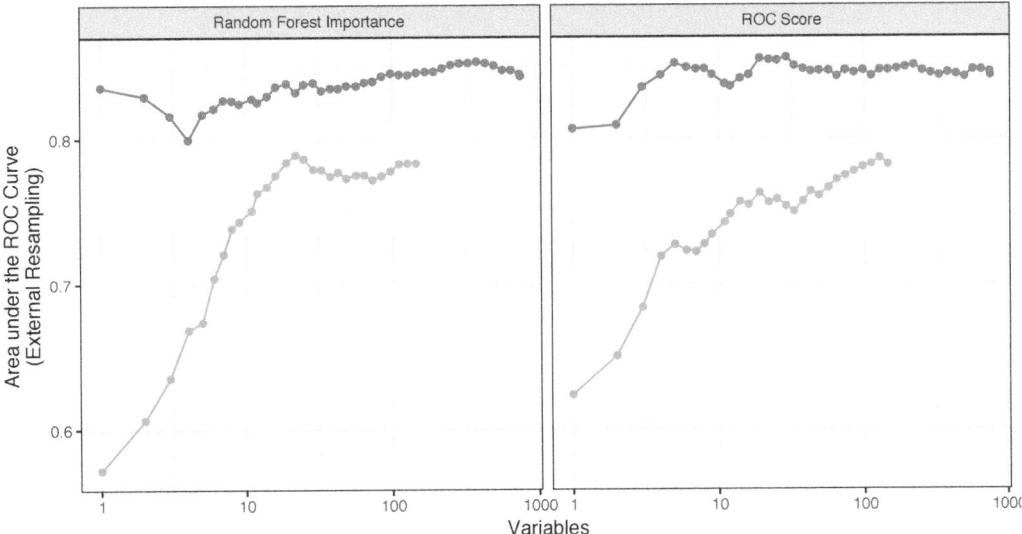

Figure 11.5: The external resampling results for RFE using random forests. The panels reflect how the predictor importances were calculated.

To test this idea, the same RFE search was used with the individual areas under the ROC curves (again, with and without a correlation filter). The results are shown in the right-hand panel of Figure 11.5. Similar to the left-hand panel, there is a loss of performance in these data when the data are pre-filtered for correlation. The unfiltered model shows solid performance up to about 30 predictors although the numerically optimal size was 128 predictors. Based on the filtered model shown here, a subset of 128 predictors could produce an area under the ROC curve of about 0.856. Using another 100 random samples of subsets of size 128, in this case with lower correlations, the optimized model has higher ROC scores than 98% of the random subsets.

How consistent were the predictor rankings in the data? Figure 11.6 shows the ROC importance scores for each predictor. The points are the average scores over resamples and the bands reflect two standard deviations of those values. The rankings appear reasonably consistent since the trend is not a completely flat horizontal line. From the confidence bands, it is apparent that some predictors were only selected once (i.e., no bands), only a few times, or had variable values across resamples.

RFE can be an effective and relatively efficient technique for reducing the model complexity by removing irrelevant predictors. Although it is a greedy approach, it is probably the most widely used method for feature selection.

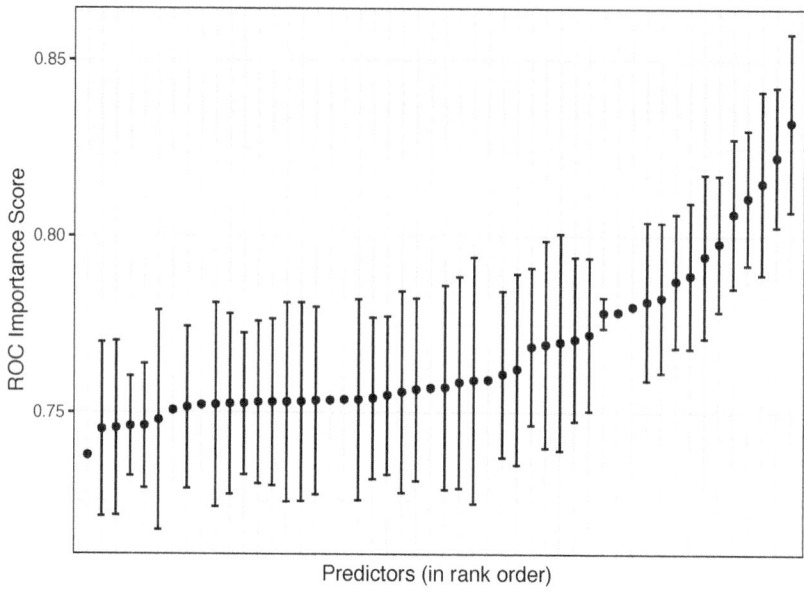

Figure 11.6: Mean and variability of predictor rankings.

11.4 | Stepwise Selection

Stepwise selection was originally developed as a feature selection technique for linear regression models. The forward stepwise regression approach uses a sequence of steps to allow features to enter or leave the regression model one-at-a-time. Often this procedure converges to a subset of features. The entry and exit criteria are commonly based on a p-value threshold. A typical entry criterion is that a p-value must be less than 0.15 for a feature to enter the model and must be greater than 0.15 for a feature to leave the model. The process begins by creating p linear regression models, each of which uses exactly one of the features.[89] The importance of the features are then ranked by their individual ability to explain variation in the outcome. The amount of variation explained can be condensed into a p-value for convenience. If no features have a p-value less than 0.15, then the process stops. However, if one or more features have p-value(s) less than 0.15, then the one with the lowest value is retained. In the next step, $p - 1$ linear regression models are built. These models consist of the feature selected in the first step as well as each of the other features individually. Each of the additional features is evaluated, and the best feature that meets the inclusion criterion is added to the selected feature set. Then the amount of variation explained by each feature in the presence of the other feature is computed and converted to a p-value. If the p-values do not exceed the exclusion criteria, then both are kept and the search procedure proceeds to search for a third feature. However, if a feature has a p-value that exceeds the exclusion criteria, then it is removed from the current set of selected features. The removed feature can still

---

[89] p, in this instance, is the number of predictors, not to be confused with p-values from statistical hypothesis tests.

enter the model at a later step. This process continues until it meets the convergence criteria.

This approach is problematic for several reasons, and a large literature exists that critiques this method (see Steyerberg et al. (1999), Whittingham et al. (2006), and Mundry and Nunn (2009)). Harrell (2015) provides a comprehensive indictment of the method that can be encapsulated by the statement:

> "... if this procedure had just been proposed as a statistical method, it would most likely be rejected because it violates every principle of statistical estimation and hypothesis testing."

Stepwise selection has two primary faults:

1. Inflation of false positive findings: Stepwise selection uses many repeated hypothesis tests to make decisions on the inclusion or exclusion of individual predictors. The corresponding p-values are unadjusted, leading to an over-selection of features (i.e., false positive findings). In addition, this problem is exacerbated when highly correlated predictors are present.

2. Model overfitting: The resulting model statistics, including the parameter estimates and associated uncertainty, are highly optimistic since they do not take the selection process into account. In fact, the final standard errors and p-values from the final model are incorrect.

It should be said that the second issue is also true of all of the search methods described in this chapter and the next. The individual model statistics cannot be taken as literal for this same reason. The one exception to this is the resampled estimates of performance. Suppose that a linear regression model was used inside RFE or a global search procedure. The internal estimates of adjusted R^2, RMSE, and others will be optimistically estimated but the external resampling estimates should more accurately reflect the predictive performance of the model on new, independent data. That is why the external resampling estimates are used to guide many of the search methods described here and to measure overall effectiveness of the *process* that results in the final model.

One modification to the process that helps mitigate the first issue is to use a statistic other than p-values to select a feature. Akaike information criterion (AIC) is a better choice (Akaike, 1974). The AIC statistic is tailored to models that use the likelihood as the objective (i.e., linear or logistic regression), and penalizes the likelihood by the number of parameters included in the model. Therefore models that optimize the likelihood and have fewer parameters are preferred. Operationally, after fitting an initial model, the AIC statistic can be computed for each submodel that includes a new feature or excludes an existing feature. The next model corresponds to the one with the best AIC statistic. The procedure repeats until the current model contains the best AIC statistic.

However, it is important to note that the AIC statistic is specifically tailored to models that are based on likelihood. To demonstrate stepwise selection with the AIC statistic, a logistic regression model was built for the OkCupid data. For illustration purposes, we are beginning with a model that contained terms for age, essay length, and an indicator for being Caucasian. At the next step, three potential features will be evaluated for inclusion in the model. These features are indicators for the keywords of `nerd`, `firefly`, and `im`.

The first model containing the initial set of three predictors has an associated binomial log-likelihood value of -18525.3. There are 4 parameters in the model, one for each term and one for the intercept. The AIC statistic is computed using the log-likelihood (denoted as ℓ) and the number of model parameters p as follows:

$$AIC = -2\ell + 2p = -2(-18525.3) + 8 = 37059.$$

In this form, the goal is to *minimize* the AIC value.

On the first iteration of stepwise selection, six speculative models are created by dropping or adding single variables to the current set and computing their AIC values. The results are, in order from best to worst model:

Term	AIC
+ nerd	36,863
+ firefly	36,994
+ im	37,041
current model	37,059
- white	37,064
- age	37,080
- essay length	37,108

Based on these values, adding any of the three keywords would improve the model, with `nerd` yielding the best model. Stepwise is a less greedy method than other search methods since it does reconsider adding terms back into the model that have been removed (and vice versa). However, all of the choices are made on the basis of the current optimal step at any given time.

Our recommendation is to avoid this procedure altogether. Regularization methods, such as the previously discussed `glmnet` model, are far better at selecting appropriate subsets in linear models. If model inference is needed, there are a number of Bayesian methods that can be used (Mallick and Yi, 2013; Piironen and Vehtari, 2017b,a).

11.5 │ Summary

Identifying the subset of features that produces an optimal model is often a goal of the modeling process. This is especially true when the data consist of a vast number of predictors.

A simple approach to identifying potentially predictively important features is to evaluate each feature individually. Statistical summaries such as a t-statistic, odds ratio, or correlation coefficient can be computed (depending on the type of predictor and type of outcome). While these values are not directly comparable, each can be converted to a p-value to enable a comparison across multiple types. To avoid finding false positive associations, a resample approach should be used in the search.

Simple filters are ideal for finding individual predictors. However, this approach does not take into account the impact of multiple features together. This can lead to a selection of more features than necessary to achieve optimal predictive performance. Recursive feature elimination is an approach that can be combined with any model to identify a good subset of features with optimal performance for the model of interest. The primary drawback of RFE is that it requires that the initial model be able to fit the entire set of predictors.

The stepwise selection approach has been a popular feature selection technique. However, this technique has some well-known drawbacks when using p-values as a criteria for adding or removing features from a model and should be avoided.

11.6 | Computing

The website `http://bit.ly/fes-greedy` contains R programs for reproducing these analyses.

12 | Global Search Methods

As previously mentioned in Chapter 10, global search methods investigate a diverse potential set of solutions. In this chapter, two methods (genetic algorithms and simulated annealing) will be discussed in the context of selecting appropriate subsets of features. There are a variety of other global search methods that can also be used, such as particle swarm optimization and simultaneous perturbation stochastic approximation. Spall (2005) and Weise (2011) are comprehensive resources for these types of optimization techniques.

Before proceeding, the naive Bayes classification model will be discussed. This classifier will be used throughout this chapter to demonstrate the global search methodologies.

12.1 | Naive Bayes Models

To illustrate the methods used in this chapter, the OkCupid data will be used in conjunction with the naive Bayes classification model. The naive Bayes model is computationally efficient and is an effective model for illustrating global search methods. This model uses Bayes' formula to predict classification probabilities.

$$Pr[Class|Predictors] = \frac{Pr[Class] \times Pr[Predictors|Class]}{Pr[Predictors]}$$
$$= \frac{Prior \ \times \ Likelihood}{Evidence}$$

In English:

Given our predictor data, what is the probability of each class?

There are three elements to this equation. The *prior* is the general probability of each class (e.g., the rate of STEM profiles) that can either be estimated from the training data or determined using expert knowledge. The *likelihood* measures the relative probability of the observed predictor data occurring for each class. And the *evidence* is a normalization factor that can be computed from the prior and likelihood. Most of the computations occur when determining the likelihood. For example, if there are multiple numeric predictors, a multivariate probability distribution can

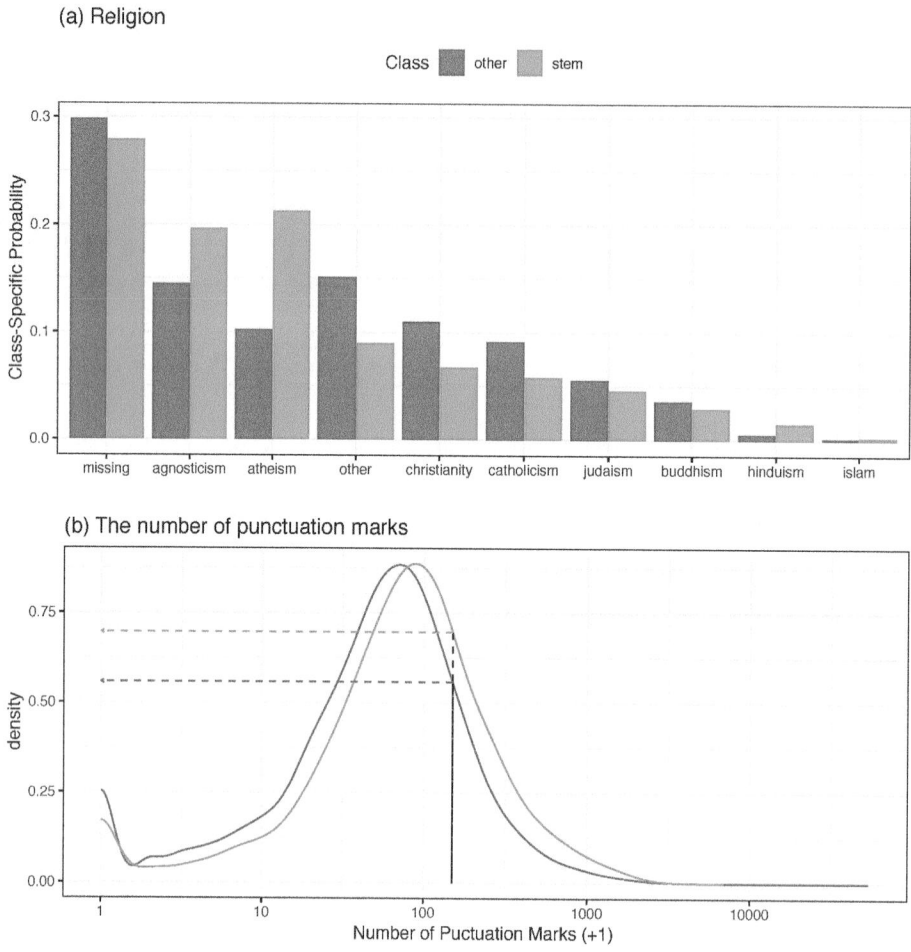

Figure 12.1: Visualizations of the conditional distributions for a continuous and a discrete predictor for the OkCupid data.

be used for the computations. However, this is very difficult to compute outside of a low-dimensional normal distribution or may require an abundance of training set data. The determination of the likelihood becomes even more complex when the features are a combination of numeric and continuous values. The *naive* aspect of this model is due to a very stringent assumption: the predictors are assumed to be *independent*. This enables the joint likelihood to be computed as a product of individual statistics for each predictor.

For example, suppose a naive Bayes model with two predictors: the number of punctuation marks in the essays as well as the stated religion. To start, consider the religion predictor (which has 10 possible values). For this predictor, a cross-tabulation is made between the values and the outcome and the probability of each religion value, within each class, is computed. Figure 12.1(a) shows the class-specific probabilities. Suppose an individual listed their religion as Atheist. The computed probability of being an atheist is 0.213 for STEM profiles and 0.103 otherwise.

For the continuous predictor, the number of punctuation marks, the distribution of the predictor is computed separately for each class. One way to accomplish this is by binning the predictor and using the histogram frequencies to estimate the probabilities. A better approach is to compute a nonparametric density function on these data (Wand and Jones, 1994). Figure 12.1(b) shows the two density curves for each class (with a $\log_{10}$ transformation on the predictor data). There is a slight trend where the non-STEM profiles trend to use less punctuation. Suppose that an individual used about 150 punctuation marks in their essay (indicated by the horizontal black line). To compute the relative density for each class, the corresponding heights of the density lines for each class are used. Here, the density for a STEM profile was 0.697 and was 0.558 for non-STEMs.

Now that the statistics have been calculated for each predictor and class value, the overall prediction can be made. First, the values for both predictors are multiplied together to form the likelihood statistic for each class. Table 12.1 shows the details for the computations. Both religion and the punctuation mark data were more consistent with the profile being STEM; the ratio of likelihood values for the classes is $0.148 \div 0.057 = 2.59$, indicating that the profile is more likely to correspond to a STEM career (based on the training data). However, Bayes' Rule also includes a term for the *prior information*, which is the overall rate that we would find STEM profiles in the population of interest.[90] In these data, the STEM rate was about 18.5%, which is most likely higher than in other parts of the country. To get the final prediction, the prior and likelihood are multiplied together and their values are normalized to sum to one to give us the *posterior probabilities* (which are just the class probability predictions). In the end, the probability that this particular profile is associated with a STEM job is only 37%; our prior information about the outcome was extreme enough to contradict what the observed data were indicating.

Considering each predictor separately makes the computations much faster and these models can be easily deployed since the information needed to compute the likelihood can be stored in efficient look-up tables. However, the assumption of independent predictors is not plausible for many data sets. As an example, the number of punctuation marks has strong rank-correlations with the number of words (correlation: 0.847), the number of commas (0.694), and so on. However, in many circumstances, the model can produce good predictions even when disregarding correlation among predictors.

One other relevant aspect of this model is how qualitative predictors are handled. For example, religion was not decomposed into indicator variables for this analysis. Alternatively, this predictor could be decomposed into the complete set of all 10 religion responses and these can be used in the model instead. However, care must be taken that the new binary variables are still treated as qualitative; otherwise a continuous density function (similar to the analysis in Figure 12.1(b)) is used to represent a predictor with only two possible values. When the naive Bayes model

---

[90]Recall that the prior probability was also discussed in Section 3.2.2 when the *prevalence* was used in some calculations.

Table 12.1: Values used in the naive Bayes model computations.

Class	Predictor Values		Likelihood	Prior	Posterior
	Religion	Punctuation			
STEM	0.213	0.697	0.148	0.185	0.37
other	0.103	0.558	0.057	0.815	0.63

used all of the feature sets described in Section 5.6, roughly 110 variables, the cross-validated area under the ROC curve was computed to be 0.798. When the qualitative variables were converted to separate indicators the AUC value was 0.799. The difference in performance was negligible, although the model with the individual indicator variables took 1.9-fold longer to compute due to the number of predictors increasing from 110 to 298.

This model can have some performance benefit from reducing the number of predictors since less noise is injected into the probability calculations. There is an ancillary benefit to using fewer features; the independence assumption tends to produce highly pathological class probability distributions as the number of predictors approaches the number of training set points. In these cases, the distributions of the class probabilities become somewhat bi-polar; very few predicted points fall in the middle of the range (say greater than 20% and less than 80%). These U-shaped distributions can be difficult to explain to consumers of the model since most incorrectly predicted samples have a high certainty of being true (despite the opposite being the case). Reducing the number of predictors can mitigate this issue.

This model will be used to demonstrate global search methods using the OkCupid data. Naive Bayes models have some advantages in that they can be trained very quickly and the prediction function can be easily implemented in a set of lookup tables that contain the likelihood and prior probabilities. The latter point is helpful if the model will be scored using database systems.

12.2 │ Simulated Annealing

Annealing is the process of heating a metal or glass to remove imperfections and improve strength in the material. When metal is hot, the particles are rapidly rearranging at random within the material. The random rearrangement helps to strengthen weak molecular connections. As the material cools, the random particle rearrangement continues, but at a slower rate. The overall process is governed by time and the material's energy at a particular time.

The annealing process that happens to particles within a material can be abstracted and applied for the purpose of global optimization. The goal in this case is to find the strongest (or most predictive) set of features for predicting the response. In the context of feature selection, a predictor is a particle, the collection of predictors is

---

Algorithm 12.1: Simulated annealing for feature selection.

---

1 Create an initial random subset of features and specify the number of iterations;

2 **for** *each iteration of SA* **do**

3   Perturb the current feature subset;

4   Fit model and estimate performance;

5   **if** *performance is better than the previous subset* **then**

6   Accept new subset;

7   **else**

8   Calculate acceptance probability;

9   **if** *random uniform variable > probability* **then**

10   Reject new subset;

11   **else**

12   Accept new subset;

13   **end**

14   **end**

15 **end**

---

the material, and the set of selected features in the current arrangement of particles. To complete the analogy, time is represented by iteration number, and the change in the material's energy is the change in the predictive performance between the current and previous iteration. This process is called *simulated annealing* (Kirkpatrick et al., 1983; Van Laarhoven and Aarts, 1987).

To initialize the simulated annealing process, a subset of features is selected at random. The number of iterations is also chosen. A model is then built and the predictive performance is calculated. Then a small percentage (1%–5%) of features is randomly included or excluded from the current feature set and predictive performance is calculated for the new set. If performance improves with the new feature set, then the new set of features is kept. However, if the new feature set has worse performance, then an acceptance probability is calculated using the formula

$$Pr[accept] = \exp\left[-\frac{i}{c}\left(\frac{old - new}{old}\right)\right]$$

in the case where better performance is associated with larger values. The probability is a function of time and change in performance, as well as a constant, c which can be used to control the rate of feature perturbation. By default, c is set to 1. After calculating the acceptance probability, a random uniform number is generated. If the random number is greater than the acceptance probability, then the new feature set is rejected and the previous feature set is used. Otherwise, the new feature set is kept. The randomness imparted by this process allows simulated annealing to escape local optima in the search for the global optimum.

To understand the acceptance probability formula, consider an example where the objective is to optimize predictive accuracy. Suppose the previous solution had an accuracy of 0.85 and the current solution has an accuracy of 0.80. The proportionate decrease of the current solution relative to the previous solution is 5.9%. If this situation occurred in the first iteration, then the acceptance probability would be 0.94. At iteration 5, the probability of accepting the inferior subset would be 0.75. At iteration 50, this the acceptance probability drops to 0.05. Therefore, the probability of accepting a worse solution decreases as the algorithm iteration increases. Algorithm 12.1 summarizes the basic simulated annealing algorithm for feature selection.

While the random nature of the search and the acceptance probability criteria help the algorithm avoid local optimums, a modification called *restarts* provides and additional layer of protection from lingering in inauspicious locales. If a new optimal solution has not been found within R iterations, then the search resets to the last known optimal solution and proceeds again with R being the number of iterations since the restart.

As an example, the table below shows the first 15 iterations of simulated annealing where the outcome is the area under the ROC curve:

Iteration	Size	ROC	Probability	Random Uniform	Status
1	122	0.776	—	—	Improved
2	120	0.781	—	—	Improved
3	122	0.770	0.958	0.767	Accepted
4	120	0.804	—	—	Improved
5	120	0.793	0.931	0.291	Accepted
6	118	0.779	0.826	0.879	Discarded
7	122	0.779	0.799	0.659	Accepted
8	124	0.776	0.756	0.475	Accepted
9	124	0.798	0.929	0.879	Accepted
10	124	0.774	0.685	0.846	Discarded
11	126	0.788	0.800	0.512	Accepted
12	124	0.783	0.732	0.191	Accepted
13	124	0.790	0.787	0.060	Accepted
14	124	0.778	—	—	Restart
15	120	0.790	0.982	0.049	Accepted

The restart threshold was set to 10 iterations. The first two iterations correspond to improvements. At iteration three, the area under the ROC curve does not improve but the loss is small enough to have a corresponding acceptance probability of 95.8%. The associated random number is smaller, so the solutions is accepted. This pays off since the next iteration results in a large performance boost but is followed by a series of losses. Note that, at iteration six, the loss is large enough to have an acceptance probability of 82.6% which is rejected based on the associated draw of a random uniform. This means that configuration seven is based on configuration number five. These losses continue until iteration 14 where the restart occurs; configuration 15 is based on the last best solution (at iteration 4).

12.2.1 | Selecting Features without Overfitting

How should the data be used for this search method? Based on the discussion in Section 10.4, it was noted that the same data should not be used for selection, modeling, and evaluation. An external resampling procedure should be used to determine how many iterations of the search are appropriate. This would average the assessment set predictions across all resamples to determine how long the search should proceed when it is directly applied to the entire training set. Basically, the number of iterations is a tuning parameter.

For example, if 10-fold cross-validation is used as the external resampling scheme, simulated annealing is conducted 10 times on 90% of the data. Each corresponding assessment set is used to estimate how well the process is working at each iteration of selection.

Within each external resample, the analysis set is available for model fitting and assessment. However, it is probably not a good idea to use it for both purposes. A nested resampling procedure would be more appropriate An *internal resample of the data* should be used to split it into a subset used for model fitting and another for evaluation. Again, this can be a full iterative resampling scheme or a simple validation set. As before, 10-fold cross-validation is used for the external resample. If the external analysis set, which is 90% of the training set, is larger enough we might use a random 80% of these data for modeling and 20% for evaluation (i.e., a single validation set). These would be the *internal analysis and assessment sets*, respectively. The latter is used to guide the selection process towards an optimal subset. A diagram of the external and internal cross-validation approach is illustrated in Figure 12.2.

Once the external resampling is finished, there are matching estimates of performance for each iteration of simulated annealing and each serves a different purpose:

- the internal resamples guide the subset selection process

- the external resamples help tune the number of search iterations to use.

For data sets with many training set samples, the internal and external performance measures may track together. In smaller data sets, it is not uncommon for the internal performance estimates to continue to show improvements while the external estimates are flat or become worse. While the use of two levels of data splitting is less straightforward and more computationally expensive, it is the only way to understand if the selection process is overfitting.

Once the optimal number of search iterations is determined, the final simulated

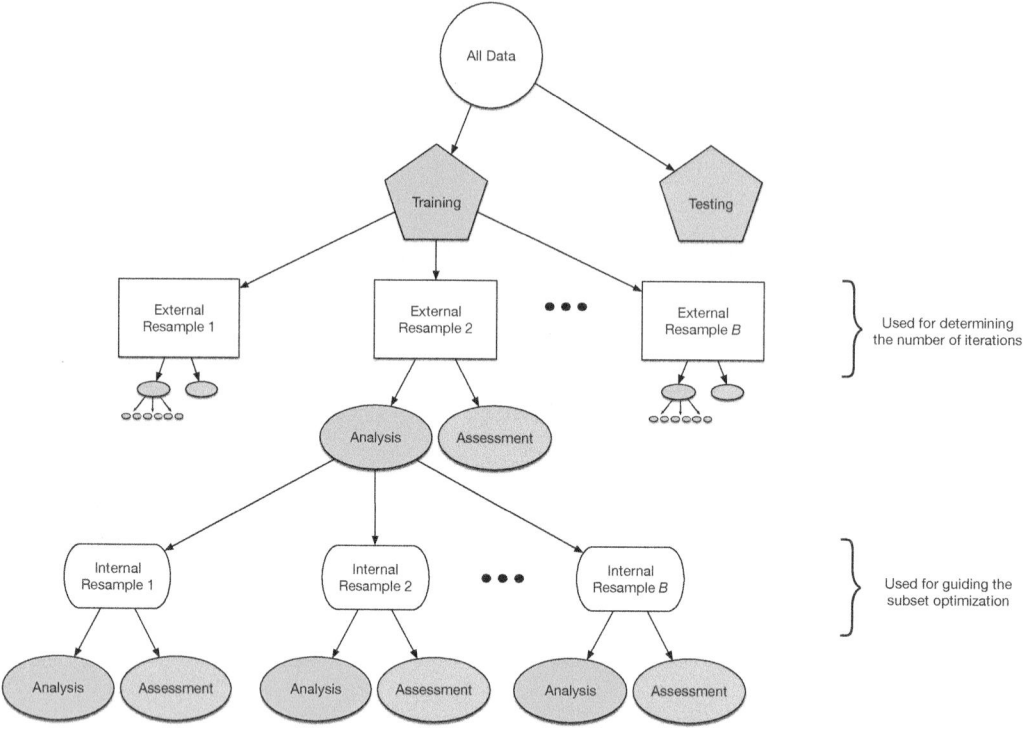

Figure 12.2: A diagram of external and internal cross-validation for global search.

annealing search is conducted on the entire training set. As before, the same data set should not be used to create the model and estimate the performance. An internal split is still required.

From this last search, the optimal predictor set is determined and the final model is fit to the entire training set using the best predictor set. This resampled selection procedure is summarized in Algorithm 12.2. It should be noted that lines 2–20 can be run in parallel (across external resamples) to reduce the computational time required.

12.2.2 | Application to Modeling the OkCupid Data

To illustrate this process, the OkCupid data are used. Recall that the training set includes 38,809 profiles, including 7,167 profiles that are reportedly in the STEM fields. As before, we will use 10-fold cross-validation as our external resampling procedure. This results in 6,451 STEM profiles and 28,478 other profiles being available for modeling and evaluation. Suppose that we use a single validation set. How much data would we allocate? If a validation set of 10% of these data were applied, there would be 646 STEM profiles and 2,848 other profiles on hand to estimate the area under the ROC curve and other metrics. Since there are fewer STEM profiles, sensitivity computed on these data would reflect the highest amount of uncertainty. Supposing that a model could achieve 80% sensitivity, the width of

---

Algorithm 12.2: A resampling scheme for simulated annealing feature selection.

```
 1  Create external resamples;
 2  for external resample do
 3  │   Create an initial random feature subset;
 4  │   for each iteration of SA do
 5  │   │   Create internal resample;
 6  │   │   Perturb the current subset;
 7  │   │   Fit model on internal analysis set;
 8  │   │   Predict internal assessment set and estimate performance;
 9  │   │   if performance is better than the previous subset then
10  │   │   │   Accept new subset;
11  │   │   else
12  │   │   │   Calculate acceptance probability;
13  │   │   │   if random uniform variable > probability then
14  │   │   │   │   Reject new subset;
15  │   │   │   else
16  │   │   │   │   Accept new subset;
17  │   │   │   end
18  │   │   end
19  │   end
20  end
21  Determine optimal number of iterations;
22  Create an initial random feature subset;
23  for optimal number of SA iterations do
24  │   Create internal resample based on training set;
25  │   Perturb the current subset;
26  │   Fit model on internal analysis set;
27  │   Predict internal assessment set and estimate performance;
28  │   if performance is better than the previous subset then
29  │   │   Accept new subset;
30  │   else
31  │   │   Calculate acceptance probability;
32  │   │   if random uniform variable > probability then
33  │   │   │   Reject new subset;
34  │   │   else
35  │   │   │   Accept new subset;
36  │   │   end
37  │   end
38  end
39  Determine optimal feature set;
40  Fit final model on best features using the training set;
```

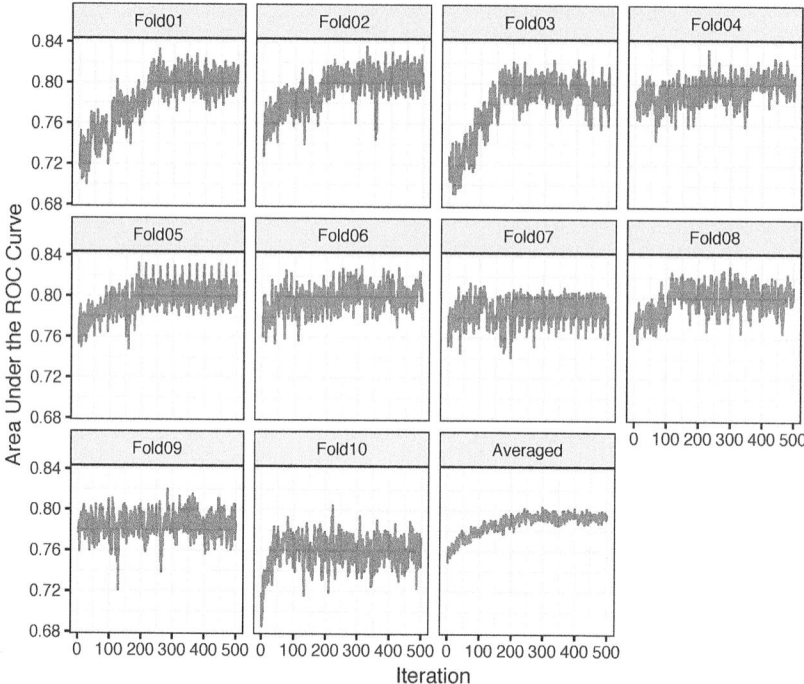

Figure 12.3: Internal performance profile for naive Bayes models selected via simulated annealing for each external resample.

the 90% confidence interval for this metric would be 6.3%. Although subjective, this seems precise enough to be used to guide the simulated annealing search reliably.

The naive Bayes model will be used to illustrate the potential utility of this search method. Five-hundred iterations of simulated annealing will be used with restarts occurring after 10 consecutive suboptimal feature sets have been found. The area under the ROC curve is used to optimize the models and to find the optimal number of SA iterations. To start, we initialize the first subset by randomly selecting half of the possible predictors (although this will be investigated in more detail below).

Let's first look at how the searches did inside each cross-validation iteration. Figure 12.3 shows how the *internal* ROC estimates changed across the 10 external resamples. The trends vary; some show an increasing trend in performance across iterations while others appear to change very little across iterations. If the initial subset shows good performance, the trend tends to remain flat over iterations. This reflects that there can be substantial variation in the selection process even for data sets with almost 35,000 data points being used within each resample. The last panel shows these values averaged across all resamples.

The external resamples also have performance estimates for each iteration. Are these consistent with the inner validations sets? The average rank correlation between the two sets of ROC statistics is 0.622 with a range of (0.43, 0.85). This indicates fairly good consistency. Figure 12.4 overlays the separate internal and external

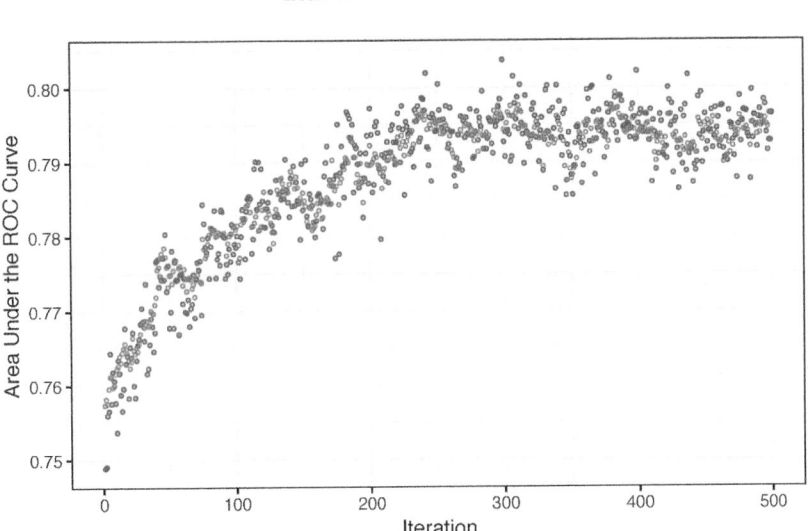

Figure 12.4: Performance profile for naive Bayes models selected via simulated annealing where a random 50 percent of the features were seeded into the initial subset.

performance metrics averaged across iterations of the search. The cyclical patterns seen in these data reflect the restart feature of the algorithm. The best performance was associated with 383 iterations but it is clear that other choices could be made without damaging performance. Looking at the individual resamples, the optimal iterations ranged from 69 to 404 with an average of 258 iterations. While there is substantial variability in these values, the number of variables selected at those points in the process are more precise; the average was 56 predictors with a range of (49, 64). As will be shown below in Figure 12.7, the size trends across iterations varied with some resamples increasing the number of predictors while others decreased.

Now that we have a sense of how many iterations of the search are appropriate, let's look at the final search. For this application of SA, we mimic what was done inside of the external resamples; a validation set of 10% is allocated from the training set and the remainder is used for modeling. Figure 12.5 shows the trends for both the ROC statistic as well as the subset size over the maximum 383 iterations. As the number of iterations increases, subset size trends upwards as does the area under the ROC curve.

In the end, 66 predictors[91] (out of a possible 110) were selected and used in the final model:

---

[91]Keyword predictors are shown in `typewriter font`.

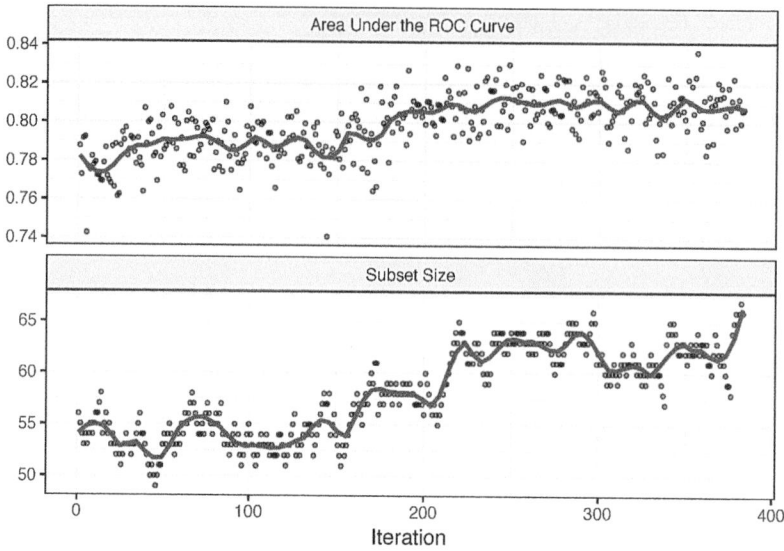

Figure 12.5: Trends for the final SA search using the entire training set and a 10 percent internal validation set.

diet, education level, height, income, time since last on-line, off-spring, pets, religion, smokes, town, 'speaks' C++ fluently, 'speaks' C++ okay, 'speaks' C++ poorly, lisp, 'speaks' lisp poorly, black, hispanic/latin, indian, middle eastern, native american, other, pacific islander, total essay length, `software`, `engineer`, `startup`, `tech`, `computers`, `engineering`, `computer`, `internet`, `science`, `technical`, `web`, `programmer`, `scientist`, `code`, `lol`, `biotech`, `matrix`, `geeky`, `solving`, `teacher`, `student`, `silicon`, `law`, `electronic`, `mobile`, `systems`, `electronics`, `futurama`, `alot`, `firefly`, `valley`, `lawyer`, the number of exclamation points, the number of extra spaces, the percentage of lower-case characters, the number of periods, the number of words, sentiment score (afinn), number of first-person words, number of first-person (possessive) words, the number of second-person (possessive) words, the number of 'to be' occurrences, and the number of prepositions

Based on external resampling, our estimate of performance for this model was an ROC AUC of 0.799. This is slightly better than the full model that was estimated in Section 12.1. The main benefit to this process was to reduce the number of predictors. While performance was maintained in this case, performance may be improved in other cases.

A randomization approach was used to investigate if the 52 selected features contained good predictive information. For these data, 100 random subsets of size 66 were generated and used as input to naive Bayes models. The performance of the model from SA was better than 95% of the random subsets. This result indicates that SA selection process approach can be used to find good predictive subsets.

12.2.3 | Examining Changes in Performance

Simulated annealing is a controlled random search; the new candidate feature subset is selected completely at random based on the current state. After a sufficient number of iterations, a data set can be created to quantify the difference in performance with and without each predictor. For example, for an SA search lasting 500 iterations, there should be roughly 250 subsets with and without each predictor and each of these has an associated area under the ROC curve computed from the external holdout set. A variety of different analyses can be conducted on these values but the simplest is a basic *t*-test for equality. Using the SA that was initialized with half of the variables, the 10 predictors with the largest absolute *t* statistics were (in order): education level, `software` keyword, `engineer` keyword, income, the number of words, `startup` keyword, `solving` keyword, the number of characters/word, height, and the number of lower-case characters. Clearly, a number of these predictors are related and measure characteristics that are connected to essay length. As an additional benefit to identifying the subset of predictive features, this *post hoc* analysis can be used to get a clear sense of the importance of the predictors.

12.2.4 | Grouped Qualitative Predictors versus Indicator Variables

Would the results be different if the categorical variables were converted to dummy variables? The same analysis was conducted where the initial models were seeded with half of the predictor set. As with the similar analyses conducted in Section 5.7, all other factors were kept equal. However, since the number of predictors increased from 110 to 298, the number of search iterations was increased from 500 to 1,000 to make sure that all predictors had enough exposure to the selection algorithm. The results are shown in Figure 12.6. The extra iterations were clearly needed to help to achieve a steady state of predictive performance. However, the model performance (AUC: 0.801) was about the same as the approach that kept the categories in a single predictor (AUC: 0.799).

12.2.5 | The Effect of the Initial Subset

One additional aspect that is worth considering: how did the subset sizes change within external resamples and would these trends be different if the initial subset had been seeded with more (or fewer) predictors? To answer this, the search was replicated with initial subsets that were either 10% or 90% of the total feature set. Figure 12.7 shows the smoothed subset sizes for each external resample colored by the different initial percentages. In many instances, the two sparser configurations converged to a subset size that was about half of the total number of possible predictors (although a few showed minimal change overall). A few of the large initial subsets did not substantially decrease. There is some consistency across the three configurations but it does raise the potential issue of how forcefully simulated annealing will try to *reduce* the number of predictors in the model. Given these results, it may make sense for many practical applications to begin with a small subset size and evaluate performance.

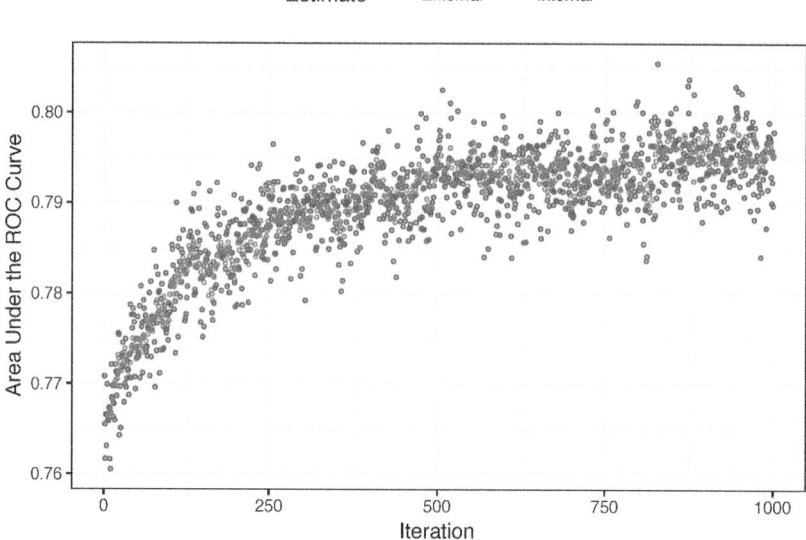

Figure 12.6: Results when qualitative variables are decomposed into individual predictors.

In terms of performance, the areas under the ROC curves for the three sets were 0.805 (10%), 0.799 (50%), and 0.804 (90%). Given the variation in these results, there was no appreciable difference in performance across the different settings.

12.3   Genetic Algorithms

A genetic algorithm (GA) is an optimization tool that is based on concepts of evolution in population biology (Mitchell, 1998; Haupt et al., 1998). These algorithms have been shown to be able to locate the optimal or near-optimal solutions of complex functions (Mandal et al., 2006). To effectively find optimal solutions the algorithm mimics the evolutionary process of a population by generating a candidate set of solutions for the optimization problem, allowing the solutions to reproduce and create new solutions (reproduction), and promoting competition to give the most evolutionarily fit solutions (i.e., optimal) the best chance to survive and populate the subsequent generation (natural selection). This process enables GAs to harness good solutions over time to make better solutions, and has been shown to converge to an optimization plateau (Holland, 1992).

The problem of identifying an optimal feature subset is also a complex optimization problem as discussed in Section 10.2. Therefore, if the feature selection problem can be framed in terms of the concepts of evolutionary biology, then the GA engine can be applied to search for optimal feature subsets. In biology, each subject in the population is represented by a chromosome, which is constructed from a sequence of genes. Each chromosome is evaluated based on a fitness criterion. The better the

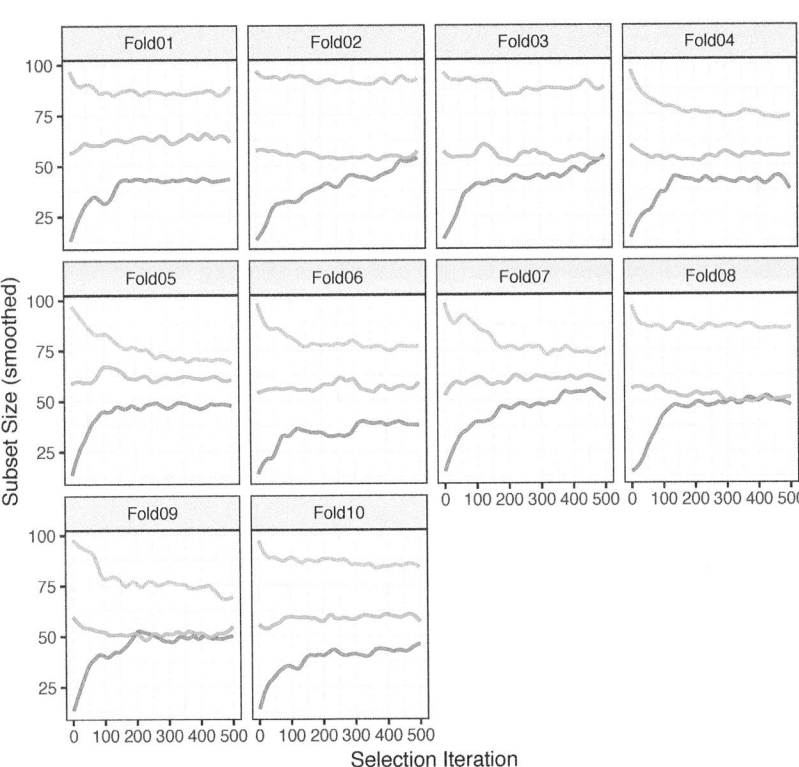

Figure 12.7: Resampling trends across iterations for different initial feature subset sizes. These results were achieved using simulated annealing applied to the OkCupid data.

fitness, the more likely a chromosome will be chosen to use in reproduction to create the next generation of chromosomes. The reproduction stage involves the process of crossover and mutation, where two chromosomes exchange a subset of genes. A small fraction of the offspring's genes are then changed due to the process of mutation.

For the purpose of feature selection, the population consists of all possible combinations of features for a particular data set. A chromosome in the population is a specific combination of features (i.e., genes), and the combination is represented by a string which has a length equal to the total number of features. The fitness criterion is the predictive ability of the selected combination of features. Unlike simulated annealing, the GA feature subsets are grouped into *generations* instead of considering one subset at a time. But a generation in a GA is similar to an iteration in simulated annealing.

In practice, the initial population of feature subsets needs to be large enough to contain a sufficient amount of diversity across the feature subset space. Often 50 or more randomly selected feature subsets are chosen to seed the initial generation.

Our approach is to create random subsets of features that contain between 10% and 90% of the total number of features.

Once the initial generation has been created, the fitness (or predictive ability) of each feature subset is estimated. As a toy example, consider a set of nine predictors (A though I) and an initial population of 12 feature subsets:

ID										Fitness	Probability (%)
1			C			F				0.54	6.4
2	A			D	E	F		H		0.55	6.5
3				D						0.51	6.0
4					E					0.53	6.2
5				D			G	H	I	0.75	8.8
6		B			E		G		I	0.64	7.5
7		B	C			F			I	0.65	7.7
8	A		C		E		G	H	I	0.95	11.2
9	A		C	D		F	G	H	I	0.81	9.6
10			C	D	E				I	0.79	9.3
11	A	B		D	E		G	H		0.85	10.0
12	A	B	C	D	E	F	G		I	0.91	10.7

Each subset has an associated performance value and, in this case, larger values are better. Next, a subset of these feature sets will be selected as parents to reproduce and form the next generation. A logical approach to selecting parents would be to choose the top-ranking feature subsets as parents. However, this greedy approach often leads to lingering in a locally optimal solution. To avoid a local optimum, the selection of parents should be a function of the fitness criteria. The most common approach to select parents is to use a weighted random sample with a probability of selection as a function of predictive performance. A simple way to compute the selection probability is to divide an individual feature subset's performance value by the sum of all the performance values. The rightmost column in the table above shows the results for this generation.

Suppose subsets 6 and 12 were selected as parents for the next generation. For these parents to reproduce, their genes will undergo a process of *crossover* and *mutation* which will result in two new feature subsets. In single-point crossover, a random location between two predictors is chosen.[92] The parents then exchange features on one side of the crossover point.[93] Suppose that the crossover point was between predictors D and E represented here:

---

[92]There are also multi-point crossover methods that create more than two regions where features are exchanged.

[93]Often a random probability is also generated to determine if crossover should occur.

ID								
6		B			E		G	I
12	A	B	C	D	E	F	G	I

The resulting children of these parents would be:

ID								
13		B			E	F	G	I
14	A	B	C	D	E		G	I

One last modification that can occur for a newly created feature subset is a *mutation*. A small random probability (1%–2%) is generated for each possible feature. If a feature is selected in the mutation process, then its state is flipped within the current feature subset. This step adds an additional protection to lingering in local optimal solutions. Continuing with the children generated above, suppose that feature I in feature subset 13 was selected for mutation. Since this feature is already present in subset 13, it would be removed since it was identified for mutation. The resulting children to be evaluated for the next generation would then be:

ID								
13		B			E	F	G	
14	A	B	C	D	E		G	I

The process of selecting two parents to generate two more offspring would continue until the new generation is populated.

The reproduction process does not guarantee that the most fit chromosomes in the current generation will appear in the subsequent generation. This means that there is a chance that the subsequent generation may have less fit chromosomes. To ensure that the subsequent generation maintains the same fitness level, the best chromosome(s) can be carried, as is, into the subsequent generation. This process is called *elitism*. When elitism is used within a GA, the most fit solution across the observed generations will always be in the final generation. However, this process can cause the GA to linger longer in local optima since the best chromosome will always have the highest probability of reproduction.

12.3.1 | External Validation

An external resampling procedure should be used to determine how many iterations of the search are appropriate. This would average the assessment set predictions across all resamples to determine how long the search should proceed when it is directly applied to the entire training set. Basically, the number of iterations is once again a tuning parameter.

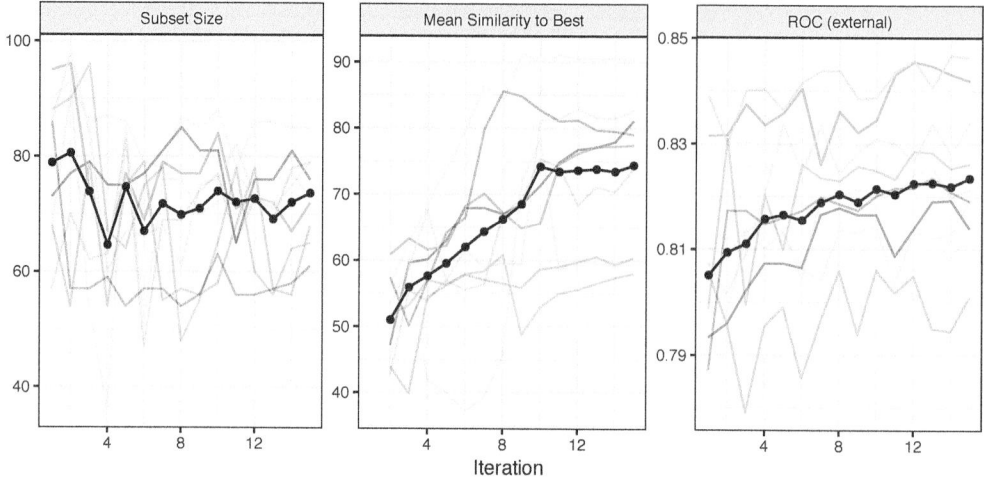

Figure 12.8: The change in subset size, similarity, and performance over time for each external resample of the genetic algorithm.

A genetic algorithm makes gradual improvements in predictive performance through changes to feature subsets over time. Similar to simulated annealing, an external resampling procedure should be used to select an optimal number of generations. Specifically, GAs are executed within each external resampling loop. Then the internal resamples are used to measure fitness so that different data are used for modeling and evaluation. This process helps prevent the genetic algorithm from finding a feature subset that overfits to the available data, which is a legitimate risk with this type of optimization tool.

Once the number of generations is established, the final selection process is performed on the training set and the best subset is used to filter the predictors for the final model.

To evaluate the performance of GAs for feature selection, we will again use the OkCupid data with a naive Bayes model. In addition, the same data splitting scheme was used as with simulated annealing; 10-fold cross-validation was used for external resampling along with a 10% internal validation set. The search parameters were mostly left to the suggested defaults:

- Generation size: 50
- Crossover probability: 80%
- Mutation probability: 1%
- Elitism: No
- Number of generations: 14[94]

---

[94]This is a low value for a genetic algorithm. For other applications, the number of generations is usually in the 100s or 1000s. However, a value this large would make the feature selection process computationally challenging due to the internal and external validation scheme.

For this genetic algorithm, it is helpful to examine how the members of each generation (i.e., the feature subsets) change across generations (i.e., search iterations). As we saw before with simulated annealing, simple metrics to understand the algorithm's characteristics across generations are the feature subset size and the predictive performance estimates. Understanding characteristics of the members *within* a generation can also be useful. For example, the diversity of the subsets within and between generations can be quantified using a measure such as Jaccard similarity (Tan et al., 2006). This metric is the proportion of the number of predictors in common between two subsets divided by total number of unique predictors in both sets. For example, if there were five possible predictors, the Jaccard similarity between subsets ABC and ABE would be 2/4 or 50%. This value provides insight on how efficiently the generic algorithm is converging towards a common solution or towards potential overfitting.

Figure 12.8 shows curves that monitor the subset size and predictive performance of the current best subset of each generation for each resample. In addition, this figure illustrates the average similarity of the subsets within a generation to the current best subset for each external resample. Several characteristics stand out from this figure. First, despite the resamples having considerable variation in the size of the first generation, they all converge to subset sizes that contain between 61 and 85 predictors. The similarity plot indicates that the subsets within a generation are becoming more similar to the best solution. At the same time, the areas under the ROC curves increase to a value of 0.843 at the last generation.

Due to an an overall increasing trend in the average area under the ROC curve, the generation with the largest AUC value was 15. Because the overall trend is still ongoing, additional generations may lead to a more optimal feature subset in a subsequent generation.

The final search was conducted on the entire training set and Figure 12.9 has the results. These are similar to the internal trends shown previously.

The final predictor set at generation 15 included more variables than the previous SA search and contains the following predictors:

> age, education level, height, income, time since last on-line, pets, astrological sign, smokes, town, religion (modifer), astrological sign (modifer), 'speaks' C++, 'speaks' C++ fluently, 'speaks' C++ poorly, lisp, 'speaks' lisp fluently, 'speaks' lisp poorly, asian, black, indian, middle eastern, native american, other, pacific islander, white, total essay length, `software`, `engineer`, `startup`, `tech`, `computers`, `engineering`, `computer`, `internet`, `technology`, `technical`, `developer`, `programmer`, `scientist`, `code`, `geek`, `lol`, `biotech`, `matrix`, `geeky`, `solving`, `problems`, `data`, `fixing`, `teacher`, `student`, `silicon`, `law`, `mechanical`, `electronic`, `mobile`, `math`, `apps`, the number of digits, sentiment score (bing), number of first-person words, the number of second-person words, and the number of 'to be' occurrences

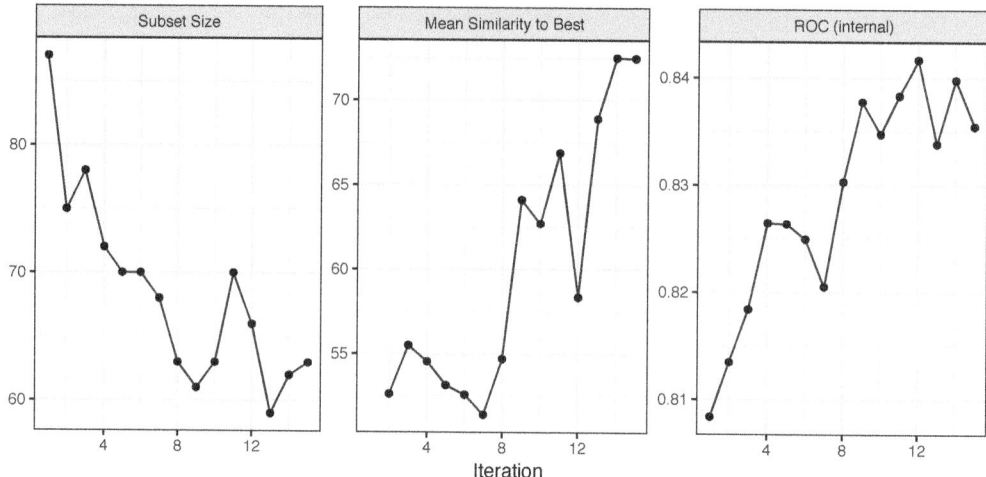

Figure 12.9: Genetic algorithm results for the final search using the entire training set.

To gauge the effectiveness of the search, 100 random subsets of size 63 were chosen and a naive Bayes model was developed for each. The GA selected subset performed better than 100% of the randomly selected subsets. This indicates that the GA did find a useful subset for predicting the response.

12.3.2 | Coercing Sparsity

Genetic algorithms often tend to select larger feature subsets than other methods. This is likely due to the construction of the algorithm in that if a feature is useful for prediction, then it will be included in the feature subset. However, there is less of a penalty for keeping a feature that has no impact on predictive performance. This is especially true when coupling a GA with a modeling technique that does not incur a substantial penalty for including irrelevant predictors (Section 10.3). In these cases, adding non-informative predictors does not coerce the selection process to strongly seek sparse solutions.

There are several benefits to a having a sparse solution. First, the model is often simpler, which may make it easier to understand. Fewer predictors also require less computational resources. Is there a way, then, that could be used to compel a search procedure to move towards a more sparse predictor subset solution?

A simple method for reducing the number of predictors is to use a surrogate measure of performance that has an explicit penalty based on the number of features. Section 11.4 introduced the Akaike information criterion (AIC) which augments the objective function with a penalty that is a function of the training set size and the number of model terms. This metric is specifically tailored to models that use the likelihood as the objective (i.e., linear or logistic regression), but cannot be applied to models whose goal is to optimize other metrics like area under the ROC curve or R^2. Also,

AIC usually involves the number of *model parameters* instead of the number of predictors. For simple parametric regression models, these two quantities are very similar. But there are many models that have no notion of the number of parameters (e.g., trees, K-NN, etc.), while others can have many more parameters than samples in the training set (e.g., neural networks). However the principle of penalizing the objective based on the number of features can be extended to other models.

Penalizing the measure of performance enables the problem to be framed in terms of *multi-parameter optimization* (MPO) or *multi-objective optimization*. There are a variety of approaches that can be used to solve this issue, including a function to combine many objective values into a final single value. For the purpose of feature selection, one objective is to optimize predictive performance, and the other objective is to include as few features as possible. There are many different options (e.g., Bentley and Wakefield (1998), Xue et al. (2013), and others) but a simple and straightforward approach is through *desirability functions* (Harrington, 1965; Del Castillo et al., 1996; Mandal et al., 2007). When using this methodology, each objective is translated to a new scale on [0, 1] where larger values are more desirable. For example, an *undesirable* area under the ROC curve would be 0.50 and a desirable area under the ROC curve would be 1.0. A simple desirability function would be a line that has zero desirability when the AUC is less than or equal to 0.50 and is 1.0 when the AUC is 1.0. Conversely, a desirability function for the number of predictors in the model would be 0.0 when all of the predictors are in the model and would be 1.0 when only one predictor is in the model. Additional desirability functions can be incorporated to guide the modeling process towards other important characteristics.

Once all of the individual desirability functions are defined, the *overall desirability statistic* is created by taking the *geometric mean* of all of the individual functions. If there are q desirability functions $d_i(i = 1 \ldots q)$, the geometric mean is defined as:

$$D = \left(\prod_{i=1}^{q} d_j \right)^{1/q} .$$

Based on this function, if any one desirability function is unacceptable (i.e., $d_i = 0$), the overall measure is also unacceptable.[95] For feature selection applications, the internal performance metric, which is the one that the genetic algorithm uses to accept or reject subsets, can use the overall desirability to encourage the selection process to favor smaller subsets while still taking performance into account.

The most commonly used desirability functions for maximizing a quantity use the piecewise linear or polynomial functions proposed by Derringer and Suich (1980):

---

[95]If we do not want desirability scores to be forced to zero due to just one undesirable characterisitc, then the individual desirability functions can be constrained such that the minimum values are greater than zero.

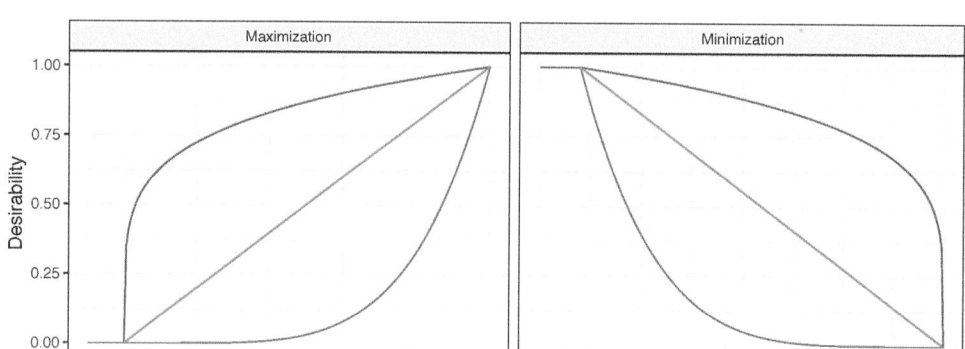

Figure 12.10: (a) Examples of two types of desirability functions and (b) the overall desirability surface when the two are combined. The contours in (b) are created using a maximization function of 0.2 and a minimization function of 1.

$$d_i = \begin{cases} 0 & \text{if } x < A \\ \left(\frac{x-A}{B-A}\right)^s & \text{if } A \leq x \leq B \\ 1 & \text{if } x > B \end{cases} \tag{12.1}$$

Figure 12.10(a) shows an example of a maximization function where $A = 0.50$, $B = 1.00$, and there are three different values of the scaling factor s. Note that values of s less than 1.0 weaken the influence of the desirability function, whereas values of $s > 1.0$ strengthen the influence on the desirability function.

When minimizing a quantity, an analogous function is used:

$$d_i = \begin{cases} 0 & \text{if } x > B \\ \left(\frac{x-B}{A-B}\right)^s & \text{if } A \leq x \leq B \\ 1 & \text{if } x < A \end{cases} \tag{12.2}$$

Also shown in Figure 12.10(a) is a minimization example of scoring the number of predictors in a subset with values $A = 10$ and $B = 100$.

When a maximization function with $s = 1/5$ and a minimization function with $s = 1$ are combined using the geometric mean, the results are shown Figure 12.10(b). In this case, the contours are not symmetric since we have given more influence to the area under the ROC curve based on the value of s.

Similar desirability functions exist for cases when a target value is best (i.e., a value between a minimum and maximum), for imposing box constraints, or for using qualitative inputs. Smoother functions could also be used in place of the piecewise equations.

For the OkCupid data, a desirability function with the following objectives was used to guide the GA:

- Maximize the area under the ROC curve between $A = 0.50$ and $B = 1.00$ with a scale factor of $s = 2.0$.
- Minimize the subset size to be within $A = 10$ and $B = 100$ with $s = 1.0$.

Overall desirability was used for internal performance while the external area under the ROC curve was still used as the primary external metric. Figure 12.11 shows a plot of the internal results similar to Figure 12.8. The overall desirability continually increased and probably would continue to do so with additional generations. The average subset size decreased to be about 28 by the last generation. Based on the averages of the internal resamples, generation 14 was associated with the largest desirability value and with a matching external estimate of the area under the ROC curve of 0.802. This is smaller than the unconstrained GA but the reduced performance is offset by the substantial reduction in the number of predictors in the model.

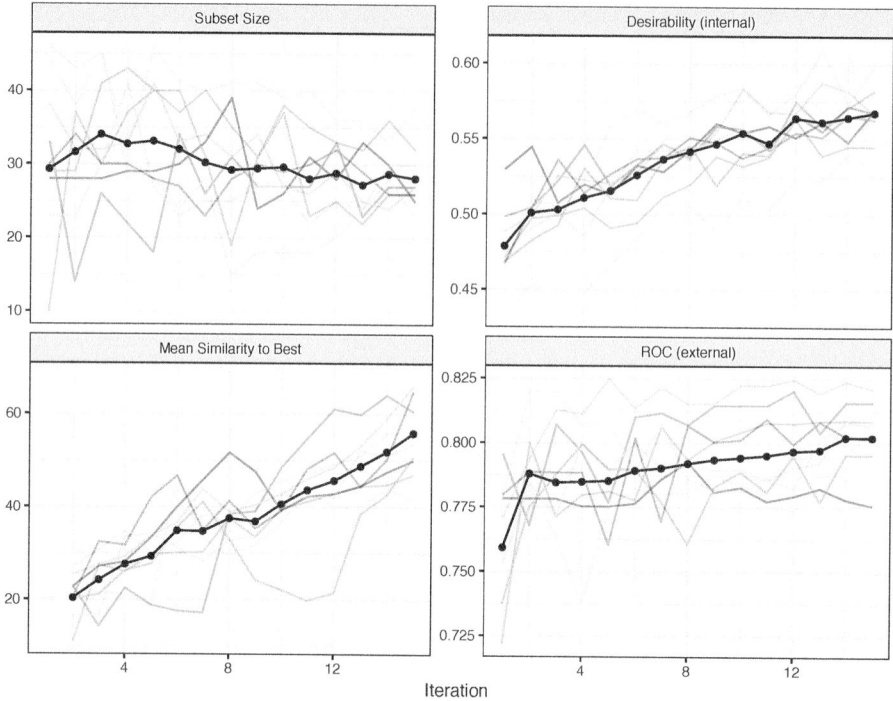

Figure 12.11: Internal performance profiles for naive Bayes models using the genetic algorithm in conjunction with desirability functions.

For the final run of the genetic algorithm, 32 predictors were chosen at generation 15. These were:

> education level, height, religion, smokes, town, 'speaks' C++ okay, lisp, 'speaks' lisp poorly, asian, native american, white, `software`, `engineer`, `computers`, `science`, `programming`, `technical`, `code`, `lol`, `matrix`, `geeky`, `fixing`, `teacher`, `student`, `electronic`, `pratchett`, `math`, `systems`, `alot`, `valley`, `lawyer`, and the number of punctuation marks

Depending on the context, a potentially strong argument can be made for a smaller predictor set even if it shows slightly worse results (but better than the full model). If naive Bayes was a better choice than the other models, it would be worth evaluating the normal and sparse GAs on the test set to make a choice.

12.4   Test Set Results

Models were built using the optimal parameter settings for the unconstrained and constrained GAs, and were applied to the test set. The area under the ROC curve for the unconstrained GA was 0.831, while the corresponding value for the constrained GA was 0.804. Both of these values are larger than their analogous resampled estimates, but the two models' performance ranks are ordered in the same way.

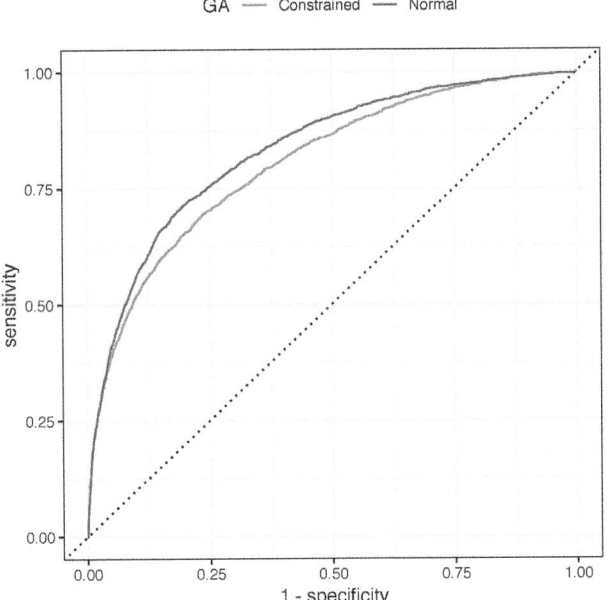

Figure 12.12: Test set results for two models derived from genetic algorithms.

Figure 12.12 shows that the unconstrained model has uniformly better sensitivity and specificity across the test set cutoffs, although it is difficult to tell if these results are practically different from one another.

12.5 │ Summary

Global search methods can be an effective tool for investigating the predictor space and identifying subsets of predictors that are optimally related to the response. Despite these methods having some tunable parameters (e.g., the number of iterations, initial subset size, mutation rate, etc.) the default values are sensible and tend to work well.

Although the global search approaches are usually effective at finding good feature sets, they are computationally taxing. For example, the complete process of conducting each of the simulated annealing analyses consisted of fitting a total of 5,501 individual models, and the genetic algorithm analyses required 8,251 models. The independent nature of the folds in external cross-validation allow these methods to be run in parallel, thus reducing the overall computation time. Even so, the GA described above with parallel processing took more than 9 hours when running each external cross-validation fold on a separate core.

In addition, it is important to point out that the naive Bayes model used to compare methods did not require optimization of any tuning parameters.[96] If the global search

---

[96]The smoothness of the kernel density estimates *could* have been tuned, but this does not typically

methods were used with a radial basis support vector machine (2 tuning parameters) or a C5.0 model (3 tuning parameters), then the computation requirement would rise significantly.

When combining a global search method with a model that has tuning parameters, we recommend that, when possible, the feature set first be winnowed down using expert knowledge about the problem. Next, it is important to identify a reasonable range of tuning parameter values. If a sufficient number of samples are available, a proportion of them can be split off and used to find a range of potentially good parameter values using all of the features. The tuning parameter values may not be the perfect choice for feature subsets, but they should be reasonably effective for finding an optimal subset.

A general feature selection strategy would be to use models that can intrinsically eliminate features from the model while the model is fit. As previously mentioned, these models may yield good results much more quickly than wrapper methods. The wrapper methods are most useful when a model without this attribute is most appealing. For example, if we had specific reasons to favor naive Bayes models, we would factor the approaches shown in this chapter over more expedient techniques that would automatically filter predictions (e.g., glmnet, MARS, etc.).

12.6 | Computing

The website `http://bit.ly/fes-global` contains R programs for reproducing these analyses.

---

have a meaningful impact on performance.

Bibliography

Abdeldayem, E., Ibrahim, A., Ahmed, A., Genedi, E., and Tantawy, W. (2015). Positive remodeling index by MSCT coronary angiography: A prognostic factor for early detection of plaque rupture and vulnerability. *The Egyptian Journal of Radiology and Nuclear Medicine*, 46(1):13–24.

Abdi, H. and Williams, L. (2010). Principal component analysis. *Wiley Interdisciplinary Reviews: Computational Statistics*, 2(4):433–459.

Agresti, A. (2012). *Categorical Data Analysis*. Wiley-Interscience.

Akaike, H. (1974). A new look at the statistical model identification. *IEEE Transactions on Automatic Control*, 19(6):716–723.

Allison, P. (2001). *Missing Data*. Sage Publications.

Altman, D. (1991). Categorising continuous variables. *British Journal of Cancer*, 64(5):975.

Altman, D. and Bland, J. (1994a). Diagnostic tests 3: Receiver operating characteristic plots. *BMJ: British Medical Journal*, 309(6948):188.

Altman, D. and Bland, J. (1994b). Statistics Notes: Diagnostic tests 2: Predictive values. *British Medical Journal*, 309(6947):102.

Altman, D., Lausen, B., Sauerbrei, W., and Schumacher, M. (1994). Dangers of using "optimal" cutpoints in the evaluation of prognostic factors. *Journal of the National Cancer Institute*, 86(11):829–835.

Amati, G. and Van R, C. J. (2002). Probabilistic models of information retrieval based on measuring the divergence from randomness. *ACM Transactions on Information Systems*, 20(4):357–389.

Ambroise, C. and McLachlan, G. (2002). Selection bias in gene extraction on the basis of microarray gene-expression data. *Proceedings of the National Academy of Sciences*, 99(10):6562–6566.

Audigier, V., Husson, F., and Josse, J. (2016). A principal component method to impute missing values for mixed data. *Advances in Data Analysis and Classification*, 10(1):5–26.

Bairey, E., Kelsic, E., and Kishony, R. (2016). High-order species interactions shape ecosystem diversity. *Nature Communications*, 7:12285.

Barker, M. and Rayens, W. (2003). Partial least squares for discrimination. *Journal of Chemometrics*, 17(3):166–173.

Basu, S., Kumbier, K., Brown, J., and Yu, B. (2018). Iterative random forests to discover predictive and stable high-order interactions. *Proceedings of the National Academy of Sciences*, 115(8):1943–1948.

Benavoli, A., Corani, G., Demsar, J., and Zaffalon, M. (2016). Time for a change: A tutorial for comparing multiple classifiers through Bayesian analysis. *arXiv.org*.

Benjamini, Y. and Hochberg, Y. (1995). Controlling the false discovery rate: A practical and powerful approach to multiple testing. *Journal of the Royal Statistical Society. Series B (Methodological)*, pages 289–300.

Bentley, P. and Wakefield, J. (1998). Finding acceptable solutions in the Pareto-optimal range using multiobjective genetic algorithms. In Chawdhry, P., Roy, R., and Pant, R., editors, *Soft Computing in Engineering Design and Manufacturing*, pages 231–240. Springer.

Bergstra, J. and Bengio, Y. (2012). Random search for hyper-parameter optimization. *Journal of Machine Learning Research*, 13:281–305.

Berry, B., Moretto, J., Matthews, T., Smelko, J., and Wiltberger, K. (2015). Cross-scale predictive modeling of CHO cell culture growth and metabolites using Raman spectroscopy and multivariate analysis. *Biotechnology Progress*, 31(2):566–577.

Berry, K. J., Mielke Jr, P. W., and Johnston, J. E. (2016). *Permutation Statistical Methods*. Springer.

Bickle, M. (2010). The beautiful cell: High-content screening in drug discovery. *Analytical and Bioanalytical Chemistry*, 398(1):219–226.

Bien, J., Taylor, J., and Tibshirani, R. (2013). A lasso for hierarchical interactions. *Annals of Statistics*, 41(3):1111.

Bishop, C. (2011). *Pattern Recognition and Machine Learning*. Springer.

Boulesteix, A. and Strobl, C. (2009). Optimal classifier selection and negative bias in error rate estimation: An empirical study on high-dimensional prediction. *BMC Medical Research Methodology*, 9(1):85.

Box, G. and Cox, D. (1964). An analysis of transformations. *Journal of the Royal Statistical Society. Series B (Methodological)*, 26(2):211–243.

Box, G., Hunter, W., and Hunter, J. (2005). *Statistics for Experimenters: An Introduction to Design, Data Analysis, and Model Building*. Wiley.

Breiman, L. (1996). Bagging predictors. *Machine Learning*, 24(2):123–140.

Breiman, L. (2001). Random forests. *Machine Learning*, 45(1):5–32.

Breiman, L., Friedman, J., Olshen, R., and Stone, C. (1984). *Classification and Regression Trees*. Chapman and Hall, New York.

Caputo, B., Sim, K., Furesjo, F., and Smola, A. (2002). Appearance-based object recognition using SVMs: Which kernel should I use? In *Proceedings of NIPS Workshop on Statistical Methods for Computational Experiments in Visual Processing and Computer Vision*, volume 2002.

Chato, L. and Latifi, S. (2017). Machine learning and deep learning techniques to predict overall survival of brain tumor patients using MRI images. In *2017 IEEE 17th International Conference on Bioinformatics and Bioengineering*, pages 9–14.

Chen, S., Sun, J., Dimitrov, L., Turner, A., Adams, T., Meyers, D., Chang, B., Zheng, S., Grönberg, H., Xu, J., et al. (2008). A support vector machine approach for detecting gene-gene interaction. *Genetic Epidemiology*, 32(2):152–167.

Chollet, F. and Allaire, J. (2018). *Deep Learning with R*. Manning.

Chong, E. and Zak, S. (2008). *Global Search Algorithms*. John Wiley & Sons, Inc.

Christopher, D., Prabhakar, R., and Hinrich, S. (2008). *Introduction to Information Retrieval*. Cambridge University Press.

Cilla, M., Pena, E., Martinez, M., and Kelly, D. (2013). Comparison of the vulnerability risk for positive versus negative atheroma plaque morphology. *Journal of Biomechanics*, 46(7):1248–1254.

Cleveland, W. (1979). Robust locally weighted regression and smoothing scatterplots. *Journal of the American Statistical Association*, 74(368):829–836.

Cleveland, W. (1993). *Visualizing Data*. Hobart Press, Summit, New Jersey.

Cover, T. and Thomas, J. (2012). *Elements of Information Theory*. John Wiley and Sons.

Davison, A. and Hinkley, D. (1997). *Bootstrap Methods and Their Application*. Cambridge University Press.

Del Castillo, E., Montgomery, D. C., and McCarville, D. R. (1996). Modified desirability functions for multiple response optimization. *Journal of Quality Technology*, 28(3):337–345.

Demsar, J. (2006). Statistical comparisons of classifiers over multiple data sets. *Journal of Machine learning Research*, 7(Jan):1–30.

Derringer, G. and Suich, R. (1980). Simultaneous optimization of several response variables. *Journal of Quality Technology*, 12(4):214–219.

Dickhaus, T. (2014). Simultaneous statistical inference. *AMC*, 10:12.

Dillon, W. and Goldstein, M. (1984). *Multivariate Analysis Methods and Applications*. Wiley.

Efron, B. (1983). Estimating the error rate of a prediction rule: Improvement on cross-validation. *Journal of the American Statistical Association*, pages 316–331.

Efron, B. and Hastie, T. (2016). *Computer Age Statistical Inference*. Cambridge University Press.

Efron, B. and Tibshirani, R. (1997). Improvements on cross-validation: The 632+ bootstrap method. *Journal of the American Statistical Association*, pages 548–560.

Eilers, P. and Marx, B. (2010). Splines, knots, and penalties. *Wiley Interdisciplinary Reviews: Computational Statistics*, 2(6):637–653.

Elith, J., Leathwick, J., and Hastie, T. (2008). A working guide to boosted regression trees. *Journal of Animal Ecology*, 77(4):802–813.

Eskelson, B., Temesgen, H., Lemay, V., Barrett, T., Crookston, N., and Hudak, A. (2009). The roles of nearest neighbor methods in imputing missing data in forest inventory and monitoring databases. *Scandinavian Journal of Forest Research*, 24(3):235–246.

Fernandez-Delgado, M., Cernadas, E., Barro, S., and Amorim, D. (2014). Do we need hundreds of classifiers to solve real world classification problems? *Journal of Machine Learning Research*, 15(1):3133–3181.

Fogel, P., Hawkins, D., Beecher, C., Luta, G., and Young, S. (2013). A tale of two matrix factorizations. *The American Statistician*, 67(4):207–218.

Friedman, J. (1991). Multivariate adaptive regression splines. *The Annals of Statistics*, 19(1):1–141.

Friedman, J. (2001). Greedy function approximation: A gradient boosting machine. *Annals of Statistics*, pages 1189–1232.

Friedman, J. (2002). Stochastic gradient boosting. *Computational Statistics & Data Analysis*, 38(4):367–378.

Friedman, J., Hastie, T., and Tibshirani, R. (2010). Regularization paths for generalized linear models via coordinate descent. *Journal of Statistical Software*, 33(1):1.

Friedman, J. and Popescu, B. (2008). Predictive learning via rule ensembles. *The Annals of Applied Statistics*, 2(3):916–954.

Friendly, M. and Meyer, D. (2015). *Discrete Data Analysis with R: Visualization and Modeling Techniques for Categorical and Count Data*. CRC Press.

Frigge, M., Hoaglin, D., and Iglewicz, B. (1989). Some implementations of the boxplot. *The American Statistician*, 43(1):50–54.

García-Magariños, M., López-de Ullibarri, I., Cao, R., and Salas, A. (2009). Evaluating the ability of tree-based methods and logistic regression for the detection of SNP-SNP interaction. *Annals of Human Genetics*, 73(3):360–369.

Geman, S., Bienenstock, E., and Doursat, R. (1992). Neural networks and the bias/variance dilemma. *Neural Computation*, 4(1):1–58.

Ghosh, A. K. and Chaudhuri, P. (2005). On data depth and distribution-free discriminant analysis using separating surfaces. *Bernoulli*, 11(1):1–27.

Gillis, N. (2017). Introduction to nonnegative matrix factorization. *arXiv preprint arXiv:1703.00663*.

Giuliano, K., DeBiasio, R., T, D., Gough, A., Volosky, J., Zock, J., Pavlakis, G., and Taylor, L. (1997). High-content screening: A new approach to easing key bottlenecks in the drug discovery process. *Journal of Biomolecular Screening*, 2(4):249–259.

Golub, G., Heath, M., and Wahba, G. (1979). Generalized cross-validation as a method for choosing a good ridge parameter. *Technometrics*, 21(2):215–223.

Good, P. (2013). *Permutation tests: A Practical Guide to Resampling Methods for Testing Hypotheses*. Springer Science & Business Media.

Goodfellow, I., Bengio, Y., and Courville, A. (2016). *Deep Learning*. MIT Press.

Gower, J. (1971). A general coefficient of similarity and some of its properties. *Biometrics*, pages 857–871.

Greenacre, M. (2010). *Biplots in Practice*. Fundacion BBVA.

Greenacre, M. (2017). *Correspondence Analysis in Practice*. CRC press.

Guo, C. and Berkhahn, F. (2016). Entity embeddings of categorical variables. *arXiv.org*.

Guyon, I., Weston, J., Barnhill, S., and Vapnik, V. (2002). Gene selection for cancer classification using support vector machines. *Machine Learning*, 46(1):389–422.

Haase, R. (2011). *Multivariate General Linear Models*. Sage.

Hammes, G. (2005). *Spectroscopy for the Biological Sciences*. John Wiley and Sons.

Hampel, D., Andrews, P., Bickel, F., Rogers, P., Huber, W., and Turkey, J. (1972). *Robust Estimates of Location*. Princeton University Press, Princeton, New Jersey.

Harrell, F. (2015). *Regression Modeling Strategies*. Springer.

Harrington, E. (1965). The desirability function. *Industrial Quality Control*, 21(10):494–498.

Hastie, T., Tibshirani, R., and Wainwright, M. (2015). *Statistical Learning with Sparsity*. CRC Press.

Haupt, R., Haupt, S., and Haupt, S. (1998). *Practical Genetic Algorithms*, volume 2. Wiley New York.

Hawkins, D. (1994). The feasible solution algorithm for least trimmed squares regression. *Computational Statistics & Data Analysis*, 17(2):185–196.

Healy, K. (2018). *Data Visualization: A Practical Introduction*. Princeton University Press.

Hill, A., LaPan, P., Li, Y., and Haney, S. (2007). Impact of image segmentation on high-content screening data quality for SK-BR-3 cells. *BMC Bioinformatics*, 8(1):340.

Hintze, J. and Nelson, R. (1998). Violin plots: A box plot-density trace synergism. *The American Statistician*, 52(2):181–184.

Hoerl, Aand Kennard, R. (1970). Ridge regression: Biased estimation for nonorthogonal problems. *Technometrics*, 12(1):55–67.

Holland, J. H. (1992). Genetic algorithms. *Scientific American*, 267(1):66–73.

Holmes, S. and Huber, W. (2019). *Modern Statistics for Modern Biology*. Cambridge University Press.

Hosmer, D. and Lemeshow, S. (2000). *Applied Logistic Regression*. John Wiley & Sons, New York.

Hothorn, T., Hornik, K., and Zeileis, A. (2006). Unbiased recursive partitioning: A conditional inference framework. *Journal of Computational and Graphical Statistics*, 15(3):651–674.

Hothorn, T., Leisch, F., Zeileis, A., and Hornik, K. (2005). The design and analysis of benchmark experiments. *Journal of Computational and Graphical Statistics*, 14(3):675–699.

Hyndman, R. and Athanasopoulos, G. (2013). *Forecasting: Principles and Practice*. OTexts.

Hyvarinen, A. and Oja, E. (2000). Independent component analysis: Algorithms and applications. *Neural Networks*, 13(4-5):411–430.

Jahani, M. and Mahdavi, M. (2016). Comparison of predictive models for the early diagnosis of diabetes. *Healthcare Informatics Research*, 22(2):95–100.

Jones, D., Schonlau, M., and Welch, W. (1998). Efficient global optimization of expensive black-box functions. *Journal of Global Optimization*, 13(4):455–492.

Karthikeyan, M., Glen, R., and Bender, A. (2005). General melting point prediction based on a diverse compound data set and artificial neural networks. *Journal of Chemical Information and Modeling*, 45(3):581–590.

Kenny, P. and Montanari, C. (2013). Inflation of correlation in the pursuit of drug-likeness. *Journal of Computer-Aided Molecular Design*, 27(1):1–13.

Kim, A. and Escobedo-Land, A. (2015). OkCupid data for introductory statistics and data science courses. *Journal of Statistics Education*, 23(2):1–25.

Kirkpatrick, S., Gelatt, D., and Vecchi, M. (1983). Optimization by simulated annealing. *Science*, 220(4598):671–680.

Kuhn, M. (2008). The caret package. *Journal of Statistical Software*, 28(5):1–26.

Kuhn, M. and Johnson, K. (2013). *Applied Predictive Modeling*. Springer.

Kvalseth, T. (1985). Cautionary note about R^2. *American Statistician*, 39(4):279–285.

Lambert, J., Gong, L., Elliot, C., Thompson, K., and Stromberg, A. (2018). rFSA: An R package for finding best subsets and interactions. *The R Journal*.

Lampa, E., Lind, L., Lind, P., and Bornefalk-Hermansson, A. (2014). The identification of complex interactions in epidemiology and toxicology: A simulation study of boosted regression trees. *Environmental Health*, 13(1):57.

Lawrence, I. and Lin, K. (1989). A concordance correlation coefficient to evaluate reproducibility. *Biometrics*, pages 255–268.

Lee, T.-W. (1998). *Independent Component Analysis*. Springer.

Levinson, M. and Rodriguez, D. (1998). Endarterectomy for preventing stroke in symptomatic and asymptomatic carotid stenosis. review of clinical trials and recommendations for surgical therapy. In *The Heart Surgery Forum*, pages 147–168.

Lewis, D., Yang, Y., Rose, T., and Li, F. (2004). Rcv1: A new benchmark collection for text categorization research. *Journal of Machine Learning Research*, 5:361–397.

Lian, K., White, J., Bartlett, E., Bharatha, A., Aviv, R., Fox, A., and Symons, S. (2012). NASCET percent stenosis semi-automated versus manual measurement on CTA. *The Canadian Journal of Neurological Sciences*, 39(03):343–346.

Little, R. and Rubin, D. (2014). *Statistical Analysis with Missing Data*. John Wiley and Sons.

Luo, G. (2016). Automatically explaining machine learning prediction results: A demonstration on Type 2 diabetes risk prediction. *Health Information Science and Systems*, 4(1):2.

MacKay, D. (2003). *Information Theory, Inference and Learning Algorithms*. Cambridge University Press.

Mallick, H. and Yi, N. (2013). Bayesian methods for high dimensional linear models. *Journal of Biometrics and Biostatistics*, 1:005.

Mandal, A., Jeff Wu, C., and Johnson, K. (2006). SELC: Sequential elimination of level combinations by means of modified genetic algorithms. *Technometrics*, 48(2):273–283.

Mandal, A., Johnson, K., Wu, C. J., and Bornemeier, D. (2007). Identifying promising compounds in drug discovery: Genetic algorithms and some new statistical techniques. *Journal of Chemical Information and Modeling*, 47(3):981–988.

Marr, B. (2017). IoT and big data at Caterpillar: How predictive maintenance saves millions of dollars. https://www.forbes.com/sites/bernardmarr/2017/02/07/iot-and-big-data-at-caterpillar-how-predictive-maintenance-saves-millions-of-dollars/#109576a27240. Accessed: 2018-10-23.

Massy, W. (1965). Principal components regression in exploratory statistical research. *Journal of the American Statistical Association*, 60(309):234–256.

McElreath, R. (2015). *Statistical Rethinking: A Bayesian Course with Examples in R and Stan*. Chapman and Hall/CRC.

Meier, P., Knapp, G., Tamhane, U., Chaturvedi, S., and Gurm, H. (2010). Short term and intermediate term comparison of endarterectomy versus stenting for carotid artery stenosis: Systematic review and meta-analysis of randomised controlled clinical trials. *BMJ*, 340:c467.

Micci-Barreca, D. (2001). A preprocessing scheme for high-cardinality categorical attributes in classification and prediction problems. *ACM SIGKDD Explorations Newsletter*, 3(1):27–32.

Miller, A. (1984). Selection of subsets of regression variables. *Journal of the Royal Statistical Society. Series A (General)*, pages 389–425.

Mitchell, M. (1998). *An Introduction to Genetic Algorithms*. MIT Press.

Mockus, J. (1994). Application of Bayesian approach to numerical methods of global and stochastic optimization. *Journal of Global Optimization*, 4(4):347–365.

Mozharovskyi, P., Mosler, K., and Lange, T. (2015). Classifying real-world data with the DDα-procedure. *Advances in Data Analysis and Classification*, 9(3):287–314.

Mundry, R. and Nunn, C. (2009). Stepwise model fitting and statistical inference: Turning noise into signal pollution. *The American Naturalist*, 173(1):119–123.

Murrell, B., Murrell, D., and Murrell, H. (2016). Discovering general multidimensional associations. *PloS one*, 11(3):e0151551.

Nair, V. and Hinton, G. (2010). Rectified linear units improve restricted Boltzmann machines. In Furnkranz, J. and Joachims, T., editors, *Proceedings of the 27th International Conference on Machine Learning*, pages 807–814. Omnipress.

Neter, J., Kutner, M., Nachtsheim, C., and Wasserman, W. (1996). *Applied Linear Statistical Models*. Irwin Chicago.

Nitish, S., Geoffrey, H., Alex, K., Ilya, S., and Ruslan, S. (2014). Dropout: A simple way to prevent neural networks from overfitting. *Journal of Machine Learning Research*, 15:1929–1958.

Opgen-Rhein, R. and Strimmer, K. (2007). Accurate ranking of differentially expressed genes by a distribution-free shrinkage approach. *Statistical Applications in Genetics and Molecular Biology*, 6(1).

Piironen, J. and Vehtari, A. (2017a). Comparison of Bayesian predictive methods for model selection. *Statistics and Computing*, 27(3):711–735.

Piironen, J. and Vehtari, A. (2017b). Sparsity information and regularization in the horseshoe and other shrinkage priors. *Electronic Journal of Statistics*, 11(2):5018–5051.

Preneel, B. (2010). Cryptographic hash functions: Theory and practice. In *ICICS*, pages 1–3.

Qi, Y. (2012). Random forest for bioinformatics. In *Ensemble Machine Learning*, pages 307–323. Springer.

Quinlan, R. (1993). *C4.5: Programs for Machine Learning*. Morgan Kaufmann Publishers.

Raimondi, C. (2010). How I won the predict HIV progression data mining competition. http://blog.kaggle.com/2010/08/09/how-i-won-the-hiv-progression-prediction-data-mining-competition/. Accessed: 2018-05-11.

Reid, R. (2015). A morphometric modeling approach to distinguishing among bobcat, coyote and gray fox scats. *Wildlife Biology*, 21(5):254–262.

Reshef, D., Reshef, Y., Finucane, H., Grossman, S., McVean, G., Turnbaugh, P., Lander, E., Mitzenmacher, M., and Sabeti, P. (2011). Detecting novel associations in large data sets. *Science*, 334(6062):1518–1524.

Rinnan, Å., Van Den Berg, F., and Engelsen, S. B. (2009). Review of the most common pre-processing techniques for near-infrared spectra. *TrAC Trends in Analytical Chemistry*, 28(10):1201–1222.

Roberts, S. and Everson, R. (2001). *Independent Component Analysis: Principles and Practice*. Cambridge University Press.

Robnik-Sikonja, M. and Kononenko, I. (2003). Theoretical and empirical analysis of ReliefF and RReliefF. *Machine Learning*, 53(1):23–69.

Rousseeuw, P. and Croux, C. (1993). Alternatives to the median absolute deviation. *Journal of the American Statistical Association*, 88(424):1273–1283.

Sakar, C., Serbes, G., Gunduz, A., Tunc, H., Nizam, H., Sakar, B., Tutuncu, M., Aydin, T., Isenkul, E., and Apaydin, H. (2019). A comparative analysis of speech signal processing algorithms for Parkinson's disease classification and the use of the tunable Q-factor wavelet transform. *Applied Soft Computing*, 74:255 – 263.

Sapp, S., van der Laan, M. J. M., and Canny, J. (2014). Subsemble: An ensemble method for combining subset-specific algorithm fits. *Journal of Applied Statistics*, 41(6):1247–1259.

Sathyanarayana, A., Joty, S., Fernandez-Luque, L., Ofli, F., Srivastava, J., Elmagarmid, A., Arora, T., and Taheri, S. (2016). Sleep quality prediction from wearable data using deep learning. *JMIR mHealth and uHealth*, 4(4).

Schoepf, U., van Assen, M., Varga-Szemes, A., Duguay, T., Hudson, H., Egorova, S., Johnson, K., St. Pierre, S., Zaki, B., Oudkerk, M., Vliegenthart, R., and Buckler, A. (in press). Automated plaque analysis for the prognostication of major adverse cardiac events. *European Journal of Radiology*.

Schofield, A., Magnusson, M., and Mimno, D. (2017). Understanding text pre-processing for latent Dirichlet allocation. In *Proceedings of the 15th Conference of the European Chapter of the Association for Computational Linguistics*, volume 2, pages 432–436.

Schofield, A. and Mimno, D. (2016). Comparing apples to apple: The effects of stemmers on topic models. *Transactions of the Association for Computational Linguistics*, 4:287–300.

Schölkopf, B., Smola, A., and Müller, K. (1998). Nonlinear component analysis as a kernel eigenvalue problem. *Neural Computation*, 10(5):1299–1319.

Serneels, S., Nolf, E. D., and Espen, P. V. (2006). Spatial sign preprocessing: A simple way to impart moderate robustness to multivariate estimators. *Journal of Chemical Information and Modeling*, 46(3):1402–1409.

Shaffer, J. (1995). Multiple hypothesis testing. *Annual Review of Psychology*, 46(1):561–584.

Shao, J. (1993). Linear model selection by cross-validation. *Journal of the American Statistical Association*, 88(422):486–494.

Shawe-Taylor, J. and Cristianini, N. (2004). *Kernel Methods for Pattern Analysis*. Cambridge University Press.

Shmueli, G. (2010). To explain or to predict? *Statistical Science*, 25(3):289–310.

Silge, J. and Robinson, D. (2017). *Text Mining with R: A Tidy Approach*. O'Reilly.

Spall, J. (2005). *Simultaneous Perturbation Stochastic Approximation*. John Wiley and Sons.

Stanković, J., Marković, I., and Stojanović, M. (2015). Investment strategy optimization using technical analysis and predictive modeling in emerging markets. *Procedia Economics and Finance*, 19:51–62.

Stekhoven, D. and Buhlmann, P. (2011). MissForest: Non-parametric missing value imputation for mixed-type data. *Bioinformatics*, 28(1):112–118.

Steyerberg, E., Eijkemans, M., and Habbema, D. (1999). Stepwise selection in small data sets: A simulation study of bias in logistic regression analysis. *Journal of Clinical Epidemiology*, 52(10):935 – 942.

Stone, M. and Brooks, R. (1990). Continuum regression: Cross-validated sequentially constructed prediction embracing ordinary least squares, partial least squares and principal components regression. *Journal of the Royal Statistical Society. Series B (Methodological)*, pages 237–269.

Strobl, C., Boulesteix, A., Zeileis, A., and Hothorn, T. (2007). Bias in random forest variable importance measures: Illustrations, sources and a solution. *BMC Bioinformatics*, 8(1):25.

Svetnik, V., Liaw, A., Tong, C., Culberson, C., Sheridan, R., and Feuston, B. (2003). Random forest: A classification and regression tool for compound classification and QSAR modeling. *Journal of Chemical Information and Computer Sciences*, 43(6):1947–1958.

Tan, P., Steinbach, M., and Kumar, V. (2006). *Introduction to Data Mining*. Pearson Education.

Thomson, J., Johnson, K., Chapin, R., Stedman, D., Kumpf, S., and Ozoliņš, T. (2011). Not a walk in the park: The ECVAM whole embryo culture model challenged with pharmaceuticals and attempted improvements with random forest design. *Birth Defects Research Part B: Developmental and Reproductive Toxicology*, 92(2):111–121.

Tibshirani, R. (1996). Regression shrinkage and selection via the lasso. *Journal of the Royal Statistical Society. Series B (Methodological)*, pages 267–288.

Timm, N. and Carlson, J. (1975). Analysis of variance through full rank models. *Multivariate Behavioral Research Monographs*.

Tufte, E. (1990). *Envisioning Information*. Graphics Press, Cheshire, Connecticut.

Tukey, J. W. (1977). *Exploratory Data Analysis*. Reading, Mass.

Tutz, G. and Ramzan, S. (2015). Improved methods for the imputation of missing data by nearest neighbor methods. *Computational Statistics and Data Analysis*, 90:84–99.

United States Census Bureau (2017). Chicago illinois population estimates. https://tinyurl.com/y8s2y4bh. Accessed: 2017-09-20.

U.S. Energy Information Administration (2017a). Weekly Chicago all grades all formulations retail gasoline prices. https://tinyurl.com/ydctltn4. Accessed: 2017-09-20.

U.S. Energy Information Administration (2017b). What drives crude oil prices? https://tinyurl.com/supply-opec. Accessed: 2017-10-16.

Van Buuren, S. (2012). *Flexible Imputation of Missing Data*. Chapman and Hall/CRC.

Van Laarhoven, P. and Aarts, E. (1987). Simulated annealing. In *Simulated Annealing: Theory and Applications*, pages 7–15. Springer.

Wand, M. and Jones, C. (1994). *Kernel Smoothing*. Chapman and Hall/CRC.

Wasserstein, R. and Lazar, N. (2016). The ASA's statement on p-values: Context, process, and purpose. *The American Statistician*, 70(2):129–133.

Weinberger, K., Dasgupta, A., Langford, J., Smola, A., and Attenberg, J. (2009). Feature hashing for large scale multitask learning. In *Proceedings of the 26th Annual International Conference on Machine Learning*, pages 1113–1120. ACM.

Weise, T. (2011). *Global Optimization Algorithms: Theory and Application*. www.it-weise.de.

West, Band Welch, K. and Galecki, A. (2014). *Linear Mixed Models: A Practical Guide using Statistical Software*. CRC Press.

Whittingham, M., Stephens, P., Bradbury, R., and Freckleton, R. (2006). Why do we still use stepwise modelling in ecology and behaviour? *Journal of Animal Ecology*, 75(5):1182–1189.

Wickham, H. and Grolemund, G. (2016). *R for Data Science: Import, Tidy, Transform, Visualize, and Model Data*. O'Reilly Media, Inc.

Willett, P. (2006). The Porter stemming algorithm: Then and now. *Program*, 40(3):219–223.

Wolpert, D. (1996). The lack of a priori distinctions between learning algorithms. *Neural Computation*, 8(7):1341–1390.

Wood, S. (2006). *Generalized Additive Models: An Introduction with R*. Chapman and Hall/CRC.

Wu, C. J. and Hamada, M. S. (2011). *Experiments: Planning, Analysis, and Optimization*. John Wiley & Sons.

Xue, B., Zhang, M., and Browne, W. (2013). Particle swarm optimization for feature selection in classification: A multi-objective approach. *IEEE Transactions on Cybernetics*, 43(6):1656–1671.

Yaddanapudi, L. (2016). The American Statistical Association statement on p-values explained. *Journal of Anaesthesiology, Clinical Pharmacology*, 32(4):421–423.

Yandell, B. (1993). Smoothing splines: A tutorial. *The Statistician*, pages 317–319.

Yeo, I.-K. and Johnson, R. (2000). A new family of power transformations to improve normality or symmetry. *Biometrika*, 87(4):954–959.

Zanella, F., Lorens, J., and Link, W. (2010). High content screening: Seeing is believing. *Trends in Biotechnology*, 28(5):237 – 245.

Zhu, X. and Goldberg, A. B. (2009). Introduction to semi-supervised learning. *Synthesis Lectures on Artificial Intelligence and Machine Learning*, 3(1):1–130.

Zou, H. and Hastie, T. (2005). Regularization and variable selection via the elastic net. *Journal of the Royal Statistical Society: Series B (Statistical Methodology)*, 67(2):301–320.

Zuber, V. and Strimmer, K. (2009). Gene ranking and biomarker discovery under correlation. *Bioinformatics*, 25(20):2700–2707.

Zumel, N. and Mount, J. (2016). vtreat: A data.frame processor for predictive modeling. *arXiv.org*.

Index